FUNDAMENTAL PHYSICS FOR PROBING AND IMAGING

Fundamental Physics for Probing and Imaging

WADE ALLISON

*Department of Physics and Keble College,
University of Oxford*

OXFORD
UNIVERSITY PRESS

OXFORD

UNIVERSITY PRESS

Great Clarendon Street, Oxford OX2 6DP

Oxford University Press is a department of the University of Oxford.
It furthers the University's objective of excellence in research, scholarship,
and education by publishing worldwide in

Oxford New York

Auckland Cape Town Dar es Salaam Hong Kong Karachi
Kuala Lumpur Madrid Melbourne Mexico City Nairobi
New Delhi Shanghai Taipei Toronto

With offices in

Argentina Austria Brazil Chile Czech Republic France Greece
Guatemala Hungary Italy Japan Poland Portugal Singapore
South Korea Switzerland Thailand Turkey Ukraine Vietnam

Oxford is a registered trade mark of Oxford University Press
in the UK and in certain other countries

Published in the United States
by Oxford University Press Inc., New York

British Library Cataloguing in Publication Data

Data available

Library of Congress Cataloging in Publication Data

Data available

Printed in Great Britain
on acid-free paper by
Antony Rowe Ltd., Chippenham, Wiltshire

ISBN 0–19–920388–1   978–0–19–920388–8 (Hbk)
ISBN 0–19–920389–X   978–0–19–920389–5 (Pbk)

10 9 8 7 6 5 4 3 2 1

For Alice, Joss and Alfie

Preface

Fear has dominated much of the experience of the human race from earliest times. Fear of death, fear of natural disaster, fear of human enemies and fear of deities: these were fused together beneath a dense shroud of the unseen and the unknown. The major impact of physics on civilisation has been to roll back this shroud. Physics explains. It enables us to see inside the Earth and inside our own bodies. It gives us ways to probe and to cure.

It has seemed to me that there are some big questions to ask, and a dearth of books that ask them. Which aspects of physics are primarily responsible for this revolution? How do they work and how are they used to provide the information and images? Are the dangers that surround applications of this physics understood? Are these safety matters overstated or understated, and is the public misinformed?

This book is written to answer these questions. It is written for all physicists who wish to understand the physics basis. Its coverage is broad but it is also quite demanding in places, for I have a deep dislike of asking the reader to take statements on trust. Anyway, there are other books that do just that, as they rush through the fundamentals in order to reach the excitement of the applications at an early stage. I skip over many experimental details of particular technical realisations but give enough examples of applications for useful comparisons between different modalities to be made. I strongly believe that the widest understanding of the basic physics is essential if future advances in technology are to exploit the possibilities to the full.

The book developed from a short optional course entitled 'Medical and Environmental Physics' that I have given in recent years to third year mainstream physics undergraduates at Oxford University, and assumes some familiarity with basic mathematical methods and the core physics of optics, electromagnetism, quantum mechanics and elementary atomic structure.

In the introduction we ask which aspects of pure physics have enabled mankind to delve into their environment by seeing into or through otherwise opaque objects. Successful solutions have centred on three areas of fundamental physics: firstly the physics of magnetism and low frequency radiation, secondly ionising radiation and the physics of nuclei, and thirdly the mechanical properties of matter and sound. Practical examples range from safe navigation to medical diagnosis, from finding minerals to border security.

The early chapters give a pedagogical development of the pure physics of these three fundamental areas. The later chapters follow how these ideas have been developed in applications. They are concerned not just with imaging, but with further questions of dating, function and provenance, and finally with intervention and therapy. The applications illustrate both the principles at work and the comparison between different possibilities.

The pure physics concerned has changed slowly compared with the recent rapid development of applications. The necessary understanding of magnetism and electromagnetic radiation began in the mid nineteenth century and was completed a century later with the theory of magnetic resonance. Similarly the relevant ionising radiation and nuclear physics was understood within 75 years of the discovery of radioactivity in the 1890s. The basic physics of sound is classical and the understanding of it dates back to the work of Lord Rayleigh, more than a century ago. In every case what has changed recently is that developments and applications using modern materials, electronics and computational power have enabled this academic understanding to escape from the pure physics laboratory into the everyday world.

I have avoided the temptation to follow, logically and immediately, the discussion of each set of fundamental ideas with examples of its application. The subject of successive chapters switches back and forth to encourage parallel thinking about the choice of methods available. Chapter 5 in the middle gives an overview of information and methods of data analysis which have been used in academic physics research for decades. In the past these were too computationally intensive to be deployed in everyday analysis. Now, as the required computational power has become available, they are used routinely in the analysis of images and data.

Inevitably from such a broad field, the applications are selected and their discussion avoids experimental detail which may be found on the Web and elsewhere. To have followed every idea raised in the early chapters would have lengthened this book beyond what could conceivably be covered in a single text. Therefore many fields of application have been omitted entirely, or have only been mentioned in passing.

The concluding chapter takes a bird's eye view of possible developments and the ideas that might emerge from the cupboard of pure physics in the future. There is much in the physical world that we do not understand, and the book ends by looking at a few such cases. For some readers the book will open many questions that it does not answer, but it will not have failed in its aim if such omissions stimulate further study. Other readers will feel the need to rebalance completely society's perception of the threats and dangers that surround it. Perhaps the book may be a beginning to the process of turning public opinion and decision making in the direction of a safer world.

Structure of the book

The chapters are written in such a way that some may be omitted without affecting all of those that follow, and shorter courses may be constructed by reading them selectively, albeit with some loss of the overview. Thus one or more of the following sub-sets of chapters might be omitted:

◇ chapters 2 and 7 on magnetism and magnetic resonance, and related imaging methods;

◇ chapters 3, 6 and 8 on interactions of ionising radiation, analysis and damage by irradiation, and medical imaging and therapy with such radiation;

◇ chapters 4 and 9 on mechanical waves and properties of matter, and ultrasound for imaging and therapy.

Every chapter is divided into a number of sections, each of which starts with a summary. Some sections are more demanding and are marked with a dagger (†). On a first reading of the book some readers may prefer to study just the summary of these, returning to pick up the detail of the derivations at a second reading.

At the end of each chapter there is a short list of recommended books and a list of references and searches for further material on the Web. These should enable the reader both to keep up to date and also to broaden programmes of study based on this book. With its basic introduction I hope that the reader will be able to appreciate in context the technical details of applications, galleries of images, and yet wider links that may be found. Included is a link to the website for the book:

www.physics.ox.ac.uk/users/allison/booksite.htm

Some colour images and video related to the grey-scale material in the book may be found there, together with later comments and news.

Because of the interdisciplinary nature of the material some clarity is needed in the use of terms, abbreviations and conventions. These are laid out for reference in two appendices. Each of the main chapters ends with a short selection of questions, and the final appendix gives hints and answers to some of these.

Acknowledgements

In writing this book I have relied heavily on others to keep my balance and perspective in a wide landscape. Those who have read large sections of the manuscript and provided exactly the combination of crisp comment and encouragement that was most helpful were Richard Tuley, Daniel McGowan, Louis Lyons, John Mulvey and Peter Jezzard. Over the years, Peter, with Stuart Clare, Steve Smith and other members of his group at the FMRIB at the John Radcliffe Hospital, has given me much time and encouragement. More recently I have enjoyed the benefit of discussions with Chris Gibson, Andrew Nisbett and Fares Mayia at the Churchill Hospital on ultrasound and therapy, and stimulating experiments with Geoff Lewis, Chris Fursdon-Davis and students, Lauren McDonald and Frances Lavender, on noise emission from the neck. In fact this book would not have been written without the interest and enthusiasm of many students, both those on the course and those who have carried out medical physics projects. I am indebted especially to Peter Jezzard and Chris Gibson among the many people who have readily provided medical and other images that have brought this story to life. I should like to thank Dieter Jaksch who shouldered my teaching responsibilities at Keble College during my sabbatical year when much of this work was done. Help from Ian Macarthur and his IT team in the Oxford Physics Department is warmly acknowledged. And I thank Sonke Adlung and his editorial team at OUP for their positive and welcoming cooperation in this venture.

I have received ideas, stimulation and correction from many people but the mistakes that remain are mine. Comments on these are welcomed.

Finally, thanks and love to Kate who has sustained and encouraged me throughout this absorbing task.

Wade Allison
Oxford, August 2006

Contents

1	**Physics for security**	**1**
	1.1 The task	1
	1.1.1 Stimulation by fear and the search for security	1
	1.1.2 Crucial physics for probing	5
	1.1.3 Basic approaches to imaging	9
	1.2 Value of images	10
	1.2.1 Information from images	11
	1.2.2 Comparing modalities	12
	1.3 Safety, risk and education	16
	1.3.1 Public apprehension of physics	16
	1.3.2 Assessing safety	17
2	**Magnetism and magnetic resonance**	**21**
	2.1 An elemental magnetic dipole	21
	2.1.1 Laws of electromagnetism	21
	2.1.2 Current loop as a magnetic dipole	22
	2.1.3 The Larmor frequency	25
	2.2 Magnetic materials	27
	2.2.1 Magnetisation and microscopic dipoles	27
	2.2.2 Hyperfine coupling in B-field	31
	2.3 Electron spin resonance	34
	2.3.1 Magnetic resonance	34
	2.3.2 Detection and application	36
	2.4 Nuclear magnetic resonance	37
	2.4.1 Characteristics	38
	2.4.2 Local field variations	39
	2.4.3 Relaxation	42
	2.4.4 Elements of an experiment	44
	2.4.5 Measurement of relaxation times	45
	2.5 Magnetic field measurement	47
	2.5.1 Earth's field	48
	2.5.2 Measurement by electromagnetic induction	48
	2.5.3 Measurement by magnetic resonance	50
3	**Interactions of ionising radiation**	**55**
	3.1 Sources and phenomenology	55
	3.1.1 Sources of radiation	55
	3.1.2 Imaging with radiation	56
	3.1.3 Single and multiple collisions	57

	3.2	Kinematics of primary collisions	58
		3.2.1	Kinematics and dynamics	59
		3.2.2	Energy and momentum transfer	59
		3.2.3	Recoil kinematics	60
		3.2.4	Applications of recoil kinematics	61
	3.3	Electromagnetic radiation in matter	65
		3.3.1	Compton scattering	65
		3.3.2	Photoabsorption	66
		3.3.3	Pair production	68
	3.4	Elastic scattering collisions of charged particles	68
		3.4.1	Dynamics of scattering by a point charge †	69
		3.4.2	Cross section for energy loss by recoil	73
	3.5	Multiple collisions of charged particles	73
		3.5.1	Cumulative energy loss of a charged particle	74
		3.5.2	Range of charged particles	77
		3.5.3	Multiple Coulomb scattering	78
	3.6	Radiative energy loss by electrons	81
		3.6.1	Classical, semi-classical and QED electromagnetism	81
		3.6.2	Weissäcker–Williams virtual photon picture	81
		3.6.3	Radiation length	82

4 | **Mechanical waves and properties of matter** | **85** |
	4.1	Stress, strain and waves in homogeneous materials	85
		4.1.1	Relative displacements and internal forces	85
		4.1.2	Elastic fluids	87
		4.1.3	Longitudinal waves in fluids	88
		4.1.4	Stress and strain in solids †	92
		4.1.5	Polarisation of waves in solids †	96
	4.2	Reflection and transmission of waves in bounded media	99
		4.2.1	Reflection and transmission at normal incidence	99
		4.2.2	Relative directions of waves at boundaries †	100
		4.2.3	Relative amplitudes of waves at boundaries †	103
	4.3	Surface waves and normal modes	111
		4.3.1	General surface waves	113
		4.3.2	Rayleigh waves on free solid surfaces	113
		4.3.3	Waves at fluid–fluid interfaces	115
		4.3.4	Normal mode oscillations	119
	4.4	Structured media	120
		4.4.1	Interatomic potential wells	121
		4.4.2	Linear absorption	126

5 | **Information and data analysis** | **131** |
	5.1	Conservation of information	131
	5.2	Linear transformations	135
		5.2.1	Fourier transforms	135
		5.2.2	Wavelet transforms	142
	5.3	Analysis of data using models	143
		5.3.1	General features	144

| | | 5.3.2 | Least squares and minimum χ^2 methods | 145 |
| | | 5.3.3 | Maximum likelihood method | 149 |

6 Analysis and damage by irradiation — **157**
	6.1	Radiation detectors	157
		6.1.1	Photons and ionisation generated by irradiation	157
		6.1.2	Task of radiation detection	159
		6.1.3	Charged particle detectors	161
		6.1.4	Electromagnetic radiation detectors	165
	6.2	Analysis methods for elements and isotopes	168
		6.2.1	Element concentration analysis	169
		6.2.2	Isotope concentration analysis	172
		6.2.3	Radiation damage analysis	177
	6.3	Radiation exposure of the population at large	179
		6.3.1	Measurement of human radiation exposure	179
		6.3.2	Sources of general radiation exposure	182
	6.4	Radiation damage to biological tissue	187
		6.4.1	Hierarchy of damage in space and time	187
		6.4.2	Survival and recovery data	189
	6.5	Nuclear energy and applications	192
		6.5.1	Fission and fusion	192
		6.5.2	Weapons and the environment	193
		6.5.3	Nuclear power and accidents	199

7 Imaging with magnetic resonance — **207**
	7.1	Magnetic resonance imaging	207
		7.1.1	Spatial encoding with gradients	207
		7.1.2	Artefacts and imperfections in the image	211
		7.1.3	Pulse sequences	213
		7.1.4	Multiple detector coils	218
	7.2	Functional magnetic resonance imaging	221
		7.2.1	Functional imaging	221
		7.2.2	Flow and diffusion	223
		7.2.3	Spectroscopic imaging	225
		7.2.4	Risks and limitations	227

8 Medical imaging and therapy with ionising radiation — **233**
	8.1	Projected X-ray absorption images	233
		8.1.1	X-ray sources and detectors	233
		8.1.2	Optimisation of images	236
		8.1.3	Use of passive contrast agents	239
	8.2	Computed tomography with X-rays	241
		8.2.1	Image reconstruction in space	241
		8.2.2	Patient exposure and image quality	245
	8.3	Functional imaging with radioisotopes	246
		8.3.1	Single photon emission computed tomography	246
		8.3.2	Resolution and radiation exposure limitations	252
		8.3.3	Positron emission tomography	252

	8.4	Radiotherapy	256
	8.4.1	Irradiation of the tumour volume	256
	8.4.2	Sources of radiotherapy	257
	8.4.3	Treatment planning and delivery of RT	259
	8.4.4	Exploitation of non-linear effects	262

9   Ultrasound for imaging and therapy   **267**

	9.1	Imaging with ultrasound	267
	9.1.1	Methods of imaging	267
	9.1.2	Material testing and medical imaging	270
	9.2	Generation of ultrasound beams	272
	9.2.1	Ultrasound transducers	272
	9.2.2	Ultrasound beams	276
	9.2.3	Beam quality and related artefacts	278
	9.3	Scattering in inhomogeneous materials	280
	9.3.1	A single small inhomogeneity	281
	9.3.2	Regions of inhomogeneity	284
	9.3.3	Measurement of motion using the Doppler effect	287
	9.4	Non-linear behaviour	290
	9.4.1	Materials under non-linear conditions	290
	9.4.2	Harmonic imaging	294
	9.4.3	Constituent model of non-linearity	297
	9.4.4	Progressive non-linear waves	300
	9.4.5	Absorption of high intensity ultrasound	301

10 Forward look and conclusions   **307**

	10.1	Developments in imaging	307
	10.2	Revolutions in cancer therapy	312
	10.3	Safety concerns in ultrasound	313
	10.4	Rethinking the safety of ionising radiation	315
	10.5	New ideas, old truths and education	318

Appendices

A Conventions, nomenclature and units   **321**

B Glossary of terms and abbreviations   **323**

C Hints and answers to selected questions   **327**

Index   **331**

Physics for security

1.1 The task	1
1.2 Value of images	10
1.3 Safety, risk and education	16
Look on the Web	20

in which are discussed the ways in which physics is used to answer questions, practical and cultural, that face human society. This sets the agenda for the rest of the book.

1.1 The task

By enabling us to know and understand what is happening in the physical world, physics has reduced our fear of it. Specifically, we consider areas of human activity where physics has made a real difference by providing additions to our ability to see by probing and imaging. We identify four possible approaches in physics to seeing into and through material objects: high energy radiation, low frequency radiation (and magnetism), sound (and mechanical probing), and gravity. We mention gravity for completeness only, for it is effective on a different scale to the other three probing fields discussed in this book. There are five ways in which to gather data to make an image using a probing field. These differ according to whether the origin of signals is internal or external, natural or applied, pulsed or continuous.

1.1.1 Stimulation by fear and the search for security

Science and the external world

Scientific progress is stimulated by the urge to understand and control the physical world. It starts quite simply. First we perceive the objects around us. Next we seek to develop an understanding of the laws that determine their behaviour and how they work. Then we learn to apply this knowledge to predict their behaviour, to modify them, to conserve them and to use them for the common good. By doing this we have learned to master the fear of the external world which oppressed previous generations.

It is no coincidence that, although the space of physics is isotropic in principle, the space of the humanities is not.[1] Upwards we can see. There is light. It is 'good' and the symbolic direction of heaven. Downwards is dark, 'bad' and the supposed direction of hell. Our language is full of a symbolic fear of an underworld, both as used figuratively and in reality. Despite the light that physics has shed in the past five centuries on the physical world in general, it was still from beneath our feet that the unpredicted tsunami struck on 26 December 2004. Physics has further

[1] This divide was the substance of the Copernican revolution.

to go to protect humanity from danger and the related fear.

Physics has generalised the idea of light and the act of seeing. We often use the word 'see' to mean 'understand', and there is a degree of confidence implicit in all seeing and imaging. We 'see' when information flows into the mind of the observer. Such information often arrives rather indirectly through the intermediary of a display screen, a recorded picture or an optical instrument.[2] In such cases, with familiarity and confidence observers come to think that they see 'as with their own eyes'. Even the process of seeing itself is quite indirect, for an eye too is a physical instrument which works on the same principle as a telescope or microscope.

[2]Or even through reading the pages of a book!

Pictorial information

Pictorial information is stimulating, partly because pictures entering the brain carry a thousand times the information content or bandwidth of other senses, such as hearing or touch. Pictures of the night sky, for example, have influenced humans from the earliest times. Their constancy, predictability and dramatic changes told them something that they could not understand. These pictures have been one of the greatest influences on our minds, and modern pictures from better telescopes enable us to see deeper and answer more questions about the Universe.

However, there are objects close at hand that we cannot see. For example, intervening opaque layers obscure the inside of our bodies and the Earth beneath our feet. Screens can hide enemies, and packaging may conceal dangerous drugs or weapons. With an understanding of physics we have found methods of seeing through these barriers for the first time.

Navigation

Navigation is a historic example of the stimulation of physics by the need to see. The Sun, Moon and stars provide an outstandingly precise basis for navigation. Before the development of the chronometer by Harrison in the eighteenth century, one dimension of information was still missing. But even after this longitude problem was solved, unpredictably overcast skies continued to frustrate navigation, and fog often rendered navigators quite blind. The resulting fear, loss of life and loss of goods seriously restricted world trade.

The penetrating power of the Earth's magnetic field and the mechanical probe of the lead line were the early solutions.[3] Sound signals in the form of fog horns were used in historic times, and the use of underwater sound (or ultrasound) for navigation by simple depth determination replaced the lead line in the mid twentieth century.

[3]In the base of the lead sounding weight was placed a core of soft tallow to pick up a sample of the seabed. Charts of the composition of the seabed gave additional navigational information. Such information may be found on nautical charts to this day.

These early techniques were all based on single channel probes, and no images were produced. With radar and the ability to see through fog came the first multi-channel image. Recently, with modern electronics and software other single channel probes have been extended to provide pictures, such as the map of a whole area of the sea floor provided by a

Fig. 1.1 An ultrasound image of a sunken boat on the sea floor. The image is taken in cylindrical coordinates, shown with the z-axis horizontal and the r-axis downward. The transducer has scanned a fan-shaped beam horizontally along the z-axis on the sea surface. The r-coordinate is the radial distance from the transducer, as determined by the ultrasound reflection time. There is no discrimination in angle. The signals at the top of the picture are from surface waves. In the centre can be seen signals scattered by fish. The sharp boundary is the sea floor immediately below. The mast and rigging are further away sideways along the sea floor. The reflection of early signals therefore generates shadows on later image elements further away. [Image reproduced by kind permission of Humminbird, a Subsidiary of Johnson Outdoors, Inc.]

[4]These images, invaluable for locating wrecks, are proving injurious to the survival of the fish population of the oceans that may now be easily located and caught by modern fishing fleets.

modern ultrasonic depth sounder.[4] Figure 1.1 shows an example of such an image.

Geophysics and geology

The problem of imaging the interior of the Earth presents a major challenge. Drilling wells and digging mines is invasive, expensive and very limited. Sounding, including the use of signals generated naturally by earthquakes, is the preferred solution, both for mineral and oil prospecting, and for larger scale geophysics. Superficially, imaging with sound signals in this way is similar to imaging with ultrasound in medicine or the oceans, although there are significant differences. The task of imaging the geological structure sufficiently well to locate minerals and predict major natural disasters is difficult.

Medicine

Medicine is the single most important field for the application of physics probes. Pictures of the inner workings of the human body are now commonplace, even in routine clinical examination. They are intriguing and exciting to everybody, and their interpretation is challenging. The first pictures came a century ago following the discovery of X-rays. An early example, Figure 1.2a, shows the bones within a hand. A modern picture using magnetic resonance imaging (MRI) is shown in Fig. 1.2b. As the ability to see more clearly has developed the need for invasive exploratory surgery has declined.

In some cases the physics used to make an image of a tumour is related to the physics required to deliver a therapeutic energy dose. The task for therapy is to confine the damage to the contours of the malignant tumour, sparing the surrounding healthy tissue as far as possible. This calls for alignment of scanned images in three dimensions with the coordinates of the treated volume and for quantitative planning of the effect of the dose. The alignment is termed registration. The mathematical physics tools of fiducial mark arrays, coordinate transformations and overdetermined sets of measurements are needed.

Archaeology

Methods of imaging developed for medicine and geophysics are also successful in archaeology. Images of buried objects can highlight where excavations would be most effective. They also stimulate popular imagination and answer questions without the need to excavate.

Archaeological discoveries are followed by further questions, on dates of activities, on origins and on uses of artefacts. In many cases the answers to these questions come from the application of fundamental physics, for instance by analysing and imaging the concentration of particular elements or isotopes as discussed in chapter 6.

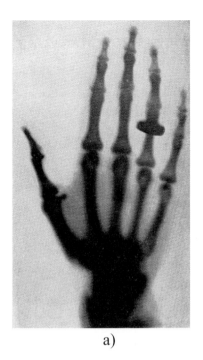

a)

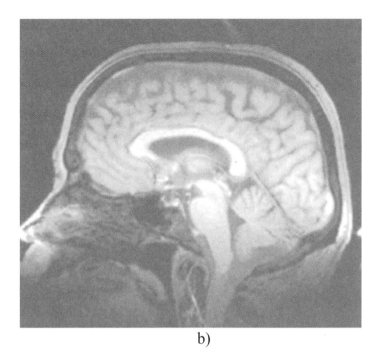

b)

Fig. 1.2 a) An X-ray of a hand with ring (printed in *McClures Magazine*, April 1896). b) A recent MRI scan of the author's head.

Transport security at railway stations, airports and ports

A challenging new problem is the task of transport security. Internal images and information are needed in order to find hidden weapons and explosives. The invasive opening of all luggage and personal stripping are time consuming, impractical and unpopular. The wider problem of searching freight for drugs, terrorists and illegal immigrants, on land or sea, is a large task. Its thoroughness is directly related to how fast, safely and conveniently it can be done. The problem is that luggage, packaging and clothes are opaque to light. How can these be examined quickly and non-invasively? What is the basis in physics for the possible probes that might be used? Can they be made specific enough to distinguish different drugs, for example? Or fast enough and definite enough to pick out a suicide bomber carrying explosives? Or sufficiently effective to help in the clearance of land-mines.

1.1.2   Crucial physics for probing

Of the fundamental interactions known to physics only gravity and electromagnetism have the necessary long range influence to provide the basis of a macroscopic probing field. In addition, as we discuss later, there is the approach based on mechanical properties and sound. This is a second order interaction from the physics viewpoint, but nonetheless effective.

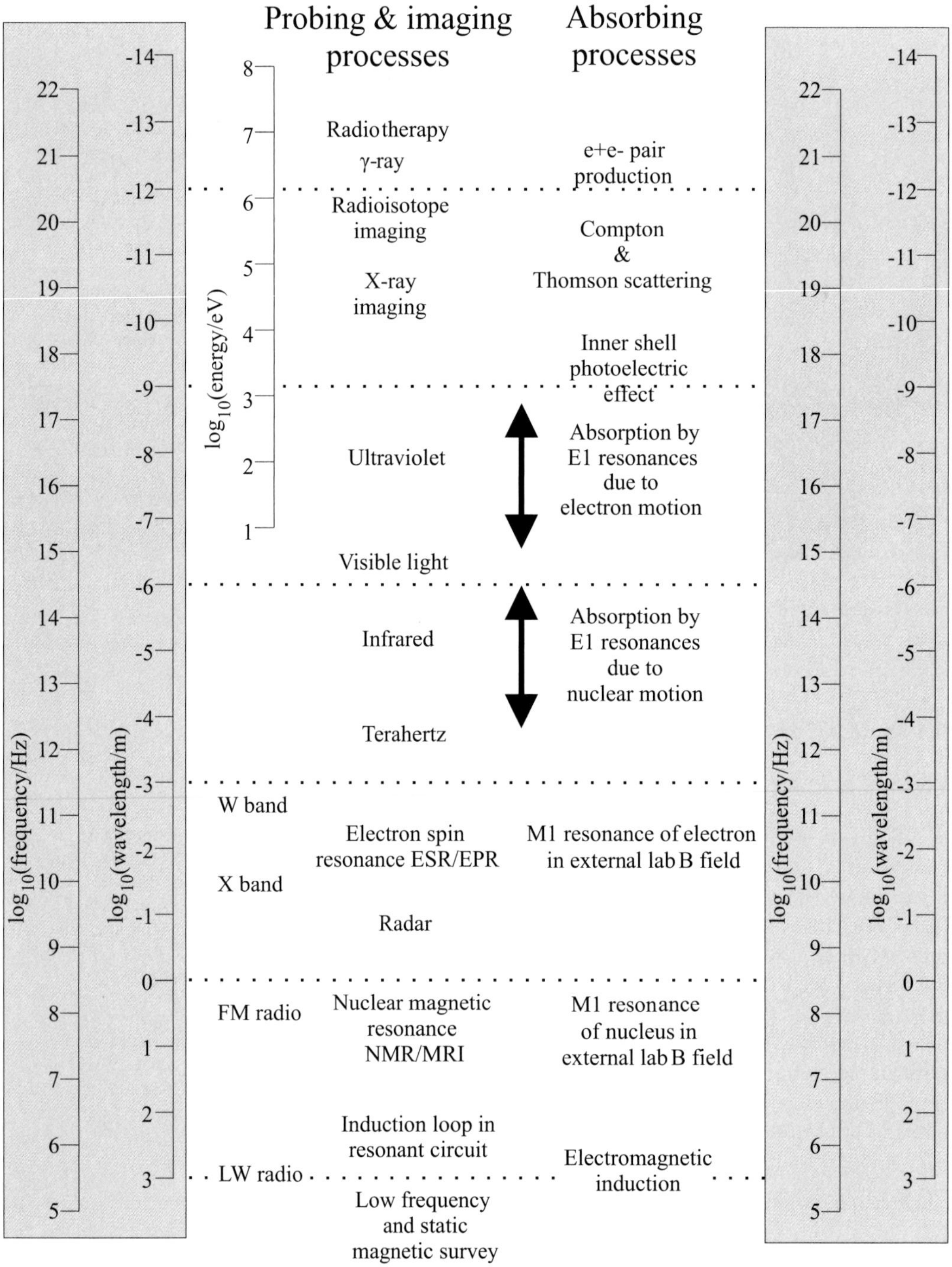

Fig. 1.3 A diagrammatic plot of the spectrum of electromagnetic radiation with frequency and wavelength shown vertically on logarithmic scales. The various phenomena shown to the right include the two main absorption regions due to electric dipole (E1) resonances. These are indicated by heavy arrows. The one in the ultraviolet region (UV) is associated with electronic motion. The other in the infrared (IR) is due to the motion of nuclei and atoms.

Gravity and dark matter

Gravity is a very weak probe and can only contribute information when the mass of the object concerned is very large. This is significant only at the geophysical scale and above.[5] In the Universe at large, gravity has an important story to tell. Among other data, in our own and other galaxies the dependence of rotation velocity on distance from the centre of the galaxy indicates that a large fraction of the mass is unobserved. In the next 50 years we may expect that the detection of this dark matter as well as gravitational waves will reveal much about the Universe of which we are unaware.[6] However, these important developments have little in common with the rest of this book, and we do not pursue them further.

X-ray and RF methods

The different regions of the electromagnetic spectrum are shown in Fig. 1.3 on a logarithmic scale. Half way up the page in the centre of the scale is visible light. This narrow range, called the optical region, lies between the ultraviolet (UV) and infrared (IR) absorption bands in matter. The UV or soft X-ray band with shorter wavelength and greater frequency than optical is associated with electric dipole resonances (E1) involving electron transitions in atoms. The IR region involves similar E1 resonances where the motion of nuclei rather than electrons is involved. Nuclei carry the inertia of whole atoms, so that the IR resonances are transitions in the rotational and vibrational states of atoms within molecules. Typically the two regions are displaced from one another by just over three orders of magnitude in energy (or frequency), arising from the mass ratio of atoms and electrons.[7]

For the task of border security, for instance, we need to see through all materials. If this is not possible with visible light, Fig. 1.3 suggests that either we should use frequencies above the UV region (3×10^{17} Hz) or below the IR absorption region (10^{12} Hz). We describe these as the X-ray and radiofrequency (RF) methods. The fundamental physics of these two options is explored in chapters 3 and 2. The X-rays extend to γ-rays with energies in the MeV range.[8] Closely related to γ-rays are the charged particle beams such as electrons (β-radiation) and protons with similar energy. All such radiation is termed ionising radiation and its general applications are discussed in chapter 6 and its use for medical imaging and therapy in chapter 8.

Although the high frequency or X-ray methods are quite distinct from the magnetic or radio frequency (RF) methods, they are competitive in many applications. Magnetic imaging applications are covered in chapter 7.

Mechanical probing

Both the X-ray and the RF methods are routinely used in airport security. Luggage is scanned with X-rays and passengers walk through

[5] Even there, the sensitivity of the gravitational field at short range to changes in mass density is reduced if the masses concerned are floating. The reason is that Archimedes' principle ensures that, for example, the mass of a floating continent is the same as the displaced mass of the mantle, where the two are in hydrostatic equilibrium. This gives some cancellation, depending on the distance at which the field is measured.

[6] These gravitational waves (concerned with G) should be clearly distinguished from water and tidal waves, discussed in chapter 4. These are sometimes called gravity waves because of their dependence on g.

[7] It is a fortunate accident that the narrow gap between the bands where many materials are transparent is also the region at which the 6000 K black body spectrum of the Sun is maximum. It is not surprising that animals with sight should have developed sensitivity to this region.

[8] We shall often refer to photons over this whole range as X-rays. Usually no distinction between the terms X-ray and γ-ray is intended.

RF magnetic induction loops. A third method of probing is also used. As soon as the RF induction loop indicates that a passenger might be carrying something magnetic or conductive, he or she is quickly taken aside and frisked by a security guard for signs of the bulk mechanical properties that might be expected of a gun or other weapon. We attempt a description in physics terms of this frisking process. It is useful to think about how this is actually done! A large quasi-static mechanical deformation of the exterior surface, in this case the clothing, is applied and variations in the magnitude and direction of the reactive force are sensed. These methods have none of the simplifying assumptions of small amplitude, linearity, isotropy or homogeneity assumed in simple physics problems. The response to the large amplitude probings are interpreted in the mind of the experienced airport security officer in terms of the possible density and elasticity variations that he or she would expect of a concealed object, such as a weapon in a pocket.

The same technique is applied in the crucially important cancer screening procedure of 'feeling for lumps', or palpation as it is termed medically. By applying large strains at the tissue surface, particularly in shear, and sensing the variation in the stress felt by the fingers as they move around, it is possible to detect a hard mass that might be a tumour, for instance in the breast. This is the way in which many people become aware of a possible tumour before alerting their doctor. Deeper tumours for which such simple early detection is not possible present a greater challenge.

The passive detection of sound emission from the body using a stethoscope also relies on the simple mechanical interactions of matter. In that case the probe is a single channel probe without imaging.

This mechanical probing field is provided by the forces between neutral atoms and molecules that determine the properties of bulk material.[9] The basic physics of these properties is discussed in chapter 4 with extensions and applications, particularly to ultrasound, in chapter 9. In the linear limit the propagation of sound and ultrasound depends solely on the density and elastic moduli of the material. When these change, the waves are reflected or scattered and an image may be formed. Such a use of ultrasound is an alternative to palpation.

At sufficient power levels and with focusing, these waves behave nonlinearly and their energy may be deposited in a small region for therapy. This is discussed at the end of chapter 9.

Weaker interactions

It is important that the items being imaged are not *completely* transparent to the probing field, otherwise the image would be blank. For example, in the application to border security, neither the case nor the gun that it might contain would be seen. So, for the two options that use electromagnetic waves, the X-ray and RF methods, we need to identify the weaker interactions that offer something short of total transparency.

For the solution using X-rays there are three such weaker phenom-

[9]In fundamental terms this is a second order electromagnetic process.

ena: inner electron ionisation by the photoelectric effect, Thomson (or Compton) scattering, and pair production. These will be discussed in chapter 3.

For the low frequency or RF region the source of the required weaker interaction is magnetic. As already stated the resonances responsible for the strong absorption at UV and IR frequencies are the electric dipole resonances. Magnetic effects involve currents or moving charges instead of the static charges of electric interactions. They are relativistic corrections to the electric interaction and are weaker by factors of the order of v^2/c^2 where v is the speed of the charge and c is the velocity of light. In atomic hydrogen, as an example, this ratio is α^2, where α, the fine structure constant, is 0.00730. Therefore typical magnetic dipole (M1) resonances are weaker in energy than E1 resonances by a factor of order 5×10^{-5}. This means that the frequency of a typical magnetic resonance is lower by the same factor. Just as for electric resonances, we expect magnetic resonances to be divided into those associated with the motion of electrons and those associated with the motion of nuclei or atoms, the two separated from one another by a factor of several thousand in frequency on account of the mass ratio. The general frequency ranges of these are shown in the right hand column in Fig. 1.3. These ideas give only rough estimates. The actual numbers depend on whether external as well as internal fields are involved and whether there are cancellation effects in the details. The important conclusion is that magnetic dipole resonances are likely to occur in the low frequency range where materials are almost transparent and which is suitable for the imaging task. These resonances are the basis of the weaker interaction response that we need for imaging to avoid complete transparency.

Very low frequency magnetic fields (below 1 MHz) usually interact with material by magnetic induction. Probing methods based on these, like the induction coil used in airport security, are important but will not be discussed in detail.

The three principal imaging methods discussed above are distinguished by their penetration. Other methods are subject to much stronger absorption in general and discussion of these is omitted from the main text. The most important are IR absorption, IR fluorescent spectroscopy and terahertz imaging. These are noted briefly in the final chapter.

1.1.3   Basic approaches to imaging

We have identified three particular carriers of imaging information. How may they be utilised to image properties in space and time? We distinguish five different ways in which this problem may be approached.

1. We can observe internal signals that are generated naturally. For example, we may attach sensors to the surface of the body and record the electric and magnetic fields emitted in the course of its normal function. Similarly an array of seismometers can be deployed to map and record earthquake vibrations as they occur. We can detect vibrations emitted

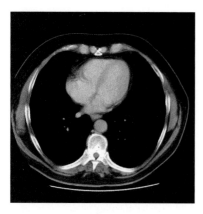

Fig. 1.4 A slice of an X-ray CT image showing lungs, heart, spine, ribs and major vessels. The grey scale has been mapped in order to increase the apparent contrast. [Image reproduced by kind permission of Medical Physics and Clinical Engineering, Oxford Radcliffe Hospitals NHS Trust.]

by the body with a classic stethoscope, or instrument it in a more modern way with accelerometers, as described in chapter 10. Such passive imaging methods are the least invasive and most benign. Any risk is confined to the attachment of sensors.

2. We can illuminate the region of interest with a steady beam of the probe field and observe what is transmitted, absorbed or reflected. This is more invasive since energy is necessarily deposited. The use of X-rays for imaging is a familiar example. Figure 1.4 shows a slice of a modern computed tomography (CT) image which uses X-rays in this way.

3. We can illuminate the region of interest with pulses of the probe field and record the transit time of the reflection. This is the basic method of radar, ultrasound imaging, and acoustic sounding in navigation, mineral prospecting and archaeology.

4. We can illuminate the region of interest and observe the frequency and width of the response of the target region. This may involve a pulsed or continuous excitation and a changed response signal. Examples here include nuclear magnetic resonance imaging (NMR/MRI), IR fluorescence, X-ray fluorescence (XRF), proton-induced X-ray emission (PIXE) and Doppler shift methods in ultrasound.

5. More invasively, active sources of the probe field may be injected, or otherwise introduced, into the target region. The imaging process then tracks the progress of these sources as a function of time. An example is the introduction of short-lived radioactive sources into the human body, single photon emission computed tomography (SPECT) and positron emission tomography (PET). These procedures are significantly more invasive. A more benign form of the same idea involves injecting contrast agents whose distinguishing but passive properties are tracked in time when illuminated by the probe field. We shall see examples of this method in connection with all three probing fields.

1.2   Value of images

The successful use of image information depends on being able to notice and highlight relevant features, a process that requires the use of a combination of symmetry, calibration and prior knowledge. In large volumes of data this may become unwieldy unless aided by software. Useful imaging and probing methods, together or separately, need to record sev-

eral properties so as to be able to optimise contrast or answer further questions. In practice, techniques vary in their resolution in time and space. Each is compromised to some extent and all fall short of what is ultimately sought. There are further criteria that influence the choice of method, namely contrast, registration, signal-to-noise ratio, artefact effects, safety and calibration. In addition there are the socio-economic criteria of availability of educated staff, patient throughput and acceptability.

1.2.1 Information from images

Imaging and noticing

Having obtained an image the task is to notice whether it contains an important message so that action can be taken. A security scan of airport luggage which produces a good image of a gun inside a suitcase or a medical scan of the brain which shows a tumour is not effective unless the features are actually picked up by the security guard or the clinician respectively. Images may be formed from measurements of several different properties in three dimensions, and the important message may be hidden in the way in which these have changed with time. Such large data sets in the four dimensions of space and time are not simple to look at, and clear evidence may be missed.[10]

In the days when the information could be laid out as a number of simple pictures in black and white, the experienced eye of the professional was a sufficiently reliable and quick way to notice and interpret the evidence. With the vast increase in the volume and complexity of data that is no longer generally true. The time required can be too long to examine all by eye. Computer software can help by searching the data very rapidly using agreed search rules, rules which emulate an initial examination by a professional with all the data available. Features that are possibly unusual can then be highlighted for the clinician's attention using colour enhancement and quantitative estimates. Ultimately a human, whether doctor or security officer, has to be alerted and action taken. This observer is often the slowest link in the chain and needs all the help that can be given.

Symmetry and calibration

It is instructive to think about how we look at pictures or plots of graphical information, even in two dimensions, and what we learn by doing so. Sometimes there is some symmetry to the image. From that symmetry, especially if it is slightly broken, we learn something, even starting from a point of complete ignorance. Thus in Fig. 1.5 we first note the symmetry between the left and right sides of the body. The eye is drawn to the black area in the region of the right lung because of the incomplete symmetry. One lung is different to the other. The human eye is good at noticing tumours and other pathologies when they violate a near symmetry. The search for symmetry has been an essential item in the physicist's tool bag in the past century, and it has proved to be crucial

[10]The reader should note that the illustrations in this book are all two dimensional. Although they may show good examples of what is being looked for in images, they obscure the difficulties of noticing or finding it.

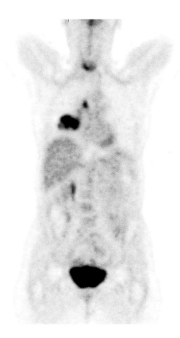

Fig. 1.5 A positron emission tomography (PET) scan of a patient diagnosed with lung cancer. Because of the asymmetry of the signals from the two lungs, the diagnosis is strongly suggestive, even with the poor image resolution in the image. (The physics of this image is discussed further in chapter 8.) [Image reproduced by kind permission of Medical Physics and Clinical Engineering, Oxford Radcliffe Hospitals NHS Trust.]

when attempting to unravel the physics of an unknown field, such as particle physics in the 1960s.

The beauty of symmetry is that it is self-calibrating. Thus, even to someone who knows nothing of medicine, the sight of someone walking with a limp indicates prima facie that they have a medical problem. But to go beyond the use of symmetry requires measurement and calibration, and a real quantitative understanding of what is going on. For example, the comparison of two medical images taken on different occasions requires calibration and position registration. Using symmetry we simply need to compare one part of the image with another.

Without symmetry clinicians compare an image qualitatively with their perception of cases they have seen previously. But to use all of the information requires a quantitative assessment of the measured values using calibration information to register and compare scans of the patient taken on other occasions. This is not a task for which the unaided human eye was designed. Computer software is better matched to do the detailed quantitative work for the clinician.

The actual detailed solutions to the technical problems of calibration and registration depend on the particular modality and lie somewhat outside the scope of this book. However, we note their importance.

Further investigation

When something unexpected does show up in an image we need to be able to answer more questions. In transport security these might be blunt questions to the traveller.[11] In archaeology the questions might be about use, provenance or date. In medical imaging the questions might be about function, blood flow or growth. Purely anatomical maps often do not provide the answer to the questions that the medical clinician is asking. Airport security is a privileged example, for the passenger is on hand to answer the questions. In clinical medicine the patient may try to answer but have limited relevant sensory information which he or she finds hard to articulate. In archaeology the witnesses are long gone, the hunter from his arrows or the mason from his stone tablets. Therefore we need to study with what success it is possible to image the answers to the follow-up questions in addition to the initial purely compositional or anatomical ones.

[11]'Where did you get this?', 'Why do you have this in your possession?', 'What is this for?' or 'How long have you had it?'

1.2.2   Comparing modalities

Space and time resolution

How should we assess different methods of imaging? There are a significant number of methods to choose from, their differences are often complementary and they are frequently used together. In the following, since it presents the most ambitious list of objectives, we have taken the example of medical imaging to discuss a basis on which to compare methods.

Although great advances have been made, there is a long way to go.

At best, current methods achieve millimetre spatial resolution in three dimensions and millisecond resolution in time, albeit with different techniques. But these fall seriously short of the ability to watch individual cells and detect the onset of pathological cell reproduction, which is the scale ultimately needed to master the medical condition.

Near-field and far-field imaging

In an imaging process a probe acts on an object through a probe field, and the reaction of the object is detected and used to form the image. The spatial resolution that may be achieved in an image of an object at a distance r from the detector using a probe at frequency ω depends on whether the influence of the probe is near-field or far-field. Qualitatively speaking, in the near-field case the interaction between the probe and object is intimate and involves no significant time delay. In the far-field case the probe field mediates an action at a distance with a delay. The significance of this delay is measured by the phase difference, $r\omega/c$ radians, where c is the phase velocity of the probe field.

In the far-field region the object–detector distance is sufficient that this phase difference is larger than unity, say. The probe field ϕ is then governed in a non-trivial way by the wave equation,

$$\nabla^2\phi - \frac{1}{c^2}\frac{\partial^2\phi}{\partial t^2} = 0. \tag{1.1}$$

This determines the wavelength, $\lambda = 2\pi c/\omega$. The spatial resolution which comes from measurement of the phase is limited to a fraction[12] of $\lambda/(2\pi)$. It follows that to improve the spatial resolution significantly, a shorter wavelength, that is to say a higher frequency probe field, must be used.

[12]The actual value of this fraction depends on the noise level but is not small.

However, this dependence of resolution on frequency does not apply in the near-field where the object–detector distance is not inferred from this phase difference. In some cases the frequency may be effectively zero. A simple example is palpation, which we already mentioned. There the spatial resolution is determined by the size of the fingers of the individual doing the examination. The dimension of time, and the wave equation, are not involved. Another near-field method is MRI, discussed in chapter 7. The wavelength of the RF used may be 3 m (100 MHz), but the spatial resolution achieved is about 1 mm. This is possible because the probing field is not the radiative electromagnetic field but the inductive magnetic near-field of NMR, discussed in chapter 2. The spatial resolution derives from measurement of field signals as a function of static B-field gradients, and the phase delay between source and object is not involved. Radar is a far-field method which uses similar frequencies to MRI but has far inferior spatial resolution. Mechanical probing, too, may be either near-field (palpation) or far-field (ultrasound). The two methods provide quite different information, as discussed in chapter 9.

A list of criteria

We draw up a list against which different modalities may be judged.

1. *The spatial resolution in three dimensions* (3-D). This may
    be described by a *point spread function* (PSF), the raw im-
    age shape generated by a point object.[13] Alternatively the
    resolution may be described in harmonic rather than impulse
    terms. This is usually expressed in terms of the image con-
    trast of an object with full sine-wave modulation at a certain
    spacing, written reciprocally as the number of *lines per mm*.
    Thus an optical system whose quality would render an ob-
    ject of black lines on a white ground with 100 μm spacing as
    an image with valley-to-peak intensity ratio of 50% can be
    described as having a spatial resolution of 10 lines per mm at
    50% contrast. In digital terms a distinguishable element of
    a 2-D image is a *pixel*, a picture element with a certain size.
    An element of a 3-D image is called a *voxel*, an elementary
    volume element.

2. *Time resolution.* If time-resolved images are available, then
    complete images may be recorded for different times, and
    movement and velocity may be derived by comparing them.
    In other modalities velocity itself can be measured and im-
    aged; an instance is the imaging of Doppler shifts in ultra-
    sound.

3. *Contrast.* The images derived from different modalities show
    up different properties of the bone and tissue, muscle and fat,
    blood and other fluids. Some tend to show fine detail with
    poor contrast; others, strong contrast but poor resolution.
    For this reason progress has been made by combining data
    from different modalities.

4. *Registration.* During imaging a patient breathes, his or her
    heart beats and he or she may move. If optimum spatial res-
    olution is to be achieved, there is therefore a general problem
    of referencing images to a coordinate system that co-moves
    with the body. Parts of the body that deform or move differ-
    entially according to the patient's attitude or muscular ten-
    sion present the most challenging problem, for example the
    breast. Registration between different examinations must
    be made so that changes over periods of weeks and months
    can be tracked. Even more important is the need to relate
    images, possibly from different modalities, to the patient co-
    ordinate system at the point of delivery of therapy. The
    better the intrinsic spatial resolution, the more demanding
    the task of registration. The development of any technique,
    such as ultrasound, which can deliver images and therapy
    within the same coordinate system in a short time interval,
    has an important advantage in this respect.

[13]We may describe the latter at the origin by a 3-D Dirac delta function, $\delta^3(\boldsymbol{r}) = \delta(x) \times \delta(y) \times \delta(z)$.

5. *Noise.* Every technique is limited to some extent by signal-to-noise ratio (SNR). In some cases we shall find that this is a compromise also involving spatial resolution, safety and the rate at which data can be taken and patients scanned.

6. *Artefacts.* In addition to random noise there are systematic effects that create features in the image which are not present in the object. The understanding, identification and minimisation of these requires the coordinated skills of clinician, engineer and physicist.

7. *Safety.* There are real risks that have to be weighed against the potential benefit of a procedure. There are additional safety concerns that arise from public perceptions. The public accepts risk in medicine, and therefore medical safety needs to be considered carefully. Regular maintenance, calibration and safety checks form a part of the disciplined way of life with imaging and therapy equipment. A more general discussion of risk follows later in this chapter.

8. *Cost and expertise.* The cost of making techniques, developed in research laboratories, rapidly available in hospitals is high. However, the limiting factor is expertise. With more qualified staff more use could be made of expensive equipment. The local education and knowledge base is important.

9. *Throughput.* The total time taken to prepare and execute a scan determines the patient throughput, but calibration and shimming procedures can be time consuming. In a clinical context costs are directly related to throughput.

10. *Acceptability.* Some scanning modalities form an unpleasant or forbidding experience for the patient. They may be noisy, claustrophobic, uncomfortable and protracted. Often patients are being scanned at a time when they are already upset or afraid, such as after an accident or as part of a diagnosis for a cancerous tumour.

11. *Calibration.* As discussed above the first level of information in an image comes from comparing one part of an image with another. To learn more from an image we must be able to calibrate the measurements. This will depend in part on measurements and images of known reference samples. In medicine these are termed 'phantoms', usually simple volumes of matter of known composition. In scientific archaeology they are called 'standards'.

To improve registration the patient may be asked to hold his or her breath, but this is only possible for short scans. Alternatively, data may be taken only during a selected phase of the respiratory cycle, but then the time required to complete a scan is correspondingly increased, patient throughput decreased, unit costs raised and the unpleasant experience for the patient lengthened. These points bear upon one another, and there is a continual need to compromise.

1.3   Safety, risk and education

The power of physics to provide information and to diminish real dangers, comes at some risk. This risk needs to be understood, monitored and openly explained to the public at large. Otherwise physics itself generates fear and apprehension. In the early days of a new application, when experience is fragmentary, monitoring poor, and there are no longterm records, safety standards need to be conservative. Later, as knowledge and experience grow, safety standards may be lowered, so that decisions are made in the light of the best information available at the time. Legislation and regulation do not make people feel safe. That only comes with the confidence born of understanding, information, explanation and public education. Unfortunately this is not what normally occurs. Consequently, wrong decisions are made, apprehensions may be increased, and future dangers incurred.

1.3.1   Public apprehension of physics

The impact of physics on society became a public concern with the advent of nuclear weapons. The threat of ionising radiation was used during the Cold War intentionally to frighten people. This fear was not forgotten, and remains to this day to be exploited as an instrument of political or terrorist blackmail, independent of the weight of risk actually involved. Rigid safety standards have not reassured, and the knowledge now available is largely ignored by the media and the general public. An objective of this book is to provide a broad look at such knowledge.

Risks due to ionising radiation should be looked at in the same way as other types of damage, such as laceration and bruising, or tissue over-heating. An important psychological difference is in the ability to sense the effect. Consider magnetic fields, for example. They carry no risks, but in the absence of education on the subject, the general population is naturally frightened by that which is apparently influential, but which cannot be seen or felt. So the popular perception of magnetism includes a fear of the unknown, unrelated to any uncertainty in physics. Magnetism is sometimes seen alongside water-divining[14] and other 'mysteries' that lack a scientific basis.

Nuclear radiation is associated in the public mind with the idea of the run-away chain reaction.[15] Then to the fear inherited from the era of the Cold War were added concerns about accidents in the nuclear power industry. Early accidents were hushed up. Later ones were exaggerated in the media despite subsequent reliable international reports. Both outcomes had perverse effects on public trust that now discourage irresolute politicians from reaching clear policy decisions on the future of nuclear power, in spite of the threat of global warming. The current state of knowledge of the effects of ionising radiation is discussed in chapters 3, 6 and 8. Some conclusions are drawn in chapter 10.

Unfortunately, it is in the earlier days of a new use of physics, when enthusiasm is high, that there is the greatest danger, the least understanding of risks and the least adequate technology available to monitor

[14]Alias dowsing or witching in the USA.

[15]In fact it is extremely difficult to generate a chain reaction, as discussed briefly in chapter 6.

exposure.[16] Later, when procedures for monitoring and control are fully developed, risks can be more precisely evaluated. This suggests that safety levels should be determined in the light of current knowledge, with the implication that normally they may be relaxed subsequently rather than raised, assuming positive experience and improved knowledge. In practice the cumbersome machinery of safety legislation is usually to be observed proceeding in the opposite direction.

There are other cultural pressures. The use of an RF induction loop for surveillance at an airport, though sensitive to the presence of an electrical conductor or magnetic material carried by a passenger, does not provide an image. Is this technically inferior solution chosen because it has been decided that it would be dangerous to expose passengers to X-rays on a routine basis? Or is it because it is thought an unacceptable invasion of privacy to take images of people through their clothes? Was this the right decision? Such broader questions arise in medical physics, archaeology and other areas where the choice between different technologies has to be made in terms of risk, benefit and wider cultural sensitivities. Physicists should know what they are talking about, so that they can contribute to such debates. The media and the public in general have great difficulty in balancing matters of benefit, risk and acceptability. Consequently those charged with making decisions too often prefer to conceal debate rather than discuss matters openly. When this becomes known, faith in the science suffers unjustly. Ultimately, our ability to survive on this planet may depend on improving the confidence and communication that scientists have with the public about what science can do for everybody. The only solution is a combination of education and open debate.

1.3.2   Assessing safety

Evolution of acceptable levels of risk

We take a fresh look at safety. This is a subject on which much is written and much assumed. What is important is what is known in principle, and what can be monitored in a particular instance. Procedures should evolve as knowledge and experience grow and as instrumentation improves. Regulation should follow such development, not lead it.

The exposure of the population to sources of energy, whether in the course of probing and imaging or otherwise, may be looked at objectively in terms of the following levels:

◇ the detectable level at which dose may be reliably measured and monitored;

◇ the background level within which humans have lived and evolved naturally;

◇ the damage level above which long-term damage is possible as informed by reliable epidemiological study or current knowledge of causative mechanisms;

[16] An early example of a poorly monitored risk was the use of X-rays in the so-called pedoscope. This was used in children's shoe shops in the 1940s and 1950s. The author recalls his mother's enthusiastic reception of the display on a fluorescent screen showing that his toes and feet were well matched in size to the newly acquired pair of shoes. It is interesting to note that:

1 the scan was applied regularly to all children, not just a minority with a serious problem;

2 the standard high street shoe shop would not have had anyone with the skill to calibrate or maintain the equipment regularly;

3 in spite of the fact that the radiation levels involved would now be considered unsafe, there does not appear to be any evidence that either he or any of his contemporaries suffered any ill effects.

◇ the lethal level at which breakdown of the functioning of cells and organisms causes their early death.

Doses at the background level carry no risk, but doses that cannot be monitored reliably are of concern. Thus diagnostic doses below the damage level can be considered safe provided that dose monitoring is reliable. Where this implies a narrow window, a significant safety problem exists. As discussed in chapters 8 and 10 cancer therapy involves the delivery of localised lethal doses. The problem is to avoid damage to peripheral healthy tissue associated with the difference between the lethal level and the damage level, a window in which some cells will be seriously affected but not all.

The damage level includes special cases, such as metal implants in MRI patients, or the role of iodine in thyroid cancer in the radiation environment. Generally establishing confidence in a value for the damage level requires continuing high quality research, record keeping and vigilance over long periods.

Thus there should be an expectation that accepted safety levels may be relaxed as knowledge increases, confidence builds and standards of monitoring and instrumentation improve. Often this is not what happens. Safety legislation is often tardy and driven by the conservative ALARA principle, 'as low as reasonably achievable'. This is ill-founded for the following reasons:

◇ it does not relax with improving knowledge;

◇ by concentrating only on achievable levels it ignores the need to balance absolute risks against one another;

◇ it does not take account of the possibility that achievable levels may be inadequate, that is to say of questionable safety;

◇ it does not take into account the possibility that achievable levels may be factors of 10 lower than background levels, and therefore irrelevant.

With ALARA, technical improvements tend to encourage ever tighter safety standards on the basis that 'you cannot be too safe'. This statement is dangerous. Responsible living depends on a continually updated balance between benefits and risks. In an overcautious safety-legislated society the largest and latest risks, for which knowledge is least developed, are incurred preferentially, and older dangers, which may be well understood and far less of a threat, are avoided. The result is that the greater risk is incurred.

The relative threats of global warming and nuclear power are the prime example. Another is the balance between the risks of ionising radiation, MRI and ultrasound in medicine. We return to these comparisons in the course of the book and again in the final chapter where conclusions are drawn.

Thermal, resonant and disruptive damage

By interacting with materials, all probing and imaging methods deposit energy to a greater or lesser extent. The most pervasive type of damage caused by this energy is unspecific and thermal. However, it is also possible that the energy absorption is specific, that is associated with a resonance in a non-thermal way. A third possibility is that the energy is absorbed in a disruptive and inelastic fashion.

An unspecific thermal dose may be characterised by the local tissue temperature rise. In time, such an increase may be dispersed, for instance by convection or evaporation. It is well established that an increase of 1°C in the local temperature of human tissue above its norm causes no serious damage, but that a rise of 2°C or more can be harmful. Temperature excursions at this level occur in tissue as a result of normal exercise or a mild fever. The corresponding tolerable specific energy absorption rate (SAR) that can be dispersed by tissue varies[17] between 2 and 10 W kg^{-1}. At the other extreme a temperature increase of 21°C for a duration of 1 s causes coagulative necrosis.[18] So for temperature changes, referring to the levels defined above, we may say that the background level is 1°C, the damage level is 2°C and the lethal level is 21°C. The monitoring level presents a problem, for it is difficult to measure localised temperature excursions *in vivo* with precision.

To the extent that the rate of energy deposition or damage in tissue is a non-linear function of the incident energy flux, monitoring and safety levels need to be more tightly defined. For example, if the damage depends on the third power of the energy flux and the energy flux was only measured to a precision of 10%, the damage would be uncertain at the level of 30%. So significant non-linearity in tissue response imposes a need for more carefully defined monitoring procedures.

There is always the question whether more serious damage might be inflicted if energy is delivered non-thermally, for instance via a particular local atomic or molecular resonance. Whether this is a problem depends on the strength of the coupling of the resonance to the probing field and to the other degrees of freedom of the tissue. These questions have to be considered in the physics of each case.

The problem of high magnetic fields is exceptional, as no energy is deposited unless the field changes or there is some movement.[19] Because no firm evidence of damage by high steady fields has been reported, the maximum field considered safe in clinical use for MRI has been raised to 4 tesla. This change has a direct benefit on the signal-to-noise ratio that may be achieved in MRI, or, equivalently, on the speed with which such scans can be made.[20] This is an unusual case in which increased knowledge and experience has lead to an appropriate relaxation of safety levels.

Distinct from either thermal or resonant energy deposition is the damage caused by disruption. An example is abrasion, or cuts and bruises. Given time such damage heals. In this case the questions are 'What is the healing time?' and 'How much abrasion can be tolerated within

[17] These figures are quite high because the body responds dynamically in various ways to reduce any temperature increase. They cover a range because of the variation in this cooling for different parts of the body. For example, cooling is poor in the eyes and the testes.

[18] This medical description means the cells are cooked or melted at 56°C, a terminal condition for functioning tissue.

[19] Strictly, this refers to movement of conductors, currents or magnetic materials, but all materials experience induced magnetisation or eddy currents to some degree.

[20] These are discussed in chapter 7.

that time without incurring permanent damage?'. The energy deposited by ionising radiation causes such rupture of atoms, molecules and cells. The debris tends to be confined to the immediate local path followed by the individual charged particles, including the secondary ones created by the absorption of photons. The extent to which such damage gets repaired, by cell and bio-molecule reproduction, is an important matter addressed in chapter 6. This gives a risk of long-term damage that depends quite non-linearly on the dose integrated over repair time. The provision of effective cancer therapy turns on an appreciation of this. The naive assumption that the dependence is linear without a threshold is called the linear no-threshold (LNT) model and is not credible.

There are also important risks of a more mundane variety. Examples are burns caused through contact with metal probes which become overheated during ultrasound scanning, and impacts by metal objects accelerated by the high fringe B-field of an MRI scanner.

The principal regulatory bodies are given below. Their websites may be consulted for more details. The facts given there are usually reliable, but, regrettably, the ALARA principle and the discredited LNT model remain deeply embedded in their application in some cases.

Look on the Web

Look at the book website at
   www.physics.ox.ac.uk/users/allison/booksite.htm

Find links to simple ideas and developments on gravitational fields
   gravitational waves LIGO LISA
   gravitational anomalies

Examine safety and other regulatory websites:

IAEA, the International Atomic Energy Agency

OECD NEA, the Nuclear Energy Agency is a specialised organisation within the international Organisation for Economic Cooperation and Development.   Find recent reports on Chernobyl at   *www.nea.fr/html/rp/chernobyl/chernobyl.html*   and links to World Health Organisation Reports

SRP The UK Society for Radiological Protection

NRPB UK National Radiological Protection Board (became part of HPA after April 2005), including the text of Report, Vol. 14, No. 2 (2003) on RF fields.

HPA UK Health Protection Agency, including ultrasound. Find *Ionising Radiation Exposure of the UK Population: 2005 Review*, and also search for references to ultrasound

ICNIRP International Commission on Non-ionising Radiation, including the text of Magnetic Resonance 2004, *Health Physics*, 87, 197

USFDA US Food and Drug Administration

NICE UK National Institute for Health and Clinical Excellence

BMUS British Medical Ultrasound Society

RERF LSS Radiation Effects Research Foundation, Life Span Study, for more information on the studies of Hiroshima and Nagasaki survivors.

HPS The Health Physics Society, for example *hps.org/documents/radiationrisk.pdf*

Magnetism and magnetic resonance

2

in which are developed the fundamental physics of the single isolated magnetic dipole, materials as ensembles of such dipoles, their resonant behaviour, and magnetic field measurements. This provides the underlying physics for chapter 7.

2.1 An elemental magnetic dipole 21

2.2 Magnetic materials 27

2.3 Electron spin resonance 34

2.4 Nuclear magnetic resonance 37

2.5 Magnetic field measurement 47

Read more in books 53

Look on the Web 53

Questions 54

2.1 An elemental magnetic dipole

Magnetic forces are weaker than electric ones to which they are a relativistic correction. Current loops, both microscopic and macroscopic, form the basic source of magnetic flux $\boldsymbol{B}$. The torque on a magnetic dipole $\boldsymbol{\mu} = I\boldsymbol{S}$ due to a current I in a loop of area $\boldsymbol{S}$ in a field $\boldsymbol{B}$ is $\boldsymbol{\Gamma} = \boldsymbol{\mu} \times \boldsymbol{B}$. A charge Q of mass m circulating in an orbit has both a magnetic moment and an angular momentum. The ratio of these is $\gamma = gQ/2m$, where the gyromagnetic ratio g in this case is unity. In a classical view such a magnetic moment in a uniform B-field precesses with the Larmor frequency, $\omega_L = \gamma B$. In a quantum view the magnetic substates m_j of an atom are split in energy from one another by $\hbar\gamma\Delta m_j B$. The dipole selection rule $\Delta m_j = \pm 1$ ensures that classical and quantum frequencies are the same. Electron spins, nuclear spins and muon spins behave in similar ways to orbital motion but with values of g different from 1. The effective value of g for an electron with both orbital and spin angular momentum is g_j given by the Landé formula.

2.1.1 Laws of electromagnetism

Maxwell's equations in the absence of materials

The basic laws of electromagnetism are known as Maxwell's equations. In their differential form in the absence of materials these link the fields $\boldsymbol{E}$ and $\boldsymbol{B}$ to their sources, the charge and current densities ρ and $\boldsymbol{J}$, through the constants ε_0 and μ_0:

$$\mathrm{div}\,\boldsymbol{E} = \rho/\varepsilon_0 \tag{2.1}$$

$$\mathrm{div}\,\boldsymbol{B} = 0 \tag{2.2}$$

$$\mathrm{curl}\,\boldsymbol{E} = -\frac{\partial \boldsymbol{B}}{\partial t} \tag{2.3}$$

$$\mathrm{curl}\,\boldsymbol{B} = \mu_0 \boldsymbol{J} + \varepsilon_0 \mu_0 \frac{\partial \boldsymbol{E}}{\partial t}. \tag{2.4}$$

Equations 2.1 and 2.2 are both known as Gauss's law. Equation 2.3 is Faraday's law of electromagnetic induction describing the electric field generated by a changing B-field. Equation 2.4 is Ampere's law for the B-field generated by a current density J and displacement current density $\varepsilon_0 \partial E / \partial t$. The conservation of charge relates the current flowing out of unit volume to the rate of change of charge density

$$\mathrm{div} J = -\frac{\mathrm{d}\rho}{\mathrm{d}t}. \tag{2.5}$$

Taking the divergence of both sides of equation 2.4 shows that the displacement current density term is necessary for the consistency of Ampere's law with the conservation of charge. As shown in books on electromagnetism, in the absence of the source terms wave solutions to these equations exist with velocity $c = 1/\sqrt{\varepsilon_0 \mu_0}$, the velocity of light in vacuum.

2.1.2   Current loop as a magnetic dipole

Origin of magnetism

In the first observations of magnetism a sample of iron ore was suspended in the magnetic field of the Earth. The torque on this ore formed a compass which indicated the direction of the Earth's B-field, that is the direction of the magnetic pole.[1] This continues to be a very important aid to navigation.[2]

Equation 2.4 shows that the source of the magnetic B-field is the current density J; that is, magnetism arises from the movement of electric charges. Magnetic interactions are relativistic corrections to electric interactions and so magnetic energies and forces are expected to be smaller by a factor of order v^2/c^2. For atomic electrons this factor is typically 10^{-4}.

This expectation is rather misleading. Electric fields achieve rather complete cancellation except on an atomic scale, while magnetic effects often do not. There are two reasons why magnetic forces do not average out as effectively as electric ones, namely thermodynamics and spin.

The density of electric charge is often balanced, although its flow may not be. Any significant imbalance of charge density is suppressed by the large energy that this would generate. This contrasts with the smaller magnetic energy associated with a significant net charge flow. So on general thermodynamic grounds magnetic forces and net current flows are not suppressed by cancellation to the same extent as electric forces and net charge densities.

The electron spin which is equivalent to a circulating charge or current does not balance out effectively in certain kinds of atoms. Because the electron spin is $\frac{1}{2}$, the net angular momentum cannot be zero unless electrons can pair their angular momenta.[3]

Thus magnetism arises from materials that contain these microscopic atomic currents and spins, as well as from currents provided by macroscopic current loops. However, these microscopic atomic magnets and

---

[1] A north (+) magnetic pole is attracted towards the North. Since opposite sign magnetic poles attract, the North magnetic pole of the Earth is actually a south (−) magnetic pole. This is consistent with the direction of the B-field which by convention points from + to −, as for an electric field.

[2] There are two conditions for a compass to work. Firstly the electron has to have spin $\frac{1}{2}$, so that the unpaired electron spin of iron gives it spontaneous magnetisation. Secondly the static B-field due to the Earth's core must have a long enough range to reach the Earth's surface. It may be shown that this requires the Compton wavelength of the photon $\hbar/m_\gamma c$ be as large as an Earth radius. Thus, for a compass to work, m_γ, the mass of the photon, must be less than about 10^{-49} kg.

It has been suggested that the discovery of America depended on the spin of the electron and such an upper limit to the mass of the photon, although Christopher Columbus did not know that!

[3] Such lack of pairing occurs in particular for the transition elements (such as iron, cobalt, nickel or manganese), the lanthanides (such as cerium, lanthanum or gadolinium) and the actinides. These have partially filled shells of electrons with high angular momentum.

the macroscopic magnets of currents are fundamentally the same from an electromagnetic viewpoint.

Macroscopic current loop

The current in such a loop arises from a combination of any directly applied voltage source, such as a battery, and the effect of changing B-fields through Faraday's law of electromagnetic induction, equation 2.3. In the absence of any applied voltage, the electromotive force V induced in a loop of area S in an externally applied changing B-field is

$$V = -\frac{\partial \boldsymbol{B}}{\partial t} \cdot \boldsymbol{S}. \tag{2.6}$$

Such a loop can be described in terms of its equivalent components shown in Fig. 2.1a, a series resistance R, capacitance C and magnetic inductance L. The current in the loop I obeys the differential equation

$$\frac{\partial V}{\partial t} = L\frac{\partial^2 I}{\partial t^2} + R\frac{\partial I}{\partial t} + CI, \tag{2.7}$$

where each term is the time derivative of the voltage across each component. Driven by a harmonic external B-field of frequency ω the current in the steady state is then given by[4]

$$I = V/Z = \frac{-\mathrm{j}\omega \boldsymbol{B} \cdot \boldsymbol{S}}{R + \mathrm{j}\omega L + 1/(\mathrm{j}\omega C)}. \tag{2.8}$$

The electrical impedance Z is given by

$$Z = R + \mathrm{j}\omega L + \frac{1}{\mathrm{j}\omega C}. \tag{2.9}$$

Resonant circuit

The dependence of the phase, the imaginary and real parts of this current on frequency are plotted out in Fig. 2.1b. They show the general characteristics of a resonance, namely a peak in the imaginary part at the resonant frequency ω_0, a phase that changes by π as the frequency passes through resonance, and a real part that changes sign at resonance.[5] The resonant frequency is

$$\omega_0 = \frac{1}{\sqrt{LC}}. \tag{2.10}$$

The full width of the peak in the imaginary part at half-maximum energy is $\gamma_r = R/L$, and the Q of the resonance[6] is ω_0/γ_r. At resonance the real part of the current amplitude passes through zero and the imaginary part is maximum. Note that below resonance the current produces a B-field that opposes the applied field within the loop and increases it outside the loop. Above resonance the situation is reversed.

This may be described as a classical example of magnetic resonance.

Next we consider the torque and force on such a small current loop in a steady B-field.

[4]The time development is assumed to be of the form $\exp(\mathrm{j}\omega t)$, see the discussion of the choice of complex sign convention in appendix A.

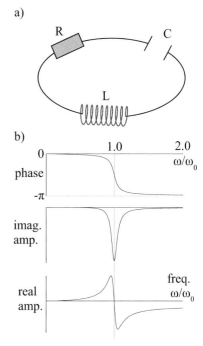

Fig. 2.1 a) An electrical circuit loop. b) The phase, real and imaginary parts of the induced current in the circuit as a function of the frequency of the external applied B-field, in terms of the resonant frequency $\omega_0 = 1/\sqrt{LC}$. The value of Q is taken as 10 for purposes of illustration.

[5]The sign of this phase depends on the phase convention used.

[6]This may be derived from the definition: $2\pi\times$ the energy stored divided by the energy lost per cycle.

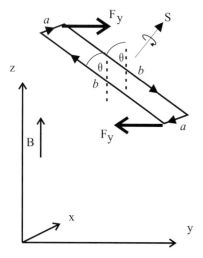

Fig. 2.2 A rectangular coil carrying a current I placed in a uniform field B.

[7]The forces F_x on the two b sides are also equal and opposite with magnitude $IB_z b \sin\theta$ but are not displaced and do not contribute a torque.

Torque on a magnetic dipole in a B-field

Consider a small rectangular coil carrying a current I and placed in a uniform field B parallel to the z-axis. The side of length a is parallel to the x-axis and the side of length b lies in the yz-plane at an angle θ to the z-axis, as shown in Fig. 2.2. We calculate the force and torque on this coil, and then express the result in general terms. The force per unit length on a conductor carrying a current I in a B-field is

$$\boldsymbol{F} = \boldsymbol{I} \times \boldsymbol{B}. \tag{2.11}$$

So the forces on the a sides of the coil are equal and opposite, $F_y = \pm IaB$. Although there is no net force along y, the two forces are displaced by a distance $b\cos\theta$ along z. They therefore form a torque in the x-direction of magnitude $IabB\cos\theta$. So there is a net torque which can be expressed with a cross product as[7]

$$\boldsymbol{\Gamma} = I\boldsymbol{S} \times \boldsymbol{B} \tag{2.12}$$

where $\boldsymbol{S}$ is equal in magnitude to the area ab and normal to it with a right hand sense determined by the direction of the current I as shown. This is a general result. The torque depends only on the vectors $\boldsymbol{S}$ and $\boldsymbol{B}$, but not on other details of the geometry.

In SI units the magnetic dipole moment of the loop is defined as

$$\boldsymbol{\mu} = I\boldsymbol{S}. \tag{2.13}$$

The torque is then

$$\boldsymbol{\Gamma} = \boldsymbol{\mu} \times \boldsymbol{B}. \tag{2.14}$$

The potential energy of the magnetic dipole in the B-field is found by integrating the torque over angle,

$$W = \int_0^\theta \mu B \sin\theta' \mathrm{d}\theta' = \boldsymbol{\mu} \cdot \boldsymbol{B}. \tag{2.15}$$

Gyromagnetic ratio

Such a current loop also possesses a mechanical angular momentum $\boldsymbol{L}$ by virtue of the rotating charge carriers which have mass m and charge Q. If the orbital velocity of the charge is v, then $\boldsymbol{L} = m\boldsymbol{v} \times \boldsymbol{r}$ where we have now assumed that the loop is circular with radius r for ease of calculation. The ratio of the magnetic moment of the rotating charge to the angular momentum of its orbiting mass is a constant as we now show.

The current I in a loop length $2\pi r$ is equivalent to a total circulating charge $Q = 2\pi r I/v$. From equation 2.13 the magnetic moment is then $\mu = Qvr/2$. The gyromagnetic ratio is then defined as

$$\gamma = \mu/L = Q/2m. \tag{2.16}$$

Since the angular momentum L is quantised in units of $\hbar$, the size of the magnetic moment is fixed to be in multiples of $\hbar\gamma$. This classical derivation gives the correct answer for charges in orbital motion as determined by quantum mechanics. The magnetic moment of an electron with $L = l\hbar$ and quantum number $l = 1$ by virtue of its orbiting motion is defined as the Bohr magneton,

$$\mu_B = \hbar\gamma = e\hbar/(2m_e) = 5.788 \times 10^{-5} \text{ eV T}^{-1}. \qquad (2.17)$$

Similarly the moment of a proton charge with $l = 1$ by virtue of its orbiting motion is defined as the nuclear magneton,

$$\mu_N = e\hbar/(2m_p) = 3.152 \times 10^{-8} \text{ eV T}^{-1}, \qquad (2.18)$$

which is three orders of magnitude smaller than the Bohr magneton.

For magnetic moments arising from quantum spins the classical picture described above fails and the magnetic moment is different. The factor by which it differs is denoted by the g-factor. For the electron spin, $g = 2$ as determined by the Dirac equation for an elementary point-like spin-$\frac{1}{2}$ fermion, ignoring a tiny correction given by quantum electrodynamics (QED). On the other hand a proton is a composite particle with an internal structure that is not described by the Dirac equation. In its case the value of $g = 5.59$ comes from experiment.

Thus with a spin $\frac{1}{2}$ and $g = 2$ the electron spin magnetic moment is μ_B. The magnitude of the proton spin magnetic moment is $\mu_p = 2.79\mu_N$.

2.1.3   The Larmor frequency

Classical precession

Consider the motion of a current-carrying coil or magnetic moment $\boldsymbol{\mu}$ in a uniform B-field. As a result of the torque, equation 2.14, the rate of change of angular momentum $\boldsymbol{L}$ of the magnetic dipole is

$$\frac{d\boldsymbol{L}}{dt} = \boldsymbol{\mu} \times \boldsymbol{B}. \qquad (2.19)$$

Using equation 2.16 this may be written more usefully as

$$\frac{d\boldsymbol{\mu}}{dt} = \gamma\boldsymbol{\mu} \times \boldsymbol{B}. \qquad (2.20)$$

If $\boldsymbol{B}$ is constant then $d\boldsymbol{\mu}/dt$ is always perpendicular to $\boldsymbol{B}$. Being also perpendicular to $\boldsymbol{\mu}$, the length of $\boldsymbol{\mu}$ is constant. So the end of the $\boldsymbol{\mu}$-vector executes circular motion in the plane perpendicular to $\boldsymbol{B}$. In other words, $\boldsymbol{\mu}$ precesses about $\boldsymbol{B}$ with angular frequency $\omega_L = \gamma B$ as shown schematically in Fig. 2.3a. This is called the Larmor precession.

Larmor precession in quantum mechanics

This picture of a magnetic moment with a well-defined direction in space is essentially classical, and incorrect. In the quantum description the direction of the angular momentum of a state, and therefore its magnetic

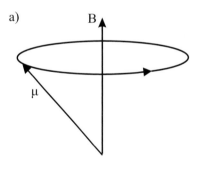

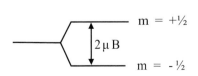

Fig. 2.3 a) A classical view of a magnetic moment μ precessing in a B-field. b) The energy of a spin-1/2 magnetic moment μ split by a B-field as seen in quantum mechanics.

moment, may not be well-defined. Only the projections parallel with the B-field are stationary. In particular, for a spin-$\frac{1}{2}$ particle, the angular momentum components $m\hbar$ with $m = \pm\frac{1}{2}$ give magnetic moments $\pm\frac{1}{2}\hbar\gamma$ aligned with the B-field. The potential energy of the moment in these two quantum states differs by $\hbar\gamma\Delta mB$, see Fig. 2.3b. In the quantum picture it is these two states with $\Delta m = 1$ and a difference of azimuthal rotational phase velocities of γB whose interference replicates the Larmor precession that provides a correspondence with the classical picture.

Spin–orbit coupling

In general an atomic electron has both orbital angular momentum $l\hbar$ and spin angular momentum $\frac{1}{2}\hbar$. In the absence of a significant external B-field, the magnetic interaction of these creates an energy difference $Al \cdot s$, the spin–orbit interaction. Then electron states are eigenstates of $l \cdot s$. Defining the total electronic angular momentum (in units of $\hbar$) as $j = l+s$, we have $j^2 = l^2+s^2+2l\cdot s$, so that states that are eigenstates of $l \cdot s$ are also eigenstates of j^2. In the classical or vector model picture in the absence of an applied field the spin magnetic moment precesses in the B-field of the orbital moment (and vice versa). Thus the z-components, m_s and m_l, are not constant, although their sum m_j is.

a)

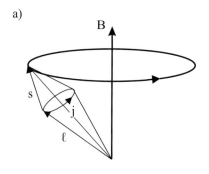

b)

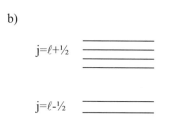

Fig. 2.4 a) A classical view of the vectors s and l precessing rapidly about j as the latter precesses slowly about the direction of a weak magnetic field B. b) A quantum view of two states,

$$j = l + 1/2 \text{ and } j = l - 1/2,$$

split apart by spin–orbit coupling. In a weak field the upper state has split into m_j substates, and the lower state likewise. A weak field means that this splitting is significantly smaller than the spin–orbit splitting.

In the presence of a weak external field B_0 the total j precesses slowly about the B-field axis with constant component m_j. This classical picture is illustrated in Fig. 2.4a. The weakness of the external field relative to the internal spin–orbit coupling field is reflected in the slow precession of j about B_0, compared with the precession of l and s about j.

In the quantum picture, see Fig. 2.4b, the perturbation energy $\mu_j \cdot B_0$ is assumed small compared with $Al \cdot s$. The eigenstates in the presence of the B-field are still the same as in the absence of B, provided that the z-axis is chosen along B. In this model the magnetic moment vector is the vector sum of its two component parts $\mu_B(2s + l)$ where $g = 2$ has been taken for the spin. However, it is only the projection of this along j that does not average quickly to zero. Therefore the effective g-factor, the magnetic moment divided by the magnitude of $j\mu_B$, is

$$g_j = \frac{\mu_B(2s + l) \cdot j}{j(j + 1)\mu_B} = \frac{3j(j+1) + s(s+1) - l(l+1)}{2j(j+1)} \quad (2.21)$$

where we have taken the eigenvalue of each angular momentum squared operator to be of the form $j(j + 1)\hbar^2$. This g_j is the Landé g-factor.

If the external field is strong the $\hbar l$ and $\hbar s$ simply precess around B_0 ignoring the weaker mutual interaction. Then m_l and m_s are separately constants of the motion, and j is not. This is the Paschen–Back effect. In a quantum picture of the states in a strong field the separation of the components in each multiplet becomes of the same order as the separation of the multiplets themselves (see Fig. 2.4b). Then the eigenstates involve the diagonalisation of the Hamiltonian involving all states with the same total angular momentum. We do not pursue this here because we shall not need it. However, for the analogous case of hyperfine structure, the strong field case will be needed, as described later.

Precession by electrons, nuclei and muons

Frequency is a parameter which is easily and accurately measured. Many technologies depend on measurements of the precession frequency of magnetic moments in B-fields. The moments of most interest are the spin moments of the electron, the nucleus (especially the proton) and the muon. Large parts of this book are concerned with the precession frequency of electron spin as measured with electron spin resonance (ESR) and of nuclear spin as measured with nuclear magnetic resonance (NMR). Investigations with the muon spin resonance (μSR) are largely of interest to academic research programmes outside the scope of this book.

2.2   Magnetic materials

Materials exhibit magnetic behaviour because of the moments of their constituent electrons and nuclei. These moments are influenced by the field and by their mutual interaction. They may exist without the field (electronic paramagnetism and nuclear magnetism) or they may be induced by the field (diamagnetism). Nuclear magnetic moments are strongly influenced by the electronic magnetic moments. Within the same atom this is termed hyperfine structure, but this influence is also important at much longer range.

2.2.1   Magnetisation and microscopic dipoles

Macroscopic magnetisation

Now we look at the magnetic behaviour of a whole ensemble of atoms and molecules in the form of a magnetic material. In this case there are many magnetic dipoles in a unit volume. These do not all point in the same direction. The net magnetic moment density $\boldsymbol{M}$, the magnetisation, is the vector sum of these. At finite temperature this vector, though composed of quantum vectors, behaves classically and may have simultaneously well-defined projections on all three cartesian axes. The magnetisation is both a source of B-field, call it $\boldsymbol{B}'$, and is itself influenced to a greater or lesser extent by the net B-field. The net B-field is made up of the externally applied field $\boldsymbol{B}_0$ and the field $\boldsymbol{B}'$. These two sources of B-field are not different in principle. However, traditionally $\boldsymbol{B}_0$ is written in terms of the H-field[8] $\boldsymbol{B}_0 = \mu_0 \boldsymbol{H}$ and then

[8]The H-field is only introduced here in order to relate the discussion to standard treatments of electromagnetism. We eliminate it and revert to a discussion in terms of $\boldsymbol{B}$, $\boldsymbol{M}$ and the magnetic susceptibility χ.

$$\boldsymbol{B} = \boldsymbol{B}_0 + \boldsymbol{B}' = \mu_0(\boldsymbol{H} + \boldsymbol{M}), \tag{2.22}$$

where $\mu_0 = 4\pi \times 10^{-7}$.

If the magnetisation of the medium is in thermal equilibrium, linear and isotropic, then all three fields ($\boldsymbol{B}$, $\boldsymbol{H}$, $\boldsymbol{M}$) are proportional and collinear. This is usually written

$$\boldsymbol{B} = \mu_r \mu_0 \boldsymbol{H} = (1 + \chi)\mu_0 \boldsymbol{H}. \tag{2.23}$$

We can eliminate (and then ignore) the H-field and express μ_r the relative magnetic permeability and χ the magnetic susceptibility of a material in terms of the magnitudes of B and M by

$$\mu_r = \left(1 - \frac{\mu_0 M}{B}\right)^{-1} \text{ and } \chi = \mu_r - 1. \tag{2.24}$$

Both χ and μ_r are properties of the material and dimensionless in SI units. Alternatively susceptibility may be given per unit density or per mole with consequential change of dimensions. Some examples of χ' the magnetic susceptibility per mole are given in table 2.1. Often these are small compared with unity.[9] Materials with positive and negative values are called paramagnetic and diamagnetic, respectively. These are static values. As a function of frequency some materials show resonant behaviour similar to the circuit shown in Fig. 2.1. We should emphasise that the use of χ and μ_r is appropriate to linear isotropic materials in thermal equilibrium.

In the following applications we shall also be concerned with non-equilibrium magnetic states. In that case $\boldsymbol{B}$ and $\boldsymbol{M}$, although influencing one another, are to be thought of as independent fields which may vary with time and may not be collinear. We may think in terms of

$$\boldsymbol{B} = \boldsymbol{B}_0 + \mu_0 \boldsymbol{M}, \tag{2.25}$$

while remembering that this is of limited use since the only B-field that may be sensed is the net $\boldsymbol{B}$.

How does this magnetic response by matter to an applied B-field come about? There are three distinguishable effects which are due, in order of decreasing magnitude, to pre-existing electronic magnetic dipoles, to induced electronic magnetic dipoles and to pre-existing nuclear magnetic dipoles.

Electronic paramagnetism and ferromagnetism

Pre-existing free electronic magnetic dipoles arise from the net angular momenta of electrons. While the angular momentum of an even number of electrons can pair up to give zero, an odd number of electrons can never do so on account of their intrinsic spin $\frac{1}{2}$. In some materials, although the number of electrons is even, pairing is not favoured. The diatomic oxygen molecule is an important example of this. Generally transition and rare earth elements with partially complete inner electron

[9] As a result μ_r is usually positive. However, this is not required. If χ is less than -1, μ_r may be negative. This unusual case is discussed in section 10.5.

Table 2.1 Values of the static magnetic susceptibility per mole $\chi'/10^{-6}$ cm^3 mol^{-1} for various materials at room temperature, where $\chi' = \mu_0 M/B \times m_w 10^3/\rho$ with m_w the molecular weight and ρ the SI density.

Al	$+16.5$	Ba	$+20.6$	Be	-9.0
He	-2.0	Ar	-19.3	Xe	-45.5
F_2	-9.6	Cl_2	-40.4	I_2	-90
C	-6.0	N_2	-12	O_2	$+3449$
Cu	-5.5	CuCl	-40	$CuSO_4$	$+1330$
Cr	$+167$	CrO_3	$+40$	Cr_2O_3	$+1960$
H_2	-4.0	NH_3	-16.3	H_2O	-12
FeO	7200	Ge	-11.6	Gd	185000

shells also have a net electronic angular momentum J, although this depends on the chemical state and degree of ionisation.[10] In table 2.1 some examples are given. The first row is a selection of metals. The second, third and fourth rows show typical elements with paired electrons and a diamagnetic susceptibility rising steadily with Z, with the exception of oxygen. The fifth and sixth rows show that the chemical form of transition elements can affect their magnetic properties. The last two rows include the exceptional behaviour of gadolinium and the large value for ferrous oxide. Transition metals that are ferromagnetic, such as iron and cobalt, cannot be given a value because their magnetic behaviour is not linear.

Unpaired electron angular momenta also occur in chemical radicals and other non-equilibrium molecular states created by irradiation and other sources of excitation. In solids at room temperature these may be metastable. The density of such free magnetic moments is important in archaeology and radiation dosimetry as a frozen record of past irradiation.

In an applied B-field any pre-existing electronic moments are preferentially orientated, causing an increase in the net B-field and a positive value of χ. Atoms and molecules with net angular momentum $J\hbar$ have $2J + 1$ states with different values of m_J. In a field B these are equally spaced in energy as $g_J\mu_B m_J B$. To make things simple we consider the case $j = \frac{1}{2}$ and $g = 2$ as an example. There are two states for each spin with energies $+\mu_B B$ and $-\mu_B B$ in the B-field, so that the net energy difference is $E_{12} = 2\mu_B B$.

According to Boltzmann's theorem of classical statistical mechanics the ratio of populations of two states, N_2 and N_1, in thermal equilibrium at temperature T with an energy difference E_{12} is[11]

$$\frac{N_2}{N_1} = \exp(-E_{12}/k_B T), \tag{2.26}$$

taking due account of any degeneracy, where k_B is the Boltzmann con-

[10] We loosely follow the convention that lower case j, l, s refer to the quantum state of single electrons and upper case J, L, S refer to the net angular momentum of several electrons in an atom. To be general in the following we frequently use upper case letters while giving simple single electron examples. More complex examples of electron coupling would be out of place in this book.

[11] With the exception of electrons in metals quantum statistics is not relevant at room temperature in our context. It only matters whether the atoms are fermions or bosons if the temperature is much lower and the density is high, as in liquid helium or in a Bose–Einstein condensate.

stant. Thus, if the spins do not interact the population ratio of these states depends on the temperature T as $\exp(-2\mu_B B/k_B T)$. The fraction of spins aligned is only $\mu_B B/k_B T = 2.3 \times 10^{-3}$ in a B-field of 1.0 T at room temperature. This shows that in the absence of cooperative effects the paramagnetic susceptibility is small and very dependent on temperature.

The mutual interaction of the moments can be significant at finite density, and this can affect the relative population very dramatically. In these circumstances Boltzmann's theorem is still applicable but the energy has a large term depending on the mutual orientation of the moments. At a sufficiently low temperature the moments become orientated just by their own field. Then all the spins line up and ferromagnetism is observed. That is, a net macroscopic magnetic moment may be observed, even in the absence of an externally imposed B-field. The temperature below which this cooperative behaviour occurs is known as the Curie temperature.

Electronic diamagnetism

If there are no such pre-existing microscopic magnetic dipoles there is no such first order response and the magnetisation is determined by the second order effect known as diamagnetism. This response of materials to a B-field comes from the current induced in the eigenstates of otherwise symmetrical atoms and molecules. By Faraday's law this induced current opposes the applied field leading to a reduction of the net B-field. Since it affects the eigenstates themselves rather than their population, it does not depend on temperature and occurs for all materials, although obscured by the larger paramagnetic effect, if present.[12] This is responsible for the negative values of χ given in table 2.1. These are usually small although bismuth is an exception.

Metals do not conform to this simple picture. For some there is a small paramagnetic response because the magnetic field generates an energy difference between the Fermi distributions for different spins. This results in a small spin asymmetry in the population of the conduction band which in turn is responsible for a small paramagnetic effect. This is of the same order as the diamagnetic response so that the net susceptibility is small but varies in sign from one metal to another.

Nuclear paramagnetism

The contribution to magnetic properties made by atomic nuclei is on a much smaller scale. In a magnetic field the magnetic moment of a nucleus with spin[13] behaves in a similar way to an unpaired electron magnetic moment. However, its gyromagnetic ratio is three orders of magnitude less, and this is reflected in its energy and Larmor frequency. But what is the value of the nuclear spin, I? All nuclei with an even number of protons and an even number of neutrons are known to have zero spin, a fact rationalised, if not explained, in the shell model of nuclear structure. A majority of stable nuclei, 155 in fact, have such even

[12] A very important exception is hydrogen in water. Because the hydrogen-bond structure of water changes continuously with temperature, the electron wavefunction varies also. Therefore the electronic susceptibility of water, and with it the NMR chemical shift, is temperature sensitive. The shift is 0.01 ppm per degree C. This causes a temperature dependent phase shift which is measurable in MRI. This is becoming important for thermal mapping in ultrasound therapy, as noted in chapter 9.

[13] By 'nuclear spin' in this context we mean the total angular momentum of a nucleus. Discussed in terms of nuclear structure this angular momentum may be composed of a combination of the orbital motion of protons within the nucleus together with the intrinsic spin angular momenta of both protons and neutrons. However, we are not concerned here with these details of nuclear physics.

numbers and zero spin. With the exception of hydrogen, this includes almost all those with a particularly high natural abundance, for example ^{16}O, ^{4}He, ^{12}C, ^{40}Ca, ^{56}Fe, ^{28}Si. The remaining 107 stable nuclei have non-zero spin[14] and several of these are given in table 2.2 with the values of their Larmor precession frequency. Hydrogen is the most important entry in this table, because it is abundant and its frequency is highest. Hence hydrogen nuclei tend to give the strongest and most useful magnetic signals.

2.2.2 Hyperfine coupling in B-field

Atomic hydrogen in zero B-field

Because nuclear magnetic moments are so small their effect on one another is negligible, and there is no nuclear equivalent of ferromagnetism. However, the behaviour of nuclear spins is heavily influenced by nearby electrons. The B-field which the nucleus sees is composed of the vector sum of the local magnetic field due to the electron angular momenta as well as the macroscopic applied field itself. This local field depends on position within the atom and also on the relative location of other atoms. As an example and to simplify the discussion, we start with the case of an isolated atom of neutral hydrogen far from any neighbours.

In the absence of an applied B-field the proton magnetic moment aligns either parallel or antiparallel to the B-field due to the electron magnetic moment. This field is proportional to J and the magnetic energy splitting between these states is proportional to $I \cdot J$, where I is the nuclear spin vector and J is the electronic angular momentum vector.[15]

This is an example of hyperfine structure. In the classical or vector model the interaction causes I and J to precess around one another, just as the l and s do in electronic fine structure. The total angular momentum $F = I + J$ is a constant of the motion. So for atomic hydrogen there are four states, a singlet $F = 0$ and a triplet $F = 1$ with $M_F = 0, \pm 1$. The three triplet states are degenerate and the singlet is separated from them by 1.5 GHz, or 6.2×10^{-6} eV, see Fig. 2.5. The transition between the $F = 1$ and $F = 0$ states in zero external field is the famous 21 cm line that makes it possible to observe the existence of clouds of low density cold atomic hydrogen in interstellar space. This is a magnetic dipole transition and the energy is very small. As a result the radiative mean life is long, 11.4×10^{6} years, but not long compared with the collision frequency in cold low density H clouds.

Atomic hydrogen in weak and strong B-fields

When a small magnetic field is applied to an isolated hydrogen atom, the net magnetic moment of the mutually coupled electron and proton precesses about the applied field. The $F=1$ state splits into three equally spaced levels, as shown to the left in Fig. 2.5. This is the Zeeman effect. At a slightly larger field the magnetic force on the electron moment due

[14]It is found that these even–even nuclei have tighter binding than those with unpaired spins and lower natural abundance. The fact that the more tightly bound nuclei are also the most abundant is witness to the fact that they were created in nuclear synthesis under conditions of near-thermodynamic equilibrium. This is consistent with the fact that the Earth is almost entirely composed of the residue of stellar thermonuclear combustion.

Table 2.2 Some common nuclei of interest with their spin I, value of precession frequency γ and relative isotopic percentage abundance.

Units	I $\hbar$	γ MHz T^{-1}	Abundance %
^{1}H	$\frac{1}{2}$	42.57	99.98
^{2}D	1	6.54	0.0156
^{13}C	$\frac{1}{2}$	10.71	1.1
^{14}N	1	3.08	100
^{19}F	$\frac{1}{2}$	40.05	100
^{23}Na	$\frac{3}{2}$	11.26	100
^{31}P	$\frac{1}{2}$	17.23	100

[15]In hydrogen we have $I = J = \frac{1}{2}$. We retain the notation for general I, J.

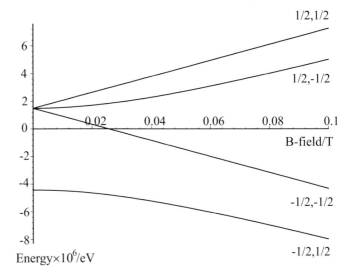

Fig. 2.5 The hyperfine energy levels of an isolated atom of hydrogen as a function of the applied B-field. As larger B-field values are considered the states become eigenstates of (M_J, M_I), as given for each curve to the right. This is called a Breit–Rabi plot.

to the nucleus becomes less important than the magnetic force on the electron due to the applied field, and decoupling occurs, at least as far as the electron is concerned. In terms of quantum mechanics there is mixing between the singlet $F=0$ state and the $F=1$, $M_F=0$ state as shown in Fig. 2.5 which is called a Breit–Rabi plot. At a field of order 0.06 T or more the eigenstates of the Hamiltonian are eigenstates separately of M_J and M_I, the separate spin-projection operators for the electron and nucleus. F is then no longer a good quantum number. This is the strong field approximation, the Paschen–Back effect.

A state described by a given wavefunction has an energy in a B-field which is proportional to B.[16] Conversely any departure from a straight line is indicative of a region where the wavefunction, not just the energy eigenvalue, is changed if the field is changed. Two of the lines on the plot are completely straight. They are the ones which are not mixed by the change of coupling.[17] The re-coupling is therefore simply the mixing of the other two states. In the higher applied B-field the strongest field felt by the nucleus is still that due to the electron. In classical terms it is still precessing round the electron spin direction, even though as far as the electron itself is concerned the coupling is off and the marriage is over! The only effect of the nuclear spin on the energy of the atom is now its orientation energy relative to the electron spin. This is independent of applied field and is seen as an energy displacement rather than a slope change within each pair of the four lines. This is seen for B at 0.1 T in Fig. 2.5.

General atomic hyperfine structure

A general atom differs quantitatively from atomic hydrogen but the basics are the same. The total electronic angular momentum J has a g-factor given by equation 2.21. The nuclear spin angular momentum I

[16] That is to say it follows a straight line on a Breit–Rabi plot.

[17] Of the four states, the ones with total angular momentum along B of ± 1 are unique. The only mixing than can occur on recoupling is between the two with component of total angular momentum zero.

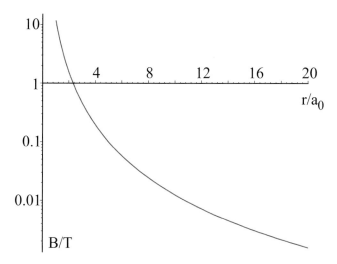

Fig. 2.6 The on-axis B-field due to an isolated unpaired electron spin as a function of radial distance in units of the Bohr radius, a_0.

may also be other than $\frac{1}{2}$ and has its own g_I-factor, so that $\mu_I = \mu_N g_I I$. The coefficient of the coupling energy $\boldsymbol{I} \cdot \boldsymbol{J}$ is an independent variable which in the absence of an external B-field will split each atomic level into $2I + 1$ or $2J + 1$ hyperfine levels, whichever is the less. Each of these hyperfine levels has its own value of F and in the presence of a weak B-field splits into $2F + 1$ components. As a result the number of states whose energies are tracked as a function of B-field in the Breit–Rabi plot is increased compared with atomic hydrogen. Qualitatively, however, the same type of behaviour occurs and we do not need to pursue details.

Range of the electronic paramagnetic B-field

The magnetic field of a non-zero electronic J has a significant influence on the precession of nuclei over a long range. The B-field of an isolated electronic magnetic dipole with moment $\boldsymbol{\mu}$ in units of μ_B at a distance r in the direction of unit vector $\boldsymbol{n}$ is given by

$$\boldsymbol{B} = \frac{\mu_0 \mu_B}{4\pi r^3} [3\boldsymbol{n}(\boldsymbol{\mu} \cdot \boldsymbol{n}) - \boldsymbol{\mu}]. \tag{2.27}$$

It is useful to consider the range at which this field would have influence on a nuclear moment. The on-axis field is plotted out in Fig. 2.6. It is evident that within a range of about three atomic radii the electronic local field is large on the scale of 1 T. A nucleus sensitive to field variations at the level of 100 ppm (parts per million) would be significantly perturbed even at distances of 20 radii. As nuclei are sensitive to field variations at the level of 1 ppm, it is clear that nuclear precession is highly sensitive to small concentrations of unpaired electronic angular momenta.

2.3   Electron spin resonance

Energy differences between electronic magnetic substates are small. Electric dipole transitions between them are forbidden and the spontaneous rate of the allowed magnetic dipole transitions is negligible. Because the energy differences are much less than thermal energies $k_\mathrm{B}T$, the population differences are small. For states with $J \neq 0$, transitions can be observed between electronic states in magnetic resonant absorption. Such states have unpaired electron angular momenta and their observation may yield information on the irradiation history or indicate the presence of chemical radicals.

2.3.1   Magnetic resonance

Absence of spontaneous radiative relaxation in magnetic resonance

Consider the states described in the Breit–Rabi plot, Fig. 2.5, as an example. The only difference between them is the direction of the nuclear and electronic spin angular momenta. They have the same orbital angular momentum and therefore the same parity. An electric dipole transition between any of them would require a change of orbital angular momentum and parity, and is therefore forbidden. Magnetic dipole transitions may be allowed but the spontaneous decay rate varies with E^3. The transition energy E, even for the electronic magnetic splittings is only of order $\mu_\mathrm{B}B$ which is very small even in a high field.

Although spontaneous transition rates are negligible, induced transitions are important. The long spontaneous mean life τ implies that the states are very narrow in resonant frequency, $\gamma_r = 1/\tau$. An applied RF field tuned to exactly the right frequency induces transitions. In fact the induced transition rate describes both absorption or emission, because the matrix elements for these two processes are the same.[18]

In electron spin resonance (ESR), also called paramagnetic resonance, the electron spin is excited by an RF field in the presence of a B-field. The coupling environment of the electron spin has a significant effect on the resonant frequency as described by the g_J-factor, discussed in section 2.2.

[18]If the RF field is sufficiently intense the populations of the upper and lower states may become equal. Then the rates of stimulated emission and absorption are the same and the transition is said to have been saturated.

ESR without orbital motion

For a simple isolated electron spin with $g = 2$ the resonant angular frequency is $2\mu_\mathrm{B}B/\hbar$ or 28 GHz T^{-1}. If a B-field of 0.34 T is chosen, the frequency is 9.5 GHz with $\lambda = 0.032$ m, in the middle of the X-band microwave region. The energy difference $E_{12} = 3.9 \times 10^{-5}$ eV is small compared with thermal energy, $k_\mathrm{B}T = 2.5 \times 10^{-2}$ eV at room temperature. In thermal equilibrium the number of electrons with spin parallel N_+ and antiparallel N_- is given by equation 2.26, so that the excess fraction of parallel over antiparallel spins is

$$\frac{N_+ - N_-}{N_+ + N_-} \approx \frac{1}{2}\left[1 - \frac{N_-}{N_+}\right] = \frac{1}{2}\left[1 - \exp\left(-\frac{2\mu_B B}{k_B T}\right)\right] \approx \frac{\mu_B B}{k_B T}. \quad (2.28)$$

At $B = 0.34$ T this fraction is 0.78×10^{-3}. So that, although the number of unpaired electrons in a sample may be large, the net macroscopic magnetisation at thermal equilibrium is small. With N unpaired electron spins per cubic metre the net magnetisation at a temperature T is

$$M \approx \frac{N\mu_B^2 B}{k_B T}. \quad (2.29)$$

In general there is a factor g_j compared with the pure electron spin $g = 2$, and there is a multiplet of states with their Boltzmann factors, all close to unity. We do not need the detail here.

In any case there is a macroscopic magnetisation M which couples to the RF field. The magnetic dipole selection rule $\Delta m_j = 1$ ensures that despite the multiplet there is only one resonant frequency, although this will be proportional to g_j and therefore depend on the atomic and molecular coupling. The amplitude of the resonant absorption signals depends on M. It should therefore increase linearly with the B-field and depend inversely on the temperature T.

This suggests that to enhance the signal a large value of B should be used. In practice there are difficulties in doing this effectively. A typical experimental arrangement is described below. The medium of interest has to be placed in a microwave resonant cavity which is necessarily no more than about one wavelength in size. The problem is that a higher field implies a shorter wavelength and thence a smaller cavity. Thus beyond a certain value the choice of a higher B-field does not result in a larger signal because the sample size is reduced.

Hyperfine structure in ESR

What role does nuclear spin play in ESR? Usually there is no role, because most nuclei are spinless.

If the nucleus has spin, the frequency of ESR is determined by the energy differences between the states at the value of the B-field on a Breit–Rabi plot. As an introductory example we consider again transitions in atomic hydrogen. In the weak B-field region to the left in the plot, M1 resonance transitions occur between members of the triplet and the singlet ($\Delta F = 1$) and also between neighbouring members of the triplet ($\Delta M_F = \pm 1$). At higher B-field to the right in the Breit–Rabi plot, the ESR transitions are those that flip the electron spin. That is, between states with $\Delta m_J = \pm 1$ and $\Delta M_I = 0$, as shown in Fig. 2.7. The nuclear spin transitions with $\Delta m_J = 0$ and $\Delta M_I = \pm 1$ are NMR transitions in the strong paramagnetic environment in which the nuclear orientation energy is determined by the electronic B-field, not the applied B-field. From Fig. 2.7 we see that there are two such ESR transitions, between the highest and lowest states and between the two middle states. The

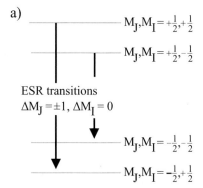

a)

$M_J, M_I = +\frac{1}{2}, +\frac{1}{2}$

$M_J, M_I = +\frac{1}{2}, -\frac{1}{2}$

ESR transitions
$\Delta M_J = \pm 1, \Delta M_I = 0$

$M_J, M_I = -\frac{1}{2}, -\frac{1}{2}$

$M_J, M_I = -\frac{1}{2}, +\frac{1}{2}$

b)

$M_J, M_I = +\frac{1}{2}, +\frac{1}{2}$

$M_J, M_I = +\frac{1}{2}, -\frac{1}{2}$

NMR transitions
$\Delta M_J = 0, \Delta M_I = \pm 1$

$M_J, M_I = -\frac{1}{2}, -\frac{1}{2}$

$M_J, M_I = -\frac{1}{2}, +\frac{1}{2}$

Fig. 2.7 The ESR and NMR transitions in an ($I = \frac{1}{2}$, $J = \frac{1}{2}$) atom, such as atomic hydrogen, in a B-field large enough to break the IJ coupling (e.g. 0.1 T in atomic hydrogen).

a)

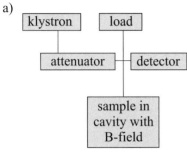

b)

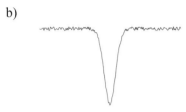

Fig. 2.8 a) A schematic diagram of the elements of an ESR experiment connected by lengths of waveguide. b) The curve shows a trace of the RF amplitude detected as a function of the modulating B-field with a minimum at resonance.

energy of the two transitions differs since $I \cdot J$ changes from parallel to antiparallel in one case, and from antiparallel to parallel in the other. So the gross dependence of the resonant frequency of the two transitions on the B-field is determined by the electronic g_j-factor. Their difference is an example of the hyperfine splitting of ESR spectra. When I and J are greater than $\frac{1}{2}$ there is an additional term in the Hamiltonian due to the electric quadrupole moment of the nucleus. We shall not explore this detail here although it affects NMR (and MRI) for nuclei with $I > \frac{1}{2}$.

2.3.2 Detection and application

ESR apparatus

Figure 2.8 shows a simplified diagram of the components of an ESR spectrometer. The klystron microwave source is connected by a rectangular waveguide to an attenuator, a cavity containing the sample of material in a uniform B-field, a diode detector to measure the RF field amplitude and a dump load. The radio frequency is constant and data are taken by modulating the amplitude of the B-field very slightly, thereby sweeping the value of the resonant frequency back and forth relative to the fixed value of the applied radio frequency. Data are recorded by observing the phase of the modulation at which maximum absorption of the RF wave occurs, as shown by the trace in Fig. 2.8b. In fact the signal is usually shown differentiated with the curve passing through zero at resonance. The amplitude and phase of the signal seen by the diode detector enables the frequency, width and amplitude of the resonant absorption by the sample to be measured. The attenuator is required because the absorption easily saturates. If a higher power level is employed, the population difference between the upper and lower states which is already small falls towards zero. As the population of the upper state rises towards 50%, the energy emitted by stimulated emission becomes equal to the energy absorbed by excitation. This is the saturation condition discussed earlier.

Applications of ESR

Studies of the frequency shifts in ESR are used extensively in research in chemistry and biochemistry. Molecular fragments or chemical radicals frequently have unpaired electron states with characteristic ESR spectra, values of g_j in fact. These stand out in the presence of ground state chemical species with paired electron angular momenta and no ESR response. ESR is therefore a prime tool in the investigation of chemical reaction mechanisms where short-lived molecular fragments with their unpaired electrons are the interesting agents.

ESR is also used in other applications which give rise to unpaired electrons. In particular the chemical debris that results from the exposure of material to ionising radiation includes a significant density of unpaired electrons. Such unpaired electrons may migrate and pair up with a relaxation time which may be fast in liquids and gases. In solids the relaxation times may be very slow indeed. Single electrons may be trapped by crystal dislocation and vacancy sites. Some of these sites may themselves have been created by irradiation. Relaxation times are reduced dramatically by heating or melting. Thus ESR has uses in measuring time-integrated radiation exposure and, thence, in archaeological dating. It is also important in the study of the effect of sterilisation of food by irradiation.

In general ESR has two big advantage over NMR at the same B-field. The larger energy splitting between the states gives a larger population difference. The magnetic moment associated with each member of this population excess is also larger. In consequence signal-to-noise ratio is not a problem in ESR. In addition the frequency shift associated with different values of g_J is a large effect. So ESR frequencies have a first order dependence on the electron environment.

But ESR also has a number of significant disadvantages compared with NMR in addition to the short wavelength and limited sample size. The strong coupling of the resonances to the crystal lattice through the electron wavefunction brings short electron spin relaxation times and therefore broader resonances. In other words the Q of the resonances is lower. If high B-fields are used the frequencies are much higher and this brings added technical problems. Advances are being made, and new applications including spatially encoded ESR may be expected. These have the prospect of fine resolution over small regions.

2.4 Nuclear magnetic resonance

NMR between nuclear states has lower frequency, higher Q-value but much smaller signal size than ESR. The motion of the net magnetisation is given by the Bloch equations. The precise value of γ (the chemical shift), and the relaxation rate caused by local variation in the magnitude (T_2) or direction (T_1) of the applied field, can all be measured. The material in a uniform steady field B_0 may be excited by an RF pulse at the resonant frequency. The amplitude, phase, frequency and polarisation of the resulting free induction decay (FID) signal is measured over the protracted period of the slow return to thermal equilibrium. By applying an RF pulse to invert the spin population during the decay, an echo signal may be generated whose amplitude is insensitive to non-uniformities in the applied B-field.

2.4.1 Characteristics

History

NMR was discovered in 1946 separately by Bloch and Purcell. From those early days the small frequency shift determined by the electronic environment of the nucleus has been exploited in investigations in chemistry and biochemistry. The extension of NMR to imaging, now described as magnetic resonance imaging (MRI), dates from a short paper by Lauterbur in 1973. In this chapter we discuss NMR. In chapter 7 we extend the treatment to MRI and spatially resolved magnetic resonance spectroscopy (MRS).

Energy and population differences

NMR depends on the relative energies of states with different nuclear spin orientation in an applied B-field. We saw in section 2.2.2 how in a paramagnetic material the local B-field seen by the nucleus is dominated by the effect of the unpaired electron moments. For diamagnetic materials the effect of electronic moments is only a small correction to the orientation energy of nuclei in an applied B-field. The orientation energy difference between states is given by $\hbar\gamma B$ where γ is given in table 2.2 for nuclei of interest. For instance, for hydrogen in water or fat where the electrons are diamagnetic, the energy splitting between the states in a 1.5 T field is $2\mu_\mathrm{p}B = 2.6 \times 10^{-7}$ eV.

According to the Boltzmann distribution, equation 2.26, it follows that in thermal equilibrium the excess of protons with spin parallel to this field compared with all protons is a fraction 5.3×10^{-6} at room temperature. Compared with ESR at the same field the fraction of aligned spins is smaller by $\mu_\mathrm{p}/\mu_\mathrm{B}$ and the magnetic moment per spin is smaller by the same factor. The net magnetic moment per unit volume is

$$M = \frac{N\mu_\mathrm{p}^2 B}{k_\mathrm{B}T}, \tag{2.30}$$

where N is now the hydrogen concentration (m^{-3}). This is six orders of magnitude smaller than the comparable figure for ESR and can only be enhanced by using the highest possible B-field and the lowest possible temperature. In the chemistry or biochemistry research laboratory such choices may be made. However, for applications in clinical medicine the maximum B-field is limited on safety grounds, and the use of cryogenic temperatures is excluded, so that NMR signals are small and the signal-to-noise ratio is a limitation.

One way around this difficulty is to use spins that are not in thermal equilibrium. Such hyperpolarised states are discussed later in this chapter in connection with the measurement of magnetic fields. Recent work has shown that the polarisation of hyperpolarised ^{129}Xe may be transferred to the state of other nuclei such as ^{13}C. This has opened up the prospect of important new applications.

Quality factor of resonance

Relaxation by spontaneous emission is effectively forbidden as discussed in section 2.3.1. Therefore once a distribution of nuclear spins becomes excited from its equilibrium Boltzmann value, the only way that it can relax back to equilibrium is through the interaction of the spins with applied fields and with the atomic and molecular environment. However, such interaction is weak and relaxation mean lives τ are typically 100–1000 ms or even longer.

To be precise, we consider the energy of the excited state with frequency ω_L to decay as $\exp(-t/\tau)$. So the amplitude of such a wave has a time dependence

$$\exp\left[-\mathrm{i}\left(\omega_L - \mathrm{i}\frac{1}{2\tau}\right)t\right].$$

Taking the Fourier transform gives the frequency dependence of this resonance amplitude $a(\omega)$ in the form of a narrow peak at ω_L. The resonance width γ_r is defined as the full width at half maximum of this peak plotted as intensity $|a(\omega)|^2$. Then $\gamma_r = 1/\tau$ as may be derived in answer to question 2.4.

The quality factor Q is the dimensionless ratio of the energy of the resonance to its width,

$$Q = E/\hbar\gamma_r = \omega_L\tau. \tag{2.31}$$

It follows that excitations with the above range of relaxation times have values of Q typically in excess of 10^8. This is a significant conclusion.[19] It is responsible for the fact that detection of NMR excitation is possible despite the weak signal strength. Within the narrow bandwidth of the signal the random noise is reduced proportionally. Furthermore, an oscillator with such a high Q can act as a very precise probe because it oscillates coherently for a long time throughout which its phase accumulates sensitivity to small perturbations. It is technically convenient that NMR resonances occur in the 100 MHz region. At such frequencies it is particularly straightforward to measure not only amplitude and frequency, but also absolute phase and signal polarisation as well. How much easier it would have been for the X-ray analysis of DNA and other molecular structures, if it had been possible to measure the phase of X-ray signals as simply!

[19]We may generalise. Any system in physics that displays a high Q is fertile ground for applications.

2.4.2   Local field variations

The Bloch equations

The equations of motion of the net macroscopic magnetisation $\boldsymbol{M}$ are known as the Bloch equations. We use a set of unit vectors $\boldsymbol{i}, \boldsymbol{j}, \boldsymbol{k}$ along the x, y, z axes. In the absence of an external RF field and with a constant applied field $\boldsymbol{B}$ along $\boldsymbol{k}$ we have

$$\frac{\mathrm{d}\boldsymbol{M}}{\mathrm{d}t} = \gamma\boldsymbol{M} \times \boldsymbol{B} - \frac{(M_x\boldsymbol{i} + M_y\boldsymbol{j})}{T_2} - \frac{(M_z - M_0)}{T_1}\boldsymbol{k} \tag{2.32}$$

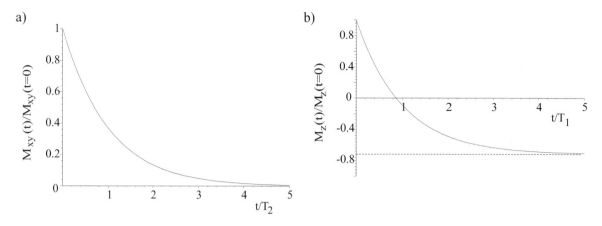

Fig. 2.9 a) The exponential decay of the transverse magnetisation with time constant T_2 relative to its value at $t = 0$.
b) The exponential recovery of the longitudinal magnetisation towards thermal equilibrium with time constant T_1 relative to its value at $t = 0$.

where $M_0\boldsymbol{k}$ is the longitudinal magnetisation at thermal equilibrium. The first term on the right hand side describes the precession of $\boldsymbol{M}$ about $\boldsymbol{B}$, as derived in section 2.1.3. The second and third terms describe the relaxation of the transverse and longitudinal components of $\boldsymbol{M}$ with time constants T_2 and T_1 respectively. These may appear similar at the macroscopic level but have a quite different microscopic interpretation.

The time T_1 describes the rate at which the relative number of up and down spins relaxes towards the thermal equilibrium value discussed above. It depends directly on the transition rate between the nuclear spin orientation states and is the relaxation connected with the actual resonance width $\gamma_r = 1/T_1$.

The relaxation time T_2 does not describe any changes in state or relaxation at the microscopic level. It arises from a loss of agreement on the phase of the Larmor precession between different protons, a loss of synchronisation, caused by the variation in the B-field that they experience at their various locations within the material.

Therefore in a steady applied B-field the Bloch equations describe a macroscopic magnetisation of two parts.

◇ A transverse magnetisation which rotates in the xy-plane at the Larmor frequency and has a magnitude that decays with time constant T_2, as illustrated in Fig. 2.9a.

◇ A longitudinal magnetisation with a magnitude that relaxes towards its non-zero thermal equilibrium value with time constant T_1, as illustrated in Fig. 2.9b.

[20]The coupling of ESR to the crystal lattice is very much stronger than in NMR and the Larmor frequency higher. Consequently the relaxation times are qualitatively shorter.

The Bloch equations of motion apply equally to ESR and NMR, although the values of the various parameters are significantly different.[20] They may also be used to describe the behaviour of spins in the B-field of an RF wave at resonance, as discussed later in section 2.4.4.

Sources of local B-field variation

The behaviour of a non-equilibrium state of proton magnetisation with time is affected by variations in the local B-field that each constituent proton experiences.[21] Such a variation may be in the magnitude or in the direction of this local field. The variation may be constant in time, depending only on the site, or it may be time dependent. In particular it may be time dependent because the proton itself moves from one place to another during the period between excitation and readout.

If the proton spin is coupled to a paramagnetic electronic J the NMR transitions are those with $\Delta M_J = 0$, $\Delta M_I = \pm 1$. For example, in atomic hydrogen these NMR transitions which are between the upper two states or between the lower two states shown in Fig. 2.5 are nearly independent of the B-field altogether except at the lowest fields. This is because the proton has an orientation energy determined by the field of the unpaired electron, not by the applied B-field. As described by equation 2.27 and plotted in Fig. 2.6, this influence will dominate within two or three atomic radii but also cause a change in the local B-field, and therefore a significant shift in NMR precession frequency, at a distance of many tens of atomic radii.

Far removed from any unpaired electron, a proton will still be influenced slightly by the medium through its diamagnetic properties. In a uniform medium to a linear approximation, if χ is the magnetic susceptibility of the macroscopic material then fractional NMR frequency shifts of order χ are to be expected. For values of χ, see table 2.1.

Chemical shifts

However, this is not a satisfactory description of the microscopic B-field seen by a proton. Within a macroscopic medium in a uniform B-field, protons at different chemical sites will see different local fields unrelated to the average macroscopic response described by χ. A proton in water, a proton in a saturated hydrocarbon chain, $-CH_2-$, and a proton in a hydroxyl group, $-OH$, all have different local electron densities and slightly different local B-fields. A molecule placed in a uniform field will have a discrete spectrum of such frequencies. Each peak is characteristic of a type of site, and the relative amplitude or peak area indicates the number of such sites. These discrete shifts, called chemical shifts, are at the part-per-million (ppm) level. This is illustrated in Fig. 2.10 for the hydrogen atoms in the organic compound, methyl propionate. Examples of proton chemical shifts which act as signatures in clinical studies of metabolism and its disorders are given in table 2.3. Zero shift is defined by the spectrum of tetra-methyl silane.[22] On this scale the shift for water is 4.7 ppm and the shift for fat, $-CH_2-$, is about 1.0 ppm.

[21] We refer to protons, but other nuclei with spin behave in a similar way.

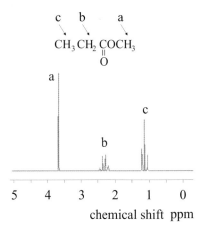

Fig. 2.10 The high resolution NMR spectrum of methyl propionate. Hydrogen atoms are in three types of positions, a, b and c. Each has its own frequency in the spectrum separated by a few ppm with normalisation 3:2:3. At high resolution each peak is split into $n+1$ sub-peaks where n is the number of hydrogen atoms on the neighbouring carbon atom. This sub-structure allows identification of radicals but it is not resolved in imaging applications. [Spectrum sourced from SDBSWeb at *www.aist.go.jp/RIODB/SDBS/* (National Institute of Advanced Industrial Science and Technology, 29 Jan 06).]

Table 2.3 The values of the chemical shift for a proton at certain molecular sites.

Molecular site	Freq. shift (ppm)
Lactate	4.1
Fat $-CH_2-$	1.32
Water	4.7
NAA (CH_3)	2.0/2.48–2.64
Choline $N(CH_3)_3$	3.2
Creatine $N(CH_3)$	3.03/3.94

[22] This molecule, $Si(CH_3)_4$, is chosen because it has 12 hydrogen atoms, all in identical chemical positions with no effects due to hydrogen bonding.

2.4.3   Relaxation

Precession dephasing by variation of the field magnitude

The variation in NMR frequency is not limited to the discrete values associated with particular chemical sites. Especially in liquids and gases there is a continuum of magnetic environments experienced by proton spins as they move about. Thus there is a spread of values of local B-field that depends in a continuous way on the position of molecules in the neighbourhood, especially if any of these should contain an unpaired electron moment. This implies that the magnetisation associated with individual protons will precess at slightly different rates and that the net macroscopic integrated transverse magnetisation will die away as contributions become progressively 'dephased', as sketched in Fig. 2.9a. The mean decay time of the transverse magnetisation is named T_2^*. Since the decay is not primarily thermal in origin it does not depend significantly on temperature. The component of magnetisation along the B-field remains unchanged because it is not affected by precession.

The variation in the magnitude of B-field responsible for T_2^* has three different contributions. These are the non-uniformity of the applied field, the position-dependent variation of the microscopic B-field experienced by hydrogen at different sites and the effect of migration of hydrogen atoms for one site to another during the measurement period. While the relaxation time T_2^* is used to describe all three, the time T_2 is used to describe the last only.

A variation of 1 part in 10^7 in the magnitude of the macroscopic applied B-field generates a phase slip at 100 MHz of 2π in 100 ms. Thus applied field uniformity at this level or better is a concern in NMR. However such dephasing is reversible by the spin echo technique, discussed later in section 2.4.5. The effect of microscopic site-dependent field variations is also reversible. Only if the proton wanders about from one location to another during the signal decay time is the dephasing not reversible. This irreversible part is T_2. The importance of the distinction between T_2 and T_2^* will become more apparent in section 2.4.5. Some sample values of T_2 are given in table 2.4 for various tissue materials.

Spin state population relaxation by variation of the field direction

The variation in the magnitude of the local B-field discussed above affects the transverse magnetisation through dephasing, but leaves the longitudinal magnetisation unchanged. The longitudinal magnetisation is determined purely by the relative population of the spin states, not by their energy splitting. In the absence of spontaneous (or induced) transitions this population can only change if the spin states are mixed. In quantum mechanics that is only possible if the quantisation axis is rotated. Only changes in the direction (not the magnitude) of the local B-field will have this effect.

Changes in the B-field direction can arise from local fluctuating trans-

Table 2.4 Values of T_1 and T_2 in milliseconds for hydrogen atoms in various tissues and at various applied B-fields. These values are given without errors for illustration.

Tissue	T_1 (ms)			T_2 (ms)
	$B = 0.2$ T	1.0 T	1.5 T	
Fat	182	241	259	85
Muscle (skeletal)	372	732	868	45
Brain – white matter	390	683	786	90
Brain – grey matter	495	813	921	100
Cerebral spinal fluid	1400	2500	3000	1400

verse magnetic moments, for instance from the thermal motion of magnetic moments and charges. In liquids and gases the fluctuation is to be understood as occurring in the form of collisions for which the matrix element includes a coupling between the nuclear moment and the transverse interacting magnetic field or electric field gradient. In consequence there is a probability that the proton spin is flipped in a collision. In solids there are no such collisions. In this case the time dependence is to be seen as coming from the thermal vibrations of the ions of the lattice, acoustic waves or phonons, and their accompanying local magnetic or electric field perturbations. It is found that the nuclear electric quadrupole moment (if present) with its coupling to electric field gradients is as important as the magnetic coupling to the spins. For both fluids and solids the response of the electron wavefunctions to the perturbation and its effect on the nucleus are complicated.

However, there are two simple conclusions.

Firstly, fluctuations in B-field direction are intrinsically thermal, so that T_1 decreases with temperature. In fluids this is because a thermally driven flux determines the relaxation rate via a spin-flip cross section. In solids a flux of phonons is thermally generated.

Secondly, T_1 increases with the value of B_0. This may be seen classically as follows. The state mixing and the relaxation depend on the angle through which the quantisation axis is rotated during the collision, and this angle is roughly equal to the transitory local B-field divided by the steady B_0. More precisely, in a perturbation theory picture the mixing that occurs during a collision depends on the perturbation energy divided by the energy separation of the spin states. Since the latter is proportional to B_0 we have the same result.

T_1 is typically much larger than T_2, while T_2^* is even shorter. Some values are included in table 2.4. In section 2.4.5 we discuss how these different time constants may be determined experimentally.

a)

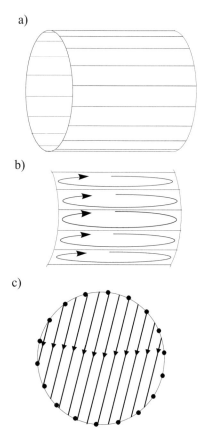

b)

c)

Fig. 2.11 a) A cylindrical birdcage coil with $N = 18$ elements.
b) A view of the current elements in the same view as a). The currents are of equal magnitude, each with a phase delay $2\pi n/N$ matching its geometrical azimuth in the structure. Thus $I_n = I_0 \sin(\omega_L t + 2\pi n/N)$.
c) A view looking along the coil axis and showing the resulting interior uniform field $\boldsymbol{B}_1$, which rotates at the radio frequency ω_L. The dots represent sections through the longitudinal conductors of the coil.

[23]This is a classical picture. In a quantum picture the matrix element may have a non-zero value only if the rate of azimuthal phase change of the product of initial and conjugate final wavefunctions $\gamma \Delta MB$ matches the rotation rate of the operator due to the RF field. The result is the same as the classical one.

2.4.4   Elements of an experiment

Solenoid for the field $\boldsymbol{B}_0$

The basic apparatus needed for an NMR experiment starts with a source of the steady uniform field B_0 which should be as high in strength as safety and resources permit. The most effective way to make a region of uniform magnetic field is to employ a large solenoidal magnet and use a relatively small volume around its centre of symmetry. In addition shimming coils can be used to reduce further the field variations throughout this region. Because of the size and stability required, this magnet is usually chosen to be superconducting and powered continuously.

Coil for the RF field

Around the uniform field volume and enclosing the sample is an RF excitation coil. This coil is fed with a short excitation pulse at the Larmor frequency. A design commonly encountered is shown in Fig. 2.11 and consists of a symmetric array of circuits forming longitudinal elements of a structure that looks like a birdcage. The axis of the birdcage is aligned with the field $\boldsymbol{B}_0$. The magnitude of the currents I_n in the n elemental circuits are all equal but the phase of each is equal to its azimuthal angle in the structure. The polarisation of the resulting RF field takes the form of a near-uniform B-field rotating at the Larmor Frequency in the plane perpendicular to $\boldsymbol{B}_0$, as illustrated in Fig. 2.11c.

This circularly polarised field is called $\boldsymbol{B}_1$. To each proton spin precessing at this same frequency, the field appears static.[23] Thus for the duration of the RF pulse the nuclear magnetisation precesses about the vector sum $\boldsymbol{B}_0 + \boldsymbol{B}_1$ instead of the vector $\boldsymbol{B}_0$ alone. As a result, at the end of such an RF pulse the magnetisation has been rotated, such that it may now have a component in the xy-plane. The angle of rotation depends on the integral of B_1 over the duration of the pulse. Thus a pulse that rotates the magnetisation through $60°$ is called 'a 60 degree pulse'. A $90°$ pulse maximises the transverse magnetisation and reduces the longitudinal magnetisation to zero. A $180°$ pulse actually inverts the magnetisation relative to its previous orientation.

Coils to detect the free induction decay signal

As shown in Fig. 2.12, following a $90°$ RF pulse the excited transverse magnetisation starts to precess at the Larmor frequency with an amplitude that decays due to dephasing. Such a rotating magnetisation generates a magnetic field which can be detected. In principle a pair of pickup coils, one in the xz-plane and the other in the yz-plane, will see the changing B_y and B_x fields respectively as an induced voltage given by Faraday's law of electromagnetic induction. So these two voltage signals are sensitive to the precessing magnetisation with a $90°$ phase difference. This signal is called the free induction decay (FID) signal. It is a near-field inductive signal, not a far-field radiative signal. In prac-

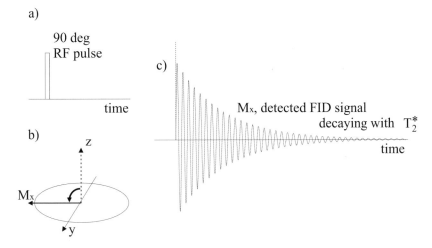

a)

90 deg
RF pulse

time

b)

c)

M_x, detected FID signal
decaying with T_2^*

time

Fig. 2.12 The effect of an RF excitation pulse on hydrogen in thermal equilibrium. a) A 90° circularly polarised RF pulse is applied. b) This tips the magnetisation from the z- into the x-direction. c) The magnetisation M_x then starts to precess about the z-direction inducing the FID signal detected by a pickup coil. A similar signal from M_y but 90° out of phase may also be detected. Both decay exponentially with the constant T_2^*.

tice such a simple pair of coils at right angles is not used. Sometimes the birdcage coil having been used to apply the RF pulse is used again to detect the FID output signals. Alternatively an array of smaller dedicated pickup coils is used.[24]

[24]See section 7.1.4.

Electronics and signal processing

The analogue FID signals lie in a frequency bandwidth of between 2 and 100 kHz centred around the nominal Larmor frequency. By mixing this signal with a reference oscillator at the nominal Larmor frequency it is converted to frequencies in the audio range. The amplitude of this audio signal is recorded by digitising it at a rate to match the bandwidth of interest. From these measurements the initial amplitude, the frequency shift and the decay rate $1/T_2^*$ can be found, and thence the number of proton spins.[25]

[25]The choice of bandwidth in NMR depends on what is being encoded. If it is chemical shift information, the bandwidth is small. If it is spatial information using gradient coils, as discussed in chapter 7, it is much larger.

2.4.5   Measurement of relaxation times

Measurement of T_1

The observed FID falls rather rapidly with time constant T_2^* which depends particularly on the uniformity of the applied B-field. There is no signal attributable to M_z because it does not precess, and therefore the FID is not sensitive to the value of T_1. What we want is to measure T_1 and T_2, which are characteristics of different media. How can this be done?

Consider the pulse sequence shown in Fig. 2.13a. With the spins initially in thermal equilibrium there is a longitudinal magnetisation but none in the transverse plane, so that no FID signal is observed. An

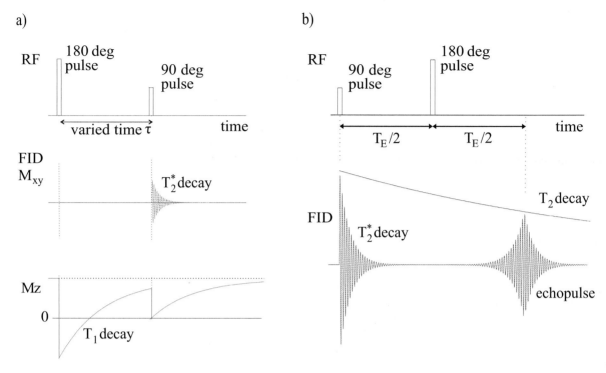

Fig. 2.13 a) A RF pulse sequence enabling T_1 to be measured by observing the FID signal amplitude as a function of the variable interval τ. The 90° pulse transfers M_z at time τ into M_{xy}, the initial value of the FID, as shown in the bottom trace. b) The spin echo RF sequence. The 180° pulse at time $T_E/2$ inverts the magnetisation. This recreates the original spin state at time T_E, except for the irreversible loss of magnetisation with time described by T_2.

initial 180° RF pulse is applied, the longitudinal magnetisation inverts, but there is still no FID signal since the transverse magnetisation is still zero. However, if, at a variable time τ after such a pulse, a RF pulse of 90° is applied, a FID signal should be seen. The initial amplitude of this FID depends on the longitudinal magnetisation at that time, τ. Thus by varying τ, the exponential relaxation of the longitudinal magnetisation towards thermal equilibrium with time constant T_1 can be plotted out and T_1 measured. Later in chapter 7 we shall find faster and better ways to measure T_1. Here we are content to see that such a measurement is possible in principle.

Measurement of T_2 using the spin echo technique

To measure T_2 we use a different sequence of two RF pulses to generate a FID with a 'spin echo'. This is shown in Fig. 2.13b. We start with a 90° pulse, for example, following which the FID falls rather quickly away with decay time T_2^*, largely due to field inhomogeneities. After a time labelled $T_E/2$ a 180° pulse is applied. This inverts all spins, not only the longitudinal magnetisation but also the sign of the precession angle of the transverse magnetic moment of each individual spin. Following the second pulse each spin then precesses as before, now unwinding its own precession angle back towards zero. Thus after a further interval $T_E/2$ every spin is back where it was at the time of the first pulse. So an echo FID signal is seen at time T_E.

However, any spin which has moved site or otherwise is seeing a different microscopic magnetic field will precess back to zero at a different rate and will have the wrong phase to contribute its full signal to the echo. This is the irreversible part of the dephasing that results in the T_2 signal loss shown in Fig. 2.13b. The time constant T_2 describes the loss of the echo signal amplitude observed at time T_E, and can be fitted to the expression $\exp\left(-T_E/T_2\right)$ when the measurement of echo amplitude is repeated for different T_E values.

Repeat time T_R

Another important time in practical NMR experiments is the repeat time T_R that describes the time at which the pulse sequence is restarted. It is important to know the state of magnetisation at the point of each new excitation. The transverse magnetisation may have decayed away but in general T_R is not many times greater than T_1 and the slower recovery of the longitudinal magnetisation towards thermal equilibrium may not be complete at T_R. This needs to be taken into account and indeed can be used to advantage, as discussed in chapter 7.

2.5   Magnetic field measurement

The B-field of the Earth is caused by its slowly moving molten core, fluctuating currents in the atmosphere and space, and the steady field of magnetised rocks. Some measurement techniques are sensitive to field direction and some to its magnitude. Some may be applied to higher fields and others only to higher fields. Amongst modern methods, the flux-gate magnetometer works only for low fields but measures direction. The Hall probe measures medium and higher fields conveniently. Measurement of the frequency of proton NMR is a higher precision method at high and medium fields but is insensitive to direction. It may be extended to lower fields if the spin population is boosted. This is done in the proton magnetometer with an auxiliary field. Other solutions are the Overhauser magnetometer and various pumped quantum magnetometers. Most sensitive are magnetometers based on the superconducting quantum interference device (SQUID). The optimum choice depends on the sensitivity, convenience, budget and ruggedness required.

2.5.1   Earth's field

Magnetic surveys

Three mechanisms contribute to the B-field of the Earth.

Firstly there is the magnetohydrodynamic effect of the moving molten core. The associated inertia of this is large enough to ensure that it changes only with a long time constant. The change in direction of the field is measured in minutes of arc per year.[26] However, on geological or even archaeological timescales such changes are fast.

Then there are the magnetic effects of currents of charged particles in space and in the atmosphere. These vary on a daily timescale being driven by the weather, fluctuations in the solar wind and charged particle emission from the Sun.

Finally there is the magnetisation of local rocks, baked bricks or the remains of a hearth. Such magnetisation was fixed at the time that the material or its ferromagnetic constituent was deposited or cooled below its Curie point. It then becomes a frozen record of the Earth's field at the time of formation. The detection and measurement of this magnetisation and its direction are of importance in geophysics, geology and archaeology.

Surveys using very sensitive devices are required to map the effect of this last signal in the presence of the other two. Some other applications requiring magnetic field maps call for even greater sensitivity, such as the measurement of the magnetic fields emitted by the brain as part of its neurological activity.

[26]This effect is large enough to require significant correction to magnetic compass readings that depend on the year.

Measurement of the Earth's field

Historically the Earth's field direction was mapped in the horizontal plane using a compass, a pivoted magnetic needle. A similar instrument in the vertical plane measured the dip angle. The magnitude of the field was determined by measuring the oscillation frequency of such a needle suspended by a torsional fibre, or by measuring the static torque with a balance. The torsion angle was read by observing the reflection of a pencil beam of light from a small mirror attached to the fibre. Work with such delicate devices was slow and laborious and belongs to a romantic but bygone age when there was something heroic about recording scientific measurements in out-of-the-way places under conditions of extreme hardship. Fortunately the days of this hair shirt school of experimental physics are passed! Today there are better and faster methods of taking accurate readings of magnetic fields in satellites and other remote locations and transmitting the data back.

2.5.2   Measurement by electromagnetic induction

Induction loop

The induction loop used for example in airport security is in the form of a resonant circuit as in Fig. 2.1. A change in the presence of the susceptible or conductive material changes the inductance of the coil L by a small fraction. This gives rise to a small shift in resonant frequency $\omega_0 = 1/\sqrt{LC}$ which may be sensed if Q is large.

As a method of sensing the presence of conductors and magnetic materials at close range this is very effective. However, its sensitivity falls off rapidly if the distance to the material d is much larger than the dimensions of the coil. The dipole field of the coil at the material falls as d^{-3}. The flux linkage of the loop due to the material therefore falls as d^{-6}. We conclude that magnetic measurements which rely on generating a magnetising field and detecting its response are not useful except at close range or with large coils.

Flux-gate magnetometer

Magnetic flux measurements are possible using Faraday's law by simply rotating a coil at constant speed in the field B and measuring the voltage generated. The orientation of the axis of the coil determines which component of $\boldsymbol{B}$ is measured.

A more sensitive device is the flux-gate magnetometer. This consists of three separate transformers mutually at right angles to one another, each consisting of a cylindrical core of high permeability metal with straight solenoidal primary and secondary windings. The primary winding is driven with a 5 kHz pure sine wave and the signal induced in the secondary winding is analysed. If there is no ambient B-field along the transformer axis, the output voltage of the secondary winding is amplified by the permeability of the core. Superimposed on the fundamental 5 kHz waves are its even harmonics which are generated as the metal core saturates. In the absence of an external field the saturation is symmetric and frequency components at 15 kHz, 25 kHz and so on are detected.

However, if there is an ambient static field along the axis the high permeability metal will saturate at lower values in one direction than the other. The magnetic response will then be asymmetric in direction, thereby generating odd harmonics, 10 kHz, 20 kHz and so on, in addition to the even ones. The B-field measurement comes from the amplitude of the odd harmonics. The flux-gate compass has a limited dynamic range of B-field (up to about 1 mT), good resolution (0.1 nT), but requires calibration for absolute field values. Its directional properties are excellent and it can be used to measure low frequency as well as static fields (to about 1 kHz). It may be used together with a quantum magnetometer which has the complementary properties of accurate absolute field measurement but lacks directional information.

Hall probe

The measurement of large B-fields is not possible with a flux-gate compass as the high permeability metal saturates too readily. However, it is possible to use a Hall probe instead.

The Hall effect is illustrated in Fig. 2.14. When a charge q moves along

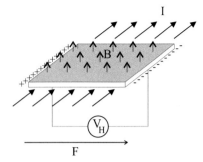

Fig. 2.14 A projected view of the 3-D configuration in which a current I flows through a conductor width w and thickness d in the presence of a transverse B-field $\boldsymbol{B}$. The current experiences a force $\boldsymbol{F}$ and this is balanced by a transverse electric field due to a build-up of charge as shown. The resulting Hall voltage V_H is generated across the width of the conductor.

a conductor in an external transverse B-field as shown, it experiences a transverse Lorentz force $\boldsymbol{F}$ perpendicular to its velocity vector $\boldsymbol{v}$ and the B-field given by

$$\boldsymbol{F} = q\boldsymbol{v} \times \boldsymbol{B}. \tag{2.33}$$

We can express this force in terms of the current formed by the moving charges. The current density, $\boldsymbol{J}$ A m^{-2}, is related to the velocity and the density of moving charges, N m^{-3}, by $\boldsymbol{J} = Nq\boldsymbol{v}$. If the conductor has width w and depth d we can rewrite this transverse Lorentz force on each charge in terms of the total current I,

$$\boldsymbol{F} = \frac{1}{Nwd}\boldsymbol{I} \times \boldsymbol{B}. \tag{2.34}$$

The direction of this force is independent of the sign of the charge. Since the net force on each charge q has to be parallel with the current and the axis of the conductor, an additional electrostatic field $\boldsymbol{E}$ is established across the wire to balance the Lorentz force, thus

$$q\boldsymbol{E} = -\frac{1}{Nwd}\boldsymbol{I} \times \boldsymbol{B}. \tag{2.35}$$

The Hall voltage across the width of the conductor is then

$$V_H = Ew = \frac{IB}{Nqd}. \tag{2.36}$$

By measuring V_H the value of B may be deduced.

In a metal N is large and the Hall voltage correspondingly low. In a semiconductor N is small and a larger effect is measured. Commercially available Hall probes cover ranges of B-field measurement from 7.5 mT up to 10 T at frequencies up to 10 kHz. Each Hall probe measures one B-field component and so in general three are required. Hall probes are compact and easy to use but they do need to be calibrated for precise measurement. Resolution is in the range 10–20 μT, so that they are not suitable for archaeology and mineral prospecting.

2.5.3 Measurement by magnetic resonance

NMR in the Earth's field

The measurement of magnetic fields with NMR is a technical application of the highest precision, especially for medium to high fields. Naturally protons are used because they have a high gyromagnetic ratio.

At low fields there are two difficulties in using NMR.

The first is the low precession frequency. The Earth's field is about 50 μT for which the proton precession frequency is 2130 Hz. The signal decays with time constant T_2, which is about 3 s in water. Field uniformity is no longer a problem but the number of periods over which the precession can be fitted is very limited.

The small signal size is an even more significant problem. The thermal excess of magnetised proton spins is tiny. The energy difference $\hbar\gamma B$ is

8.8×10^{-12} eV. The excess fraction of aligned proton spins is 1.8×10^{-10} at ambient temperature. To see how small this is, consider 1 cm^3 of water. It contains $n = 6 \times 10^{22}$ protons. The root mean square (RMS) of the statistical fluctuation of the number of these protons with spin up rather than down[27] is not much smaller than the number oriented by the Earth's field.

[27] This is about $\sqrt{n/2} \approx 2 \times 10^{11}$.

The conclusion is clear. A method has to be found to increase the magnetisation of the proton spins. The population difference must be raised from its thermal Boltzmann value. There are three different ways in which this may be done. The first uses a macroscopic auxiliary field, as in the proton magnetometer described below. The second uses the Overhauser effect in which a similar role is played by a microscopic auxiliary field. The third way is to use quantum pumping.

Proton magnetometer

The proton magnetometer is the simplest method of measuring a small B-field by NMR since no RF generator is required. Protons are supplied in the form of a bottle of water or paraffin. This is placed inside a simple coil that can be fed with a steady current to provide a magnetic field. Measurement with the proton magnetometer follows four sequential steps.

⋄ In the first the coil is powered to provide an auxiliary field B' for a time of the order of T_1.[28] Thus with B' of order 0.01 to 0.05 T and applied for several hundred milliseconds, the proton spins will be aligned along the axis of the coil with 2–3 orders of magnitude greater magnetisation than the equilibrium alignment in the Earth's field.

[28] To work, the orientation of B' should be at some angle to the Earth's field, otherwise the coil will not give rise to a large transverse magnetisation signal.

⋄ Next B' is reduced to a value several times the Earth's field, 300 μT say. This does not have to be done very quickly because the net field direction does not change.

⋄ In the third step the residual value of B' is reduced rapidly to zero in a time short compared with the precession period. At 300μ T the latter is 80 μs. This step is done quickly so that the magnetisation does not have time to track the change in field direction. As a result a significantly large transverse magnetisation is available to precess in the Earth's field.

⋄ In the fourth step the FID signal is read out and its frequency analysed. The precession can be observed for several times T_1 or the sequence can be repeated for another measurement.

Sensitivities of about 1 in 10^5 are reasonable. Commercially available proton magnetometers quote sensitivities better than a nanotesla.

Overhauser magnetometer

An improvement on the proton magnetometer relies on the Overhauser effect to increase the magnetisation further and to provide a continuous signal.

A fluid medium is required containing protons and a low concentration of paramagnetic ions which are not bound to the protons. At any time, with the fluid in thermal equilibrium, a proton may be either in a location magnetically influenced by the paramagnetic ion, or far from such an ion. In a coupled location it is close enough for its spin to form a singlet or triplet state $F = 0, 1$ with the unpaired electron.[29] As a result of thermal motion in the liquid protons may diffuse from being in coupled sites (under the influence of an ion in this way) to uncoupled sites (where they are influenced only by the Earth's field) and then may diffuse back again.

In the Earth's field the $F = 1$ energy levels in a coupled location are split. This magnetic splitting is dominated by the large effect of the electron moment. Then the Boltzmann factors that determine the population densities in thermal equilibrium are characterised by the magnetic moment of the electron, to which the proton spin is aligned by the hyperfine splitting.[30] Therefore in thermal equilibrium protons in a coupled location have a polarisation enhanced by a factor of order the ratio of magnetic moments, μ_B/μ_p. The medium is a fluid and as the protons diffuse away from the paramagnetic ion they take their polarisation with them. The result is that the free protons in the uncoupled state have a much higher magnetisation than they would have in thermal equilibrium. This is the Overhauser effect and is an example of dynamic nuclear polarisation (DNP).

To generate a precession signal the liquid is excited by a very broad band 90° RF pulse at around 2 kHz. This excites the magnetisation of the enlarged population of protons in uncoupled locations into the transverse plane. An enhanced FID signal may be measured.

Pumped quantum and SQUID magnetometers

More sensitive magnetometers are available. Most are based on pumped spin population states in potassium, rubidium or caesium vapour.

The most sensitive depend on the magnetic flux linkage of a superconducting ring and the Josephson effect. These are called SQUID devices. Their sensitivity is required to detect the transient magnetic fields emitted by the brain. Such a procedure is called magnetic encephalography (MEG). These signals, though very weak, $\sim 10^{-13}$ T, are not attenuated by the conductivity of the skull. They have good time resolution of order milliseconds, but very poor spatial resolution of order 10 cm. As such they are complementary to MRI and other modalities discussed in chapters 7 and 8.

[29] In reality it is more complicated, but we have an atomic hydrogen coupling model in our mind for illustration.

[30] The hyperfine coupling is weak enough that the $F = 1$ level is populated as well as the $F = 0$ one.

Read more in books

Classical Electrodynamics, JD Jackson, Wiley, 3rd edn (1998), the authoritative text on electricity and magnetism

Electricity and Magnetism, WJ Duffin, McGraw-Hill, 3rd edn (1980), a standard text on electromagnetism

Electricity and Magnetism, BI Bleaney and B Bleaney, Oxford, 3rd edn (1976), a standard text on electromagnetism

Electromagnetic Fields and Waves, P Lorrain, DR Corson, and F Lorrain, Freeman, 3rd edn (1988), a standard text on electromagnetism

Principles of Nuclear Magnetism, A Abragam, Oxford (1961), the classic reference work on magnetic resonance

Magnetism in Condensed Matter, SJ Blundell, Oxford (2001), a modern advanced text on magnetism

Look on the Web

Look at the book website at
  www.physics.ox.ac.uk/users/allison/booksite.htm

Watch for further developments and experimental details in ESR
  electron spin resonance

Read about the related behaviour of muons
  muon spin resonance

Follow the story of the astrophysics revealed by observations of the M1 transition in atomic hydrogen
  neutral H 21 cm

Find out more about the information given by chemical shifts and their splitting
  chemical shift NMR

Read about developments in quantum magnetometers and their application
  caesium, potassium
  or *rubidium quantum magnetometers*

Follow current developments in hyperpolarisation
  NMR hyperpolarisation or *DNP enhanced NMR*

Learn about SQUID magnetometers and their uses
  SQUID magnetometers

Questions

2.1 Calculate the numerical value of the nuclear magneton in SI units from fundamental constants and thence the Larmor precession frequency in Hz for the proton in hydrogen in a 3 T field using $g = 5.59$.

2.2 The radiative decay mean lifetime of the $F = 1$ state of atomic hydrogen is 11.4×10^6 years. Calculate the corresponding resonance width in eV. What determines the actual lifetime and measured width, a) in the laboratory, and b) in interstellar space.

2.3 A birdcage coil is formed from a ladder of 12 circuits each of self-inductance L with parallel capacitance C, of which 4 are shown:

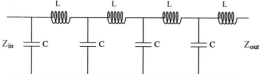

The coil is to be driven at 120 MHz from a power supply with output impedance Z. Suggest how the values of L and C should be chosen in first approximation.

Explain in principle how a more realistic solution to the problem that takes into account the interaction between the coils might be achieved.

2.4 Show that the full width at half maximum of the excitation spectrum for a single resonance plotted as a function of frequency is the reciprocal of the mean life of the resonant state, as suggested on section 2.4.1.

2.5 Calculate the magnetisation of pure water placed in a steady magnetic induction of $B_0 = 3$ T in equilibrium at a temperature of 20°C.

Following a 90° RF pulse the maximum FID signal amplitude that can be detected in a single turn coil from a container of water under the above conditions is 1.0 mV. Calculate the frequency of the signal and the mass of the water. Discuss the geometrical position of the coil and the water that is required for this maximum signal.

2.6 Calculate the frequency of ESR for an isolated electron in the Earth's B-field, assumed to be 50 μT. Explain why ESR is not a practicable way accurately to measure a value for this field.

2.7 Explain carefully the distinction between the NMR relaxation times T_1, T_2 and T_2^*. How are the phenomena behind these related to the chemical shift, the shift in NMR frequency determined by the chemical environment of a hydrogen atom?

2.8 A water phantom is at equilibrium at 20°C in a B_0 field of 1.5 T along the z-axis. A B_1 field rotating in the xy-plane with constant magnitude 78 μT is applied for 50 μs. What is the condition for the proton magnetization vector to rotate? What is its direction immediately after the B_1 field is turned off?

2.9 What are the longitudinal and transverse components of the magnetisation vector of the phantom in the previous question after a further delay of 100 ms if the values of T_1 and T_2^* are 500 ms and 50 ms respectively?

Interactions of ionising radiation

<table>
<tr><td></td><td style="text-align:center; font-size:2em;">**3**</td></tr>
</table>

in which the fundamental physics of ionising radiation is developed in preparation for general applications in chapter 6 and for the physics of medical imaging and therapy in chapter 8.

3.1 Sources and phenomenology

Ionising radiation comprises beams of charged particles, and photons with energies above about 10 keV (X-rays and γ-rays). Primary sources of such radiation include natural radioactivity, cosmic radiation from space, and beams of particles accelerated in the laboratory. Beams of secondary photons may be produced by the transverse acceleration of electron beams in the atomic electric field of a heavy nucleus (bremsstrahlung) or in the macroscopic B-field of a magnet (synchrotron radiation). Radiation in any of these forms interacts with matter through stochastic collision events. These collisions vary from single catastrophic events to multiple collisions that progressively reduce the energy of the incident particle. The probability of a collision is described by its cross section.

3.1	Sources and phenomenology 55
3.2	Kinematics of primary collisions 58
3.3	Electromagnetic radiation in matter 65
3.4	Elastic scattering collisions of charged particles 68
3.5	Multiple collisions of charged particles 73
3.6	Radiative energy loss by electrons 81
Read more in books	83
Look on the Web	83
Questions	84

3.1.1 Sources of radiation

History of ionising radiation

The history of ionising radiation starts in the four extraordinary years, 1895–1898. In 1895 Wilhelm Röntgen discovered X-rays, in 1896 Henri Becquerel discovered natural radioactivity, in 1897 J.J. Thompson discovered the electron and in 1898 Marie Curie discovered Radium. The book by Mould gives a brief account supported by pictures of the pioneers, their apparatus, data and contemporary reports in the press. The fact that these discoveries were made in such a short time generated a euphoria such that subsequently all sorts of claims were made and many believed, culminating in the fraudulent discovery of 'N rays' in 1903–1904. Radiation attracted immediate publicity as an extension of the fairground peepshow by recording pictures of otherwise hidden objects. However, its more serious use for medical imaging was also understood and exploited from the earliest days. The damaging effects of radiation were found when those working with radiation sources recorded burns that did not clear up. On the other hand in 1896 already ionising radiation was used to treat skin tumours experimentally.

Ionising radiation is normally understood to include beams of charged particles and photons of sufficient energy to create significant ionisation, about 10 keV and above. We discuss three primary sources of such radiation. Natural sources of radioactivity are concentrated in certain minerals, but also present in most unrefined materials at the level of 10^{-9} to 10^{-12}. Then there is cosmic radiation incident on the Earth from space. We discuss both of these further in the context of the natural radiation environment in chapter 6. Finally there are accelerated beams of charged particles, both electrons and nuclei, derived from ion sources in the laboratory.

The ion source for an accelerated electron beam traditionally starts with a heated metal filament providing thermionic emission. However, the current that can be obtained in this way is limited. Modern electron sources for pulsed beams use photoelectric emission from metals illuminated by lasers.

X-ray sources

X-rays and γ-rays are usually generated by the acceleration of high energy electrons as a secondary process. These electrons may be given energy by a DC voltage or a linear accelerator with an RF energy source, see section 8.1. In either case the resulting electron beam is incident on a high Z metal target where the electrons suffer large transverse accelerations in the nuclear electric fields of the atoms. The resulting electromagnetic radiation is known as bremsstrahlung and is peaked in the forward direction. Additional X-rays are emitted by the target when atoms radiate following the ejection of an inner electron.[1] The bremsstrahlung spectrum when plotted as a function of energy is quite constant[2] but cuts off at the energy of the incident electron energy. The higher the electron beam energy, the higher the efficiency of photon production. The physics of bremsstrahlung is discussed further in section 3.6.2.

In the production of bremsstrahlung much energy is lost by th electrons through ionisation and excitation of the material. For the most efficient generation of γ-rays the electron beam is accelerated transversely by macroscopic magnetic fields in vacuum, instead of by the atomic nuclear electric fields in matter. Such sources are called synchrotron light sources and have only recently become available for use other than as research tools.

3.1.2 Imaging with radiation

Ionising radiation may be used to image structure in two modes, coherent and incoherent. Probing coherently, the scattered Huygens wavelets interfere to reveal structure on scales down to $\lambda/2\pi$ where λ is the wavelength of the radiation. The de Broglie wavelength λ is given in terms of the momentum P as $\lambda = 2\pi\hbar/P$. By using radiation with enough momentum it is possible to 'see' down to the atomic scale at 10^{-10} m, or to the nuclear and quark scale at 10^{-15} m and below.

[1] Such photons are isotropic and have a spectrum characteristic of the Z of the atom. They may also be initiated by a higher energy photon (X-ray fluorescence or XRF) or by a proton beam (proton-induced X-ray emission or PIXE).

[2] The flatness of the spectrum is related to the (near) point-like charges of the nucleus and electron, the frequency spectrum of a delta function being flat.

Practical applications of ionising radiation as discussed in this book are not concerned with these scales, and radiation is used incoherently. Imaging structures on a macroscopic scale in this way, the wavelength of the radiation does not determine the spatial resolution. The radiation is merely used to measure along penetrating straight lines. No use is made of refraction, reflection or diffraction. It is the crude optics of the point source, the collimator, and the pinhole camera.

In addition to its use for imaging, X-ray and γ-radiation may be used to modify the medium through which it passes. The most important applications of interest are in the areas of electronics, cancer therapy, sterilisation and the polymerisation of organic molecules.

3.1.3   Single and multiple collisions

Range and cross section

How does radiation interact with matter? At a microscopic level the radiation, whether particles or photons, suffers collisions with the ions, atoms or molecules of which the matter is composed. There are two extreme kinds of such collision. An incident particle may suffer a catastrophic collision such that it is effectively removed from the beam. Alternatively it may simply suffer a slight loss of energy but otherwise continue with a rather small change in direction.

An example of the first kind is the behaviour of photons in matter due to the photoelectric effect. When a photoelectric collision occurs the photon is absorbed completely and an atomic electron is expelled from the atom. Therefore the range of photons under these conditions follows an exponential distribution (see Fig. 3.1) with mean range R, given by $1/R = N\sigma$, where N is the number of targets per m^3 and σ is the cross section in m^2 per target.[3]

An example of the second kind is the range of charged particles.[4] A particle finally stops after a large number of small collisions, n, each of which reduces its kinetic energy by a small amount. The mean rate of this energy loss with distance is called $\mathrm{d}E/\mathrm{d}x$. The range distribution is a sharp spike, broadened only by the effect of relatively small statistical fluctuations in the occurrence of the n collisions (see Fig. 3.1).

The range distribution of electrons in matter is described neither by a spike nor by an exponential. Because of its small mass an electron accelerates, and therefore radiates, in the nuclear electric field of atoms. Thus in addition to the gradual attrition of energy experienced by other charged particles, electrons suffer the additional process of radiative energy loss. The spectrum of this includes a significant cross section for emission of a photon carrying much of the electron energy. As a result, the electron range distribution includes the effect of a strong exponential component due to this bremsstrahlung process in addition to the usual $\mathrm{d}E/\mathrm{d}x$.

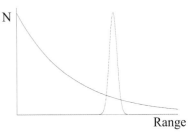

Fig. 3.1 Sketches of two examples of range distributions. An exponential distribution applies when the end of the range coincides with a single catastrophic collision. A narrow peak applies when the range is reached following attrition by a very large number of small collisions.

[3]Observe that it does not matter on what scale the 'target' is defined. For example, if N refers to the density of atoms and σ refers to the cross section per atom, the result will be the same as when N denotes the density of molecules and σ the cross section per molecule.

[4]This includes any charged particles, except electrons which are covered in the next paragraph.

Differential cross section

Whether the phenomenology of the interaction involves a large number of small collisions or a smaller number of significant collisions, the behaviour is determined solely by the cross section and its dependence on, for example, energy transfer[5] $\mathcal{E}$ and scattering angle θ. The probability per unit path of a given class of collision is

[5]In this chapter E denotes the energy of the incident charge and $\mathcal{E}$ the change to this in a single collision.

$$\mathrm{d}P = N\mathrm{d}\sigma. \tag{3.1}$$

So for a charged particle, the probability $\mathrm{d}P$ of a collision per unit path with an energy transfer between $\mathcal{E}$ and $\mathcal{E} + \mathrm{d}\mathcal{E}$ is

$$P(\mathcal{E})\,\mathrm{d}\mathcal{E} = \frac{\mathrm{d}\sigma}{\mathrm{d}\mathcal{E}}N\,\mathrm{d}\mathcal{E}, \tag{3.2}$$

where $\mathrm{d}\sigma/\mathrm{d}\mathcal{E}$ is called the differential cross section with respect to energy.

The dependence of the scattering on angle is described by the differential cross section with respect to angle, $\mathrm{d}\sigma/\mathrm{d}\theta$, or solid angle, $\mathrm{d}\sigma/\mathrm{d}\Omega$. An element of solid angle

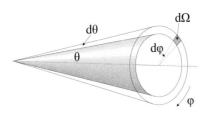

Fig. 3.2 The relationship between an element of solid angle $\mathrm{d}\Omega$ and the polar scattering angle θ and the azimuthal scattering angle ϕ.

$$\mathrm{d}\Omega = \sin\theta\,\mathrm{d}\theta\,\mathrm{d}\phi \tag{3.3}$$

in polar coordinates is shown in Fig. 3.2 with the incident beam as polar axis. If the scattering is independent of azimuth ϕ, we may integrate over it[6] so that $\mathrm{d}\Omega = 2\pi\sin\theta\,\mathrm{d}\theta$. Then the probability per unit path length for a collision to occur with a scattering angle between θ and $\theta + \mathrm{d}\theta$ is

[6]This will be the case provided neither beam nor target is polarised, and any final state polarisation is averaged. This will always be so in the following.

$$P(\theta)\,\mathrm{d}\theta = N\frac{\mathrm{d}\sigma}{\mathrm{d}\Omega}\,2\pi\sin\theta\,\mathrm{d}\theta. \tag{3.4}$$

3.2   Kinematics of primary collisions

In quantum mechanics the cross section is given by first order time-dependent perturbation theory as a product of two factors. These may be described as a kinematics factor and a dynamics factor. Kinematics involves the features that follow from energy–momentum conservation. Dynamics concerns the features arising from the matrix element of the interaction itself. Many processes can be described rather simply by the kinematics of an elastic collision with a stationary constituent mass in the medium. This gives a useful quantitative description of a diverse range of applications.

3.2.1 Kinematics and dynamics

A collision between the incident particle and the medium is usually localised to a particular constituent of that medium. The constituent
involved may be an atom, a whole molecule or just an individual electron. The rest of the medium is then not affected.[7] Such a collision
process is subject to the kinematic constraints of energy and momentum conservation. Kinematics is unaffected by the detailed mechanism
or interaction type and depends only on the masses involved. Other
factors in the cross section are described as dynamics and depend on
the interaction type.

Reaction rate and cross section

How is the cross section related to the rate of the interaction W that
may be calculated by quantum mechanics? W is the rate of the reaction
per second when the wavefunctions of the target and incident charge are
normalised in a volume $\mathcal{V}$. We imagine a single incident charge bouncing
around in this volume with velocity v and tracing out a volume σv per
second, in the sense that, if the target lies inside that volume σv, the
interaction happens in that second.[8] So the average interaction rate is
$W = \sigma v / \mathcal{V}$.

In first order time-dependent perturbation theory the interaction rate
W is given by Fermi's golden rule (FGR),[9]

$$W = \frac{2\pi}{\hbar} |\mathcal{M}|^2 \rho_f. \tag{3.5}$$

Thence the cross section,

$$\sigma = \frac{2\pi \mathcal{V}}{\hbar v} |\mathcal{M}|^2 \rho_f. \tag{3.6}$$

This comprises two factors, ρ_f the phase space factor which expresses the
kinematics and the square of the matrix element $|\mathcal{M}|^2$ which expresses
the dynamics.[10]

The value of a cross section is unchanged by a Lorentz transformation
along the incident momentum. It may therefore be evaluated in the
centre-of-mass frame and transformed into the rest frame of the target.
This is a useful picture. However, the transformation is not so trivial
for a differential cross section such as $d\sigma/d\Omega$, because the solid angle Ω
is not invariant under such a transformation.

3.2.2 Energy and momentum transfer

In a collision the material gets a kick and gains energy. The incident
particle gets an opposite kick and loses energy. In Fig. 3.3 we illustrate
this for a transfer of energy $\mathcal{E}$ and 3-vector momentum $\boldsymbol{Q}$ from the
particle to the target or medium. The $\mathcal{E}$ and Q form a 4-vector. The
length of the vector is called the 4-momentum transfer Q_4 and is given by
$Q_4^2 = Q^2 - \mathcal{E}^2/c^2$. As such, it is a relativistic invariant.[11] The differential

[7] A counter example is the emission of
Cherenkov radiation where the interaction is with the material as a whole. In
that case the emission is still linearly
dependent on path length and nothing
is lost by pretending that it is more or
less localised.

[8] To include the possibility that the target is moving, we simply note that it is
the relative velocity that is relevant.

[9] As derived in a standard text on quantum mechanics, such as one of those
given in the list for further reading at
the end of the chapter.

[10] We shall derive what we can using
the kinematics of energy and momentum conservation. Later in section 3.4
we shall tackle the dynamics. We shall
find that the volume $\mathcal{V}$ cancels out and
therefore can be taken as large and irrelevant.

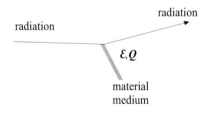

Fig. 3.3 The energy and 3-momentum
transfer between the radiation and the
medium as a result of a single collision.

[11] We have chosen a space-like definition for the 4-momentum transfer.

cross section

$$\frac{\mathrm{d}^2\sigma}{\mathrm{d}\mathcal{E}\,\mathrm{d}\boldsymbol{Q}} \tag{3.7}$$

is a function of $\mathcal{E}$, $\boldsymbol{Q}$ and the incident particle energy E. We may effectively ignore other variables.[12] We write $\mathcal{E} = \hbar\omega$ and the 3-vector momentum $\boldsymbol{Q} = \hbar\boldsymbol{k}$.

Before considering the general case involving both energy and momentum transfer, we look at examples of collisions that are dominated by one or the other.

Example of momentum transfer

Consider the case of the elastic scattering of low energy photons in a crystal. This is Bragg scattering. The photon energy is unchanged and there is no energy transfer to the medium, so that $\omega = 0$. In scattering the incident momentum vector of the X-ray is simply rotated through a significant angle, so that $\boldsymbol{k} \neq 0$.

The analysis of the experimentally observed cross section as a function of $\boldsymbol{k}$ for all sorts of crystals and molecules has enabled their structure to be deduced. The determination of the structure of DNA, the double helix, is the most famous historical example. The scattering depends on spatial structure, the electron density to be precise, and is not concerned with the resonant or frequency response of the medium.

Example of energy transfer

In photoabsorption (photoelectric effect) there is no outgoing photon and the whole incident energy and momentum are absorbed with $|\boldsymbol{k}| = \omega/c$. The energy transfer $\hbar\omega$ is now relatively large and the momentum transfer $\hbar\boldsymbol{k}$ relatively small. For example, in the photoelectric effect at optical frequencies the value of $\boldsymbol{k}$ corresponds to a wavelength of 0.5 μm, far larger than any atom. Therefore the EM wave is very uniform within the spatial dimension of any target atom or molecule.[13] The interaction is not sensitive to where the electron density is within the atom or molecule, only to what its frequency response is. Consequently the cross section depends on ω rather than on the spatial structure.

3.2.3   Recoil kinematics

Elastic recoil

Generally collisions depend in a significant way separately on both $\boldsymbol{Q}$ and $\mathcal{E}$ (or $\boldsymbol{k}$ and ω). The exception is elastic scattering for which they are related. Consider the kinematics of an elastic scatter of an incident particle by a constituent of the medium, mass M, assumed to be initially stationary.[14] On these assumptions, the final kinetic energy of the constituent is equal to the energy transfer, and its final momentum must be the momentum transfer.[15] But the constituent kinetic energy and momentum are related to its mass M. In the case of non-relativistic

[12]The particles may have spin upon which the cross section may also depend. But, provided the incident particles are not polarised and the polarisation of the outgoing particles is not measured, the effect of spin on the cross section is at most a constant, an overall statistical factor.

[13]The assumption that the electric field is completely uniform over the atom or molecule is called the dipole approximation.

[14]In an elastic scatter the identities of the incident particle and the struck particle survive intact. This is the meaning of the term 'elastic' as used here. Their energies and momenta may be exchanged but no new particles are produced. We ignore the binding energy of any constituent involved in a hard collision.

[15]We consider any kind of incident particle, including a photon.

motion by the constituent we have

$$\mathcal{E} = \frac{Q^2}{2M},$$ (3.8)

by elementary kinematics. In the relativistic case the relationship can be shown to be[16]

$$\mathcal{E} = \frac{Q_4^2}{2M}.$$ (3.9)

[16]See question 3.4.

In either case, for a given momentum transfer the energy loss to the incident particle is inversely proportional to the mass of the struck constituent.

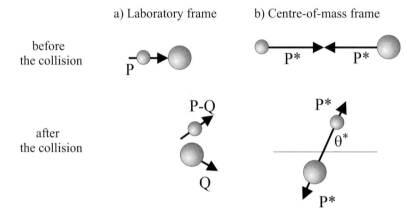

Fig. **3.4** Elastic scattering of incident radiation of momentum $\boldsymbol{P}$ by a stationary free constituent of a medium, a) as seen in the laboratory frame where the 3-momentum transfer is $\boldsymbol{Q}$, and b) as seen in the centre-of-mass frame where the length of the momentum vector P^* is constant and the only variable is θ^*.

Transformation from the centre of mass

As shown in Fig. 3.4 the process looks particularly simple in the centre-of-mass frame, where the only variable is the scattering angle θ^*. In this frame the energy and modulus of the momentum are unchanged in the scattering. In the laboratory frame there is seen to be an energy transfer. This is just a result of the Lorentz transformation from the centre of mass. The smaller the constituent-to-projectile mass ratio, the bigger this transformation and the larger the energy transfer. If the scatterer is heavy, the laboratory frame is almost the same as the centre-of-mass frame, and consequently the energy loss is small.

Equation 3.8 (or 3.9) is a useful result with a number of interesting and important practical applications.

3.2.4 Applications of recoil kinematics

Energy loss by charged particles

In their passage through materials charged particles encounter electrons as well as nuclei. For small momentum transfers $\boldsymbol{Q}$, electrons and nuclei are not resolved from one another. But if the momentum transfer $\boldsymbol{Q} = \hbar\boldsymbol{k}$ is such that $\lambda/2\pi = 1/|\boldsymbol{k}|$ is significantly smaller than the atom, electrons and nuclei may be seen as independent constituents by the incident charge. This value is about 25 keV/c for the whole atom. So separate constituent collisions with atomic electrons and nuclei are expected to occur with values of momentum transfer $Q > 25$ keV/c.

In this range of Q the energy transfers $\mathcal{E}$ for collisions with electrons and with nuclei are $Q^2/2m_e$ and $Q^2/2M_N$, respectively, where m_e is the mass of the electron and M_N is the mass of the nucleus. It follows that the energy loss in collisions with electrons is more than three orders of magnitude larger than with nuclei at the same Q. The conclusion is that constituent scattering by nuclei does not contribute to energy loss, which is dominated by scattering off atomic electrons.[17]

We have made two simplifications that can be justified at large momentum transfer. Firstly the electron binding energy was ignored. Implicitly we assumed the recoil energy to be well above this binding energy. Secondly we assumed the target atomic electron to be stationary, which is not quite true. In atomic hydrogen the RMS speed of the ground state electron $v = \alpha c$ where $\alpha = 0.0073$. This is typical for an outer electron and is negligible compared with the minimum electron recoil velocity discussed above, $Q/m_e = 0.05c$. We note that an inner atomic electron in an atom of higher Z is in a similar situation. It is bound in a smaller volume, has a greater RMS speed and a larger momentum transfer is required to resolve it.

An example of data is shown in Fig. 3.5. The calculation that shows that the peak in the data represents elastic scattering by atomic electrons is left to question 3.2. The effect of the moving electron is a broadening of the exact relation $\mathcal{E} = Q^2/2m_e$ with some atomic electrons moving towards the incident charge and some away from it.

[17] For incident electrons only, radiative scattering by nuclei is a source of energy loss, as discussed in section 3.6.2.

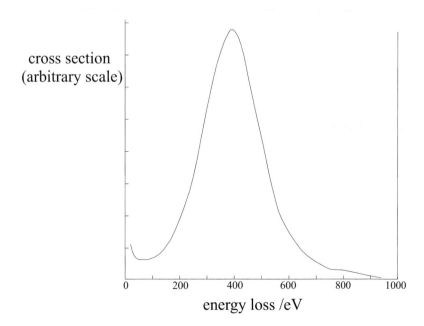

Fig. 3.5 The cross section for 25 keV electrons incident on helium, observed at 7° scattering angle and plotted against energy loss.

Rutherford back scattering

The scattering of an incident charged particle by the electric charge of a stationary nucleus is known as Rutherford scattering. The actual cross section will be discussed in section 3.4. However, using the kinematic relation 3.8, we can discuss the consequences of the recoiling nucleus without knowing the cross section. If a collision is observed in which the incident particle suffers a change in momentum Q and a change in energy $\mathcal{E}$, then the mass of the struck nucleus is

$$M_{\mathrm{N}} = \frac{Q^2}{2\mathcal{E}}. \tag{3.10}$$

This means that measurements of momentum and angle on the incident and scattered particles alone give a measurement of the struck nuclear mass, even though this is not itself detected. This is an important technique which can be used to analyse the concentration of nuclear masses in a small thin sample of material without chemical or physical preparation. To reduce the measurement error, data are taken in the backward direction where the changes in energy and momentum are largest. Hence the technique is called Rutherford back scattering (RBS). As discussed in chapter 6 this has been developed into a 3-D analytic imaging capability for thin samples.

Compton effect

The scattering of photons by free constituent electrons also shows a recoil effect. This is the famous Compton effect which helped historically to establish the semi-classical quantisation of the electromagnetic field. Because a photon carries momentum $2\pi\hbar/\lambda$, the momentum of the recoil electron depends on Planck's constant and its observation is a confirmation of quantum theory. The calculation of the kinematics is left to question 3.3.

Quark structure of hadrons

Often scattering data as a function of $\mathcal{E}$ and Q in an apparently inelastic process can be interpreted consistently in terms of elastic scattering from a constituent within the target. Indeed such an analysis may show evidence for constituents which are not otherwise observable. This was one of the principal pieces of evidence used to establish the existence of quarks inside hadrons, such as protons and neutrons.[18]

Neutron moderator in a nuclear fission reactor

In a reactor the fission process results in two heavy, slowly moving nuclear fragments and several free neutrons carrying away most of the energy released.[19] For this energy to reach the cooling circuits that power the turbines it needs to be thermalised, that is transmitted from the neutrons to the material of the reactor. This must be done quickly

[18] This is known as deep inelastic scattering (DIS); 'deep' means that it is at high Q, just as was the case for scattering from atoms.

[19] The fact that the neutrons get the lion's share of the kinetic energy is itself interesting. In a reaction the share-out between products is largely on the basis of momentum. It follows that the neutrons which are light carry away two orders of magnitude more kinetic energy than the fission fragments.

[20]This is the binding energy difference between a nucleus with an even number of neutrons and an odd number of neutrons. It is 0.5–1.0 MeV for high A nuclei, as given roughly by the Weissäcker semi-empirical mass formula.

[21]Neutron collisions with electrons could only be magnetic. Electrons have no strong nuclear interactions and neutrons, with zero charge, no electric ones. Consequently the cross section is negligible.

by elastic collisions that do not absorb the neutrons, as some of these at least are needed to maintain the chain reaction by inducing further fission processes. When successfully slowed to thermal energies the neutrons encounter large cross sections for absorption by the fuel, ^{235}U or ^{239}Pu. The resulting states of ^{236}U or ^{240}Pu are excited relative to their ground states by an energy equal to the pairing energy.[20] Such excited states have a high rate of spontaneous fission, and the chain reaction can then be maintained. Some other isotopes and impurities present, such as ^{238}U, absorb neutrons and do not lead to fission. This absorption is concentrated at a number of sharp resonances at certain energies. If the reduction in neutron energy takes place in large steps, the absorption will not be important because many neutrons simply never have the right energy for resonant absorption.

A fission reactor therefore contains a neutron moderator whose function is to reduce the neutron kinetic energy elastically in large steps. Its constituent nuclei suffer elastic collisions with high energy neutrons and large energy transfer.[21] For this to be true, our discussion has shown that these moderator nuclei must have low mass. It is the ratio of nuclear mass to neutron mass that is important. The usual choices are hydrogen (water), deuterium (heavy water), or carbon (graphite).

In the case of a solid moderator such as graphite, there is an important secondary effect of the large nuclear recoil energy. The struck nucleus may be kicked out of its location in the crystal lattice. The potential energy stored in the crystal in the form of vacant sites and interstitial occupation rises as the neutron irradiation continues. If the temperature is raised, atoms may get enough kinetic energy to hop from interstitial sites back into vacancies, thereby releasing the potential energy involved. This means that in such a temperature range a heavily irradiated neutron moderator may exhibit a negative lattice heat capacity and give a spontaneous increase in temperature. Management of this problem calls for frequent small excursions of temperature to anneal the lattice and to avoid the build-up of stored energy that can result in a runaway 'fire'. The worst nuclear radiation accident in the UK, the Windscale fire of 1956, was caused by mismanagement of this problem.

Radiation damage in materials

All solid materials are prone to radiation damage due to displaced atoms in the lattice, as discussed above. The effect is large when the momentum transfer is high and the nuclear mass is low. Such lattice dislocations and distortions give rise to donor and acceptor levels in the bandgaps of the electronic structure. If occupied, such states can absorb or scatter light for energies at which, prior to irradiation, the crystal was transparent. As a result, after low levels of irradiation transparent materials are seen to be coloured. At higher levels they become cloudy and opaque. This may be caused by the radiation populating existing metastable electron sites or by creating new ones as discussed above. In general there is coupling between the problems of electrons being excited and the lattice being excited by atoms being misplaced. In some cases transparency can be annealed by heating.

For the same reason the electronic behaviour of materials is directly affected by irradiation. Semiconductors are particularly sensitive because the density of carriers is low anyway. Radiation increases the carrier density, either through the creation of extra interstitial and vacancy sites in the lattice or through populating the resulting donor or acceptor levels with electrons. Neutron or high energy charged particle radiation are expected to have a larger effect than X-rays because of the higher momentum transfers. Generally after high levels of irradiation the characteristics of semiconductor components change and leakage currents rise because of the increased number of carriers.

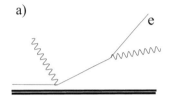

3.3   Electromagnetic radiation in matter

Photons have three significant interactions in matter. Compton scattering is a reaction with a small cross section, almost independent of energy and Z, in which the incident photon is scattered from an electron with reduced energy. The nucleus is a spectator. Photoabsorption (photoelectric effect) has a large energy-dependent and Z-dependent cross section in which the photon is completely absorbed by an atom. Pair production has a smaller cross section rising with both energy and Z from a threshold at $2m_ec^2$. The photon converts to an electron–positron pair in the nuclear Coulomb field and the atomic electron plays no role. .

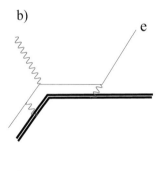

3.3.1   Compton scattering

Compton scattering is constituent scattering by atomic electrons, as shown in Fig. 3.6a. Because there are two simple electromagnetic vertices involving charge e, the interaction is second order. Therefore it is small compared with photoabsorption, a first order processes. The binding energy of the electrons can always be ignored in Compton scattering compared with the incident photon energy in the wavelength range where the electron can be resolved within the atom. In the low energy limit Compton scattering is called Thomson scattering.

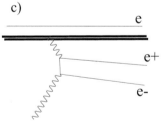

Classical Thomson scattering

If $\hbar\omega$ is much smaller than mc^2, the energy change of the photon due to the electron recoil is small and a classical picture of the interaction provides an adequate description. The electric field of the incident photon forces the electron to oscillate, and this oscillating charge then emits a dipole radiation field in its centre of mass. Electric dipole radiation is not isotropic but has a $\sin^2\theta$ angular distribution where θ is the angle

Fig. 3.6 The three processes for the interaction of electromagnetic radiation with an atom, with time shown running from left to right. The nucleus and electron are shown as a double heavy and light line, and photons as wavy lines.
a) Compton or Thomson scattering.
b) Photoabsorption, where the internal photon lines indicate significant internal momentum transfer by Coulomb fields.
c) Electron–positron pair production.

between the induced electron motion and the direction of the emitted photon, that is between the electric vector of the incident wave and the momentum vector of the outgoing wave.[22] Integrated over angle the Thomson cross section is,

$$\sigma_T = \frac{8\pi}{3} \left(\frac{\alpha\hbar}{m_e c} \right)^2. \tag{3.11}$$

This is a constant, 0.665 barn, independent of photon energy. It depends on the inverse square of the electron mass because the radiation amplitude is proportional to the electron acceleration. The equivalent cross section due to scattering by nuclear charges is therefore smaller by six or more orders of magnitude and can be ignored.

Relativistic corrections

Compton scattering is the name given to this process when it is no longer true that $\hbar\omega$ is much less than mc^2. The required relativistic corrections to Thomson scattering arise for two reasons.

The first is purely kinematic. The centre of mass of the scattering is moving in the laboratory frame. The Thomson scattering in the centre of mass, when transformed to the laboratory frame, gives an angular distribution which peaks at an angle $1/\gamma$ with respect to the incident photon direction. The same Lorentz transformation creates a Doppler shift in the photon energy. The decrease in photon energy in the laboratory frame appears as electron recoil energy.

The second relativistic correction is dynamic. There is a magnetic contribution in addition to the electric term, not only because v/c is now significant, but also because the magnetic moment associated with the spin of the electron plays a role. Further details of the relativistic field theory calculation need not concern us here. The resulting cross section falls slowly with energy from σ_T as $\hbar\omega$ approaches mc^2 and becomes[23]

$$\sigma(\omega) = \sigma_T \times \frac{3m_e c^2}{4\hbar\omega} \tag{3.12}$$

in the extreme relativistic region, $\hbar\omega \gg m_e c^2$.

3.3.2   Photoabsorption

Energy and Z-dependence

At low energy the largest contribution to the interaction of photons in materials comes from resonant interaction with the atom as a whole, resulting either in atomic excitation or in photoelectric ionisation. As shown in Fig. 3.6b there is only one external electromagnetic vertex in the excitation of the resonant state (the internal vertices just represent the potential that binds the resonant state itself), so this is a first order process. The cross section is typically three orders of magnitude larger than the Compton cross section and shows a sharp edge at the ionisation threshold of each electron energy shell. There are discrete bound

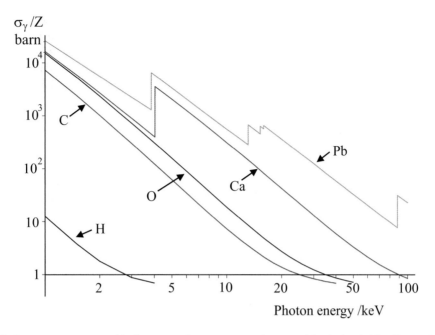

Fig. 3.7 The total photon cross section of hydrogen, carbon, oxygen, calcium and lead, divided by Z, as a function of photon energy. At high energies the cross sections fall to the Thomson cross section, 0.665 barn per electron.

state resonances just below the edge.[24] Above the edge the cross section falls roughly as $E^{-5/2}$. This fall may be envisaged as arising from a progressive loss of coherence between emitted Huygens wavelet amplitudes coming from different parts of the finite atomic volume involved. The pattern in energy of these thresholds or edges is a fingerprint of the atomic number of the element concerned. Important analytical applications are explored in chapter 6.

Figure 3.7 shows the photon cross section per electron (the atomic cross section divided by Z) for some typical elements of interest between hydrogen and lead. It is dominated by the photoabsorption process. The Thomson cross section, the same for all elements and independent of energy, is 0.665 barn, just below the axis as drawn on this log–log plot.[25] The edge for K-shell electrons in lead can be seen at 90 keV. The K-shell edge for calcium at 4 keV is at about the same energy as the M-shell edge for lead. The edges for the lower Z elements are below the left margin of this plot.

[24]These resonances when excited may decay by photon emission. This is known as resonance fluorescence.

[25]There is a difference between the photoabsorption process and Compton scattering. One gives rise to a scattered photon, whereas the other does not. They remain distinct. What is shown in the plot is the total cross section, the sum of the two.

Sum rule

The photoabsorption cross section per electron (σ_γ/Z), when integrated over frequency, is a constant for all elements,

$$\int_0^\infty \frac{\sigma}{Z}\,\mathrm{d}\omega = \frac{\pi e^2}{\varepsilon_0 m_e c}. \qquad (3.13)$$

So, although not apparent in the log–log plot, Fig. 3.7, the area under each curve is the same for all elements when plotted on a linear–linear scale and integrated down to low energy. This is known as the Thomas–Reiche–Kuhn sum rule. The photoabsorption spectra for all elements can be found on the NIST website.

3.3.3   Pair production

[26] The atomic electrons provide a much smaller source of field because it is not boosted by the Z^2-factor and its centre-of-mass energy is much lower, if not actually below pair threshold.

At photon energies above $2m_ec^2$ pair production (e$^+$e$^-$) can occur. This second order process involves the incoming photon interacting with the Coulomb field of the nucleus,[26] as shown in Fig. 3.6c. As a result the cross section involves a factor $Z^2\alpha^2$. Since particle creation is involved, the derivation of the cross section depends on quantum electrodynamics (QED).

The QED matrix element involves the same second order process as both Compton scattering

$$\gamma + e^- \rightarrow \gamma + e^- \tag{3.14}$$

and the annihilation process

$$e^+ + e^- \rightarrow \gamma + \gamma. \tag{3.15}$$

[27] The field is reduced at large distances by the electron screening of the nucleus, and at short distances by the charge distribution within the nucleus.

[28] This is defined in section 3.6.3.

Pair creation is the inverse process to annihilation, with the incoming γ interacting with a virtual γ, a component of the static nuclear Coulomb field. The nuclear Coulomb field and its Fourier components are modified[27] in the same way as in Coulomb scattering and bremsstrahlung, described later in this chapter. The effect of this modified Coulomb field may be parametrised in terms of the radiation length of the material,[28] X_0, usually expressed as the length times the density in g cm^{-2}. If material thickness is expressed as a multiple of X_0, all materials of the same thickness have the same properties as regards e$^+$e$^-$ pair creation, bremsstrahlung and multiple scattering, ignoring small threshold effects. At photon energies well above $2m_ec^2$ the mean range for a photon to create a pair is roughly equal to X_0. Values of X_0 for different materials are given in table 3.1.

3.4   Elastic scattering collisions of charged particles

Charged particles are scattered by point charges in a medium according to the Rutherford scattering formula. This differential cross section may be derived in quantum mechanics and in classical mechanics with the same result. The cross section rises as $\mathrm{cosec}^4(\theta/2)$, limited in practice at the *smallest angles by the screening of the nuclear charge by atomic electrons. Energy loss of the incident charge occurs by Rutherford scattering on target electrons.*

3.4.1   Dynamics of scattering by a point charge †

History

We derive the cross section for Rutherford scattering, that is the scattering of a point charge by a stationary fixed point charge, both by quantum mechanics and by classical mechanics.[29] The quantum derivation is straightforward. The classical derivation takes a number of lines and a good diagram.

We express the cross section in terms of the fine structure constant, $\alpha = e^2/(4\pi\varepsilon_0\hbar c)$.

Quantum derivation of cross section. The matrix element

Consider the incident charge ze of mass μ. Its wavefunction ψ describes a plane wave of momentum $\boldsymbol{P}$. The scattered wavefunction ψ' describes an outgoing wave of momentum $\boldsymbol{P}'$. Let the two waves be normalised in a large box of volume $\mathcal{V}$.[30] The wavefunctions are then

$$\psi = \frac{\exp(\mathrm{i}\boldsymbol{P}\cdot\boldsymbol{r}/\hbar)}{\sqrt{\mathcal{V}}} \text{ and } \psi' = \frac{\exp(\mathrm{i}\boldsymbol{P}'\cdot\boldsymbol{r}/\hbar)}{\sqrt{\mathcal{V}}} \tag{3.16}$$

To use Fermi's golden rule, equation 3.6, we need to evaluate the matrix element of the Coulomb potential between these states,

$$\mathcal{M} = \int\!\!\int\!\!\int_{\mathcal{V}} \psi'^*(\boldsymbol{r})V(\boldsymbol{r})\psi(\boldsymbol{r})\,\mathrm{d}^3\boldsymbol{r}. \tag{3.17}$$

We take the nucleus to be the origin and the Coulomb potential

$$V(r) = \frac{zZe^2}{4\pi\varepsilon_0 r} = \frac{zZ\alpha\hbar c}{r}. \tag{3.18}$$

We neglect the magnetic interaction of the charges and assume that the only force between them is electrostatic. We also neglect the recoil energy and assume that $M \gg \mu$. The rest frame of M is then the centre-of-mass frame. Let P be the magnitude of $\boldsymbol{P}$ (and also of $\boldsymbol{P}'$). Then the momentum transfer Q is given in terms of the scattering angle θ by

$$Q = 2P\sin(\theta/2), \tag{3.19}$$

as illustrated in Fig. 3.8.

Now we can evaluate the integral, equation 3.17. Writing the element of volume $\mathrm{d}^3\boldsymbol{r} = r^2\,\mathrm{d}\Omega\,\mathrm{d}r = 2\pi r^2\,\mathrm{d}\eta\,\mathrm{d}r$, where η is the cosine of the polar angle[31] between the momentum transfer $\boldsymbol{Q}$ and the vector $\boldsymbol{r}$,

$$\mathcal{M} = \frac{2\pi z Z \alpha\hbar c}{\mathcal{V}} \int_0^\infty \left[\int_{-1}^1 \exp\left(-\mathrm{i}Qr\eta/\hbar\right)\,\mathrm{d}\eta\right] r\,\mathrm{d}r. \tag{3.20}$$

[29] It is a matter of extraordinary historical importance that they give the same result. If the potential were other than a $1/r$, the classical calculation of the scattering cross section would give the wrong result. In that case, historically, the early shaky steps in the application of quantum ideas in the form of the Bohr atomic model would not have been made, and the theory of quantum mechanics as we know it today might not have been established, certainly at the time and in the way that it was.

[30] Later this volume cancels, so then formally we can let it go to infinity. It is the same box that was introduced already in equation 3.6.

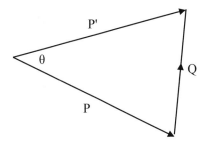

Fig. 3.8 The incident and scattered momentum vectors, P and P' with the 3-momentum transfer Q.

[31] The angles involved in this integration process should not be confused with the scattering angle θ.

Evaluating the η integral,

$$\mathcal{M} = -\frac{4\pi z Z \alpha \hbar c}{\mathcal{V}} \int_0^\infty \mathrm{d}r \, \frac{\sin(Qr/\hbar)}{Q/\hbar}. \tag{3.21}$$

The potential will be screened such that, for values of r exceeding atomic dimensions, the potential falls to zero significantly faster than the $1/r$ of the pure Coulomb potential. Therefore the r integration converges,[32]

$$\mathcal{M} = -\frac{4\pi z Z \alpha \hbar^3 c}{\mathcal{V} Q^2}. \tag{3.22}$$

Quantum derivation of cross section. The phase space factor

Next we evaluate the density of states factor in Fermi's golden rule, $\rho_f = \mathrm{d}N/\mathrm{d}E$. According to statistical mechanics there are $(2\pi\hbar)^{-3}$ states per unit volume of phase space, ignoring spin. So the number of final states in the scattering solid angle $\mathrm{d}\Omega$ and momentum range $\mathrm{d}P$ is

$$\mathrm{d}N = \frac{\mathcal{V} P^2 \, \mathrm{d}P \, \mathrm{d}\Omega}{(2\pi\hbar)^3}. \tag{3.23}$$

The phase space factor is the density of states per unit energy $\mathrm{d}E$, where $E^2 = P^2 c^2 + \mu^2 c^4$ and $E\mathrm{d}E = c^2 P \mathrm{d}P$. Thus

$$\rho_f = \frac{\mathrm{d}N}{\mathrm{d}E} = \frac{\mathcal{V} P E \, \mathrm{d}\Omega}{c^2 (2\pi\hbar)^3}. \tag{3.24}$$

Substituting in equation 3.6, the dependence on the volume $\mathcal{V}$ cancels, then

$$\frac{\mathrm{d}\sigma}{\mathrm{d}\Omega} = \left(\frac{2zZ\alpha\hbar}{\beta Q^2} P \right)^2, \tag{3.25}$$

where we have also used $v = \beta c = Pc^2/E$. The only factor depending on the scattering angle θ is $Q = 2P\sin(\theta/2)$. The Rutherford scattering formula with the explicit angular dependence is

$$\frac{\mathrm{d}\sigma}{\mathrm{d}\Omega} = \left(\frac{zZ\alpha\hbar}{2P\beta} \right)^2 \mathrm{cosec}^4(\theta/2). \tag{3.26}$$

Classical derivation of cross section

In classical mechanics (including the effects of relativity) we may follow the trajectory of the charge as it is deflected by the Coulomb field. This depends on knowing the position and momentum of the charge at all times, a state of knowledge forbidden in quantum mechanics. The following derivation is therefore unacceptable, although it gives the correct answer.

Figure 3.9 shows such a classical trajectory in the xy-plane with the y-axis chosen as the axis of symmetry. The incoming charge has an initial velocity βc at an angle $\theta/2$ with respect to the x-axis and an impact parameter b with respect to the origin, that is the distance of

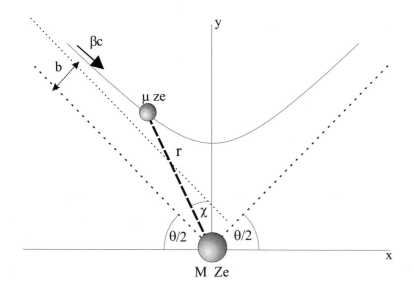

Fig. 3.9 Variables describing the trajectory of a classical particle deflected through angle θ by a Coulomb $1/r$ potential fixed at the origin. The y-axis is chosen as the axis of symmetry of the trajectory.

closest approach of the line of the initial trajectory. At any point on its path a distance r from the origin, its angle with respect to the y-axis is χ, as shown. The incident linear momentum $\boldsymbol{P}$ is $\beta\gamma\mu c$ where $\gamma = 1/\sqrt{(1 - \beta^2)}$. The initial angular momentum of the charge about the origin is $\beta\gamma\mu cb$ and at a later time t is $\mu\gamma r^2\,\mathrm{d}\chi/\mathrm{d}t$. Since these must be equal,[33]

$$r^2\,\mathrm{d}\chi/\mathrm{d}t = \beta cb. \tag{3.27}$$

The axes are chosen such that the path is symmetric about the y-axis and therefore $\boldsymbol{Q}$ is along the y-axis. $\boldsymbol{Q}$ is also equal to the time integral of the y-component of the Coulomb force $F = zZe^2/(4\pi\varepsilon_0 r^2)$; thus

$$Q = \int_{-\infty}^{\infty} F\cos\chi\,\mathrm{d}t = \int_{-(\pi-\theta)/2}^{(\pi-\theta)/2} \frac{zZe^2}{4\pi\varepsilon_0 r^2}\cos\chi\,\frac{\mathrm{d}t}{\mathrm{d}\chi}\,\mathrm{d}\chi. \tag{3.28}$$

Substituting for $r^2\,\mathrm{d}\chi/\mathrm{d}t$ from equation 3.27 gives

$$Q = \frac{2zZe^2\cos(\theta/2)}{4\pi\varepsilon_0\beta cb}. \tag{3.29}$$

Combining equations 3.19 and 3.29 gives a unique relation between the scattering angle θ and the impact parameter b,

$$b = \frac{zZe^2}{4\pi\varepsilon_0 P\beta c}\cot(\theta/2). \tag{3.30}$$

The cross section between b and $b+\mathrm{d}b$ is simply $2\pi b\,\mathrm{d}b$. Then, using equation 3.30, we can write down the cross section between θ and $\theta+\mathrm{d}\theta$,

$$\mathrm{d}\sigma = \pi\left(\frac{zZe^2}{4\pi\varepsilon_0 P\beta c}\right)^2 \frac{\cos(\theta/2)}{\sin^3(\theta/2)}\,\mathrm{d}\theta. \tag{3.31}$$

[33]This depends on an assumption that γ is constant.

Using the element of solid angle, $\mathrm{d}\Omega = 2\pi\sin\theta\,\mathrm{d}\theta$, the differential cross section is then

$$\frac{\mathrm{d}\sigma}{\mathrm{d}\Omega} = \frac{1}{2\pi\sin\theta}\frac{\mathrm{d}\sigma}{\mathrm{d}\theta}. \tag{3.32}$$

Thus

$$\frac{\mathrm{d}\sigma}{\mathrm{d}\Omega} = \left(\frac{zZe^2}{8\pi\varepsilon_0 P\beta c}\right)^2 \operatorname{cosec}^4(\theta/2). \tag{3.33}$$

Expressed in terms of α we get the same expression as we found by the quantum route,

$$\frac{\mathrm{d}\sigma}{\mathrm{d}\Omega} = \left(\frac{zZ\alpha\hbar}{2\beta P}\right)^2 \operatorname{cosec}^4(\theta/2). \tag{3.34}$$

Features of the Rutherford cross section

The cross section depends on the square of both charges and therefore the scattering is independent of their signs. In quantum mechanics this follows from the use of first order perturbation theory. The cross section could only depend on the sign of the charge if there were interference between different orders. With the obvious exception of the sense of deflection in a magnetic field, it is very unusual for the sign of the charge to affect a scattering experiment.[34] Because it depends on Z^2 the scattering cross section by heavy nuclei is largest. For the same reason the contribution to scattering by atomic electrons with charge 1 is much smaller, although there are Z of them. Approximately their contribution may be taken into account by replacing the factor Z^2 by $Z(Z+1)$.

The cross section contains the factor $1/\beta^2$. This means that the cross section gets very large as charged particles slow down towards the end of their range. This feature is inherited by all results that depend on Rutherford scattering, with important technical consequences.

The angular distribution is rather flat in the backward hemisphere, Fig. 3.10b, but a very steep function of angle near the forward direction, Fig. 3.10a. The singularity at $\theta = 0$ and zero momentum transfer does not actually happen because the nuclear Coulomb field is screened by atomic electrons at small momentum transfer (or large impact parameters). Still, in passing through even a thin sample of material, a charged particle suffers a very large number of very small scatters. Thus most observations relate to the effects of multiple scattering, rather than to the single encounters described by the formula. Multiple scattering is described in section 3.5.

In the scattering process the incident charge is accelerated, and such an accelerating electric charge radiates. The acceleration is inversely proportional to the mass μ and the radiated energy varies as $1/\mu^2$. Consequently this is important for electrons and unimportant for heavier charged particles. Radiative Rutherford scattering or bremsstrahlung is described in section 3.6.2.

[34]There are two other cases:

- If the incident particle moves at a velocity comparable with the atomic electrons ($v \sim \alpha c$);

- if it interacts coherently with crystal planes (so-called channelling), then first order perturbation theory breaks down and a dependence on the sign of charge is observed.

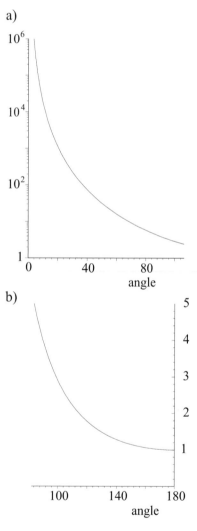

a)

b)

Fig. 3.10 The Rutherford scattering cross section as a function of angle shown relative to its value at 180°. a) At forward angles (note the logarithmic scale). b) At backward angles (note the linear scale).

3.4.2   Cross section for energy loss by recoil

Relativistic invariant cross section

We express the cross section in terms of the relativistic invariant Q_4^2 instead of θ and Ω. Since $M \gg \mu$ the cross section 3.26 refers to the centre-of-mass frame. In this frame, there is no energy transfer and so the 4-momentum transfer Q_4 is equal to the 3-momentum transfer Q,

$$Q_4^2 = Q^2 = (2P \sin(\theta/2))^2 \,, \tag{3.35}$$

and also

$$dQ_4^2 = 2P^2 \sin\theta \, d\theta = \frac{P^2}{\pi} \, d\Omega. \tag{3.36}$$

Hence the Rutherford cross section 3.25 expressed in terms of relativistic invariants is

$$\frac{d\sigma}{dQ_4^2} = \frac{4\pi z^2 Z^2 \alpha^2 \hbar^2}{\beta^2 Q_4^4}. \tag{3.37}$$

This cross section is the same in any Lorentz frame as Q_4 and β, the relative speed of beam and target, are both Lorentz scalars.

Differential energy loss cross section

If the target mass M is finite then, as seen in the frame in which M is initially stationary, it will recoil with a consequential loss of energy $\mathcal{E}$ for the incident particle. To re-express the invariant cross section in terms of energy loss, we change variables using equation 3.9, $Q_4^2 = 2M\mathcal{E}$. Thus

$$\frac{d\sigma}{d\mathcal{E}} = \frac{2\pi z^2 Z^2 \alpha^2 \hbar^2}{M \beta^2 \mathcal{E}^2}. \tag{3.38}$$

Note the inverse dependence on the target mass M. We expected this from our study of kinematics and now we see it in the formula for the actual cross section. This factor shows why the most important application of this formula is to collisions with atomic electron targets with $M = m_e$ and target charge $Z = 1$, rather than with nuclei for which the energy transfer is 3–4 orders of magnitude smaller. This is in contrast to the relative importance of nuclei and electrons to scattering in Q^2 or θ, where nuclei dominate because of the factor Z^2 and the absence of an inverse mass dependence.

3.5   Multiple collisions of charged particles

Because of the large size of the Rutherford cross section, most practical observations relate to the effect of many collisions. The energy loss of an incident charged particle by Rutherford scattering on electrons may be integrated to get the mean accumulated energy loss in a finite thickness of material. The formula so derived is close to the standard Bethe–Bloch formula. This shows that energy loss is determined by the incident charge, its velocity and the target electron density, but is otherwise insensitive to the medium. Thence universal relations between range, expressed in mass per unit area, and energy may be derived which apply to all charged particles incident on a material. Further, even the variation between different materials is quite small. The cumulative effect of scattering (in angle) in finite thickness of material depends on the Rutherford scattering from nuclei. The resulting formula for multiple scattering is characterised by the radiation length of the material, a measure of the effectiveness of the screened Coulomb field of the constituent nuclei and their density. Fluctuations in both the energy loss and the scattering angle may be significant.

3.5.1   Cumulative energy loss of a charged particle

Mean energy loss in thickness dx

We calculate the rate of energy loss of a charge due to the many collisions that it suffers as it passes through material containing N electrons m^{-3}. From equation 3.38 the differential cross section due to constituent scattering by atomic electrons is

$$\frac{d\sigma}{d\mathcal{E}} = \frac{2\pi z^2 \alpha^2 \hbar^2}{m_e \beta^2 \mathcal{E}^2}. \tag{3.39}$$

To determine the average total loss dE of the incident energy E due to all the collisions experienced by a charged particle in traversing a layer of material with thickness dx,[35] we integrate the contribution of all collision energies from $\mathcal{E}_{\min}$ to $\mathcal{E}_{\max}$. Thus

[35] So $N dx$ electrons m^{-2}.

$$
\begin{aligned}
dE &= -N\,dx \int_{\mathcal{E}_{\min}}^{\mathcal{E}_{\max}} \mathcal{E}\frac{d\sigma}{d\mathcal{E}}\,d\mathcal{E} \\
&= -N\,dx \frac{2\pi z^2 \alpha^2 \hbar^2}{m_e \beta^2} \int_{\mathcal{E}_{\min}}^{\mathcal{E}_{\max}} \frac{d\mathcal{E}}{\mathcal{E}}.
\end{aligned}
\tag{3.40}
$$

The mean rate of energy loss with distance travelled x is then

$$\frac{dE}{dx} = -\frac{2\pi z^2 \alpha^2 \hbar^2 N}{m_e \beta^2} \ln\left(\frac{\mathcal{E}_{\max}}{\mathcal{E}_{\min}}\right). \tag{3.41}$$

That leaves the problem of how to handle this logarithmic factor.

Maximum and minimum energy transfers

The minimum energy transfer that can be given to each atomic electron is its binding energy and the maximum energy must be closely related to the incident energy. We refine these two statements later. But first we note that their ratio is large, often 10^6 and frequently more. So its logarithm changes by about 1%, if either $\mathcal{E}_{\min}$ or $\mathcal{E}_{\max}$ is changed by 20%. Consequently we only need to estimate their dependence on atomic number Z and the parameters of the beam quite roughly.

Since an atom has a spectrum of binding energies and we are interested in the average of a logarithm, we define the mean ionisation potential, I,

$$\ln I = \frac{\int \ln E \, \sigma_\gamma(E) \, \mathrm{d}E}{\int \sigma_\gamma(E) \, \mathrm{d}E}, \tag{3.42}$$

where $\sigma_\gamma(E)$ is the photoabsorption cross section for photon energy E. This can be calculated from data for each atom. The result is nearly proportional to atomic number Z, such that $I/Z \approx 10$ to 10% above $Z = 12$, as shown in Fig. 3.11. We make the approximation $\mathcal{E}_{\min} = I$.

The maximum energy transfer to an atomic electron in a single collision corresponds to a $180°$ collision in the centre of mass. Assuming that $\mu \gg m_e$ the centre-of-mass momentum is $m_e c \beta \gamma$ and the maximum Q in that frame is $2 m_e c \beta \gamma$. Lorentz-transformed back to the initial electron rest frame, this gives a maximum energy transfer of order $\mathcal{E}_{\max} = 2 m_e c^2 \beta^2 \gamma^2$.

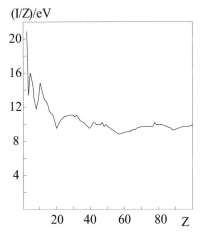

Fig. 3.11 The value of I/Z plotted as a function of atomic number Z where the mean ionisation potential I is defined by equation 3.42.

The Bethe–Bloch formula

Putting these results into the log factor we get

$$\frac{\mathrm{d}E}{\mathrm{d}x} = -\frac{2\pi z^2 \alpha^2 \hbar^2 N}{m_e \beta^2} \ln\left(\frac{2 m_e c^2 \beta^2 \gamma^2}{I}\right). \tag{3.43}$$

Except for a factor 2 and a small extra term, this is the same as the Bethe–Bloch formula for $\mathrm{d}E/\mathrm{d}x$,

$$\frac{\mathrm{d}E}{\mathrm{d}x} = -\frac{4\pi z^2 \alpha^2 \hbar^2 N}{m_e \beta^2} \left[\ln\left(\frac{2 m_e c^2 \beta^2 \gamma^2}{I}\right) - \beta^2\right]. \tag{3.44}$$

The factor 2 relates to the electron binding energy that we ignored.[36] We have succeeded in deriving the functional form of the rate of charged particle energy loss. Importantly for electrons, it does not include the effect of radiative energy loss, see section 3.6.2.

This is a strange formula because, effectively, it only depends on N, z^2 and $1/\beta^2$. In particular it only depends on the velocity of the incident charge but not on its mass μ. It is only dependent on the material which the charged particle is traversing, through the overall electron density N and through the logarithm of I, which does not vary much. Electron density and mass density are roughly proportional for most materials except hydrogen. So measurements of energy loss by charged particles are only indicative of material density. This is an important conclusion. Figures for a number of elements and materials are given in table 3.1 in terms of density. Compared with photon beams, charged particle beams have very little discrimination between materials.

The rate of energy loss depends on z^2 and is the same for positive and negative particles.[37] An important practical consequence is that energy loss is four times larger for alpha particles with charge $2e$ than for protons of the same velocity. The effect on the rate of energy loss and the range of the highly charged daughter nuclei created in the process

[36] The origin of the extra term is in part an effect of medium polarisation in the limit of low density and in part a spin–orbit effect that assumes that the incident charge is spinless. Neither of these effects is important to us here, and neither of them is well described by the Bethe–Bloch formula either. For further discussion see the review by Allison and Cobb.

[37] As previously noted, at the very lowest charge velocities (of the same order of speed as the atomic electrons) first order perturbation theory breaks down and positive and negative charges start to behave differently. In particular positive charges pick up atomic electrons and negative charges do not. However, this effect is completely negligible until the last few microns before the particle stops altogether.

Table 3.1 The energy loss coefficient and radiation length, $1/\rho \, dE/dx$ and X_0, for some materials. The energy loss coefficient should be multiplied by $z^2\rho/\beta^2$ to get the rate of energy loss with distance for a charge ze moving at velocity βc. The value of X_0 should be divided by ρ to get the radiation length in cm. The critical electron energy E_c at which dE/dx equals the radiative energy loss is given in the final column.

Material	Z	Z/A	$1/\rho \, dE/dx$ MeV cm^2 g^{-1}	X_0 g cm^{-2}	Density g cm^{-3}	E_c MeV
H$_2$ gas	1	0.992	4.103	61.28	0.089×10^{-3}	251
Carbon	6	0.500	1.745	42.70	2.265	75
N$_2$ gas	7	0.500	1.825	37.99	1.250×10^{-3}	68
O$_2$ gas	8	0.500	1.801	34.24	1.428×10^{-3}	62
Aluminium	13	0.482	1.615	24.01	2.70	38
Silicon	14	0.498	1.664	21.82	2.33	36
Argon gas	18	0.451	1.519	19.55	1.78×10^{-3}	30
Iron	26	0.466	1.451	13.84	7.87	20
Copper	29	0.456	1.371	12.25	8.96	17
Germanium	32	0.441	1.264	8.82	5.32	11
Xenon gas	54	0.411	1.255	8.48	5.86×10^{-3}	11
Tungsten	74	0.403	1.145	6.76	19.3	8
Lead	82	0.396	1.123	6.37	11.35	7
Air		0.500	1.815	36.66	1.205	66
Water		0.555	1.991	36.08	1.000	72
Concrete		0.503	1.711	26.7	2.50	46
SiO$_2$		0.499	1.699	27.1	2.20	46
Polystyrene scint.		0.538	1.936	43.7	1.03	85
NaI		0.427	1.305	9.49	3.67	12
BaF$_2$		0.422	1.303	9.91	4.89	13
BGO		0.421	1.251	7.97	7.1	10
PbWO$_4$		0.412	1.57	7.04	8.28	11

[38]The apparent dependence on $\ln(\beta\gamma)$ is usually suppressed in practice by the density effect, a material polarisation effect not accounted for by the Bethe–Bloch formula.

[39]With a computer it is quite simple to calculate these fluctuations from a knowledge of the cross section. When attempting to derive results in a pre-computer age, Landau was obliged to make approximations which are rarely valid. There is no need to use his approximations in any calculation today. See the review by Allison and Cobb.

of nuclear fission is very pronounced. We return to this when we discuss radiation damage by alpha particles and more highly charged ions in chapter 6 and again in the discussion of radiotherapy in chapter 8.

The energy loss varies as $1/\beta^2$ universally for all materials.[38] As a charge passes through a material it loses energy and slows down. As β falls, the factor $1/\beta^2$ rises sharply. Thus energy is deposited ever more densely until the charge finally stops.

Because of the stochastic nature of the energy loss process, individual charges experience different values of energy loss in the same absorber under the same circumstances. These variations are called Landau fluctuations.[39] The effect of these fluctuations can be important in very thin absorbers, and they generate variations in range, termed straggling.

The mean rate of energy loss with distance, dE/dx, with its factor

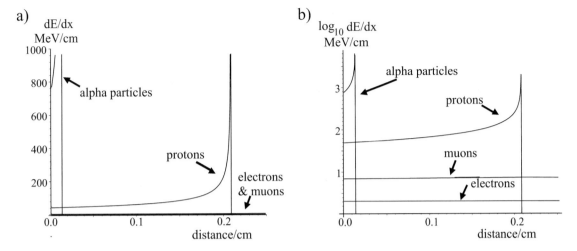

Fig. 3.12 The calculated rate of energy loss as a function of depth in water for various particles with initial kinetic energy 20 MeV, ignoring straggling. a) On a linear scale; b) on a logarithmic scale.

z^2/β^2 and linear dependence on electron density is called the linear energy transfer (LET) in the context of radiation damage. It is important in determining the localisation of radiation damage in living tissue. As discussed in chapters 6 and 8 this plays a major role in determining the probability that living tissue recovers from a given radiation dose.

3.5.2   Range of charged particles

Dependence of range on mass for incident particles of known energy

The effect of the z^2/β^2 dependence of the energy loss is illustrated in Fig. 3.12, neglecting Landau fluctuations. The rate of energy deposition with depth is calculated for various charged particles of 20 MeV kinetic energy passing into water. The alpha particles lose energy so rapidly that they stop within 150 μm. The protons go further and stop in about 2.1 mm. There is a very sharp peak in the rate of energy deposition just before they stop. In Fig. 3.12b the same calculations are shown on a logarithmic scale. In practice fluctuations spread the values of β at each distance and have a blurring effect on the peaks of these distributions when averaged over a number of particles. Such distributions are known as Bragg curves and the broadened peaks known as Bragg peaks.[40]

How far do charged particles of a given energy E go? There is a certain mean range and because of straggling there is a narrow spread in the distribution. Straggling is a relatively small effect because the number of collisions involved is large. In fact the number of collisions that occur is on the scale of 10^5 to 10^6 for the 20 MeV particles considered in Fig. 3.12. However, the fluctuations are dominated by a relatively small number of rather energetic collisions.

If a proton beam of the right energy is used for radiotherapy, a large

[40] In addition to fluctuations in its energy loss, each particle scatters so that its path is not a straight line. Strictly speaking, the mean range is the track length along its path, rather than the straight line to its stopping point. However, the effect on the range is small, except for electrons.

highly localised dose can be delivered at the point where the protons stop. Thus by locating the Bragg peak in a small target volume in a tumour within healthy tissue the energy may be deposited preferentially just where it is required as discussed in section 8.4.

Range–energy relations

In a given material the mean range R of any incident charged particle obeys the scaling relation

$$R = \frac{\mu}{z^2} \, \Phi\left(\frac{E}{\mu}\right) \tag{3.45}$$

where E is the initial kinetic energy, μ is the mass and ze the charge. The function $\Phi(E/\mu)$ is the same for all charges and is determined only by the medium, primarily through its electron density. This relation can be proved from the Bethe–Bloch formula by observing that the range is the integral along the particle path from the point at which it has kinetic energy E to the point where this falls to zero:

$$R = \int \mathrm{d}x = \int_E^0 \frac{1}{\mathrm{d}E'/\mathrm{d}x} \, \mathrm{d}E' \tag{3.46}$$

The energy loss is a function of the velocity of the charge only, that is there exists a function F characteristic of the medium such that

$$\frac{\mathrm{d}E}{\mathrm{d}x} = -z^2 F\left(\frac{E}{\mu}\right) \tag{3.47}$$

Then

$$R = -\frac{\mu}{z^2} \int_{E/\mu}^0 \frac{\mathrm{d}(E'/\mu)}{F(E'/\mu)} = \frac{\mu}{z^2} \, \Phi(E/\mu) \tag{3.48}$$

where the integral Φ depends only on the medium and the ratio (E/μ).

This is called the range–energy relation, sometimes expressed equivalently as the range–momentum relation. It says that there is a universal curve for a given material that applies for all particles. It is plotted for hydrogen, carbon and lead in Fig. 3.13. Actually, because of the relative insensitivity to parameters describing the medium, the universal curves for different materials are all rather similar when expressed in terms of (electron) density.

3.5.3   Multiple Coulomb scattering

Calculation of RMS scattering angle on passing through a material

When a parallel beam of charged particles passes through a slice of material of thickness ℓ, it exits with a symmetric distribution in net scattering angle due to the cumulative effect of Rutherford scattering. Scattering collisions at large impact parameters with small angular deviations occur many times while those at small impact parameters occur infrequently

Proton range
× density
kg m⁻²

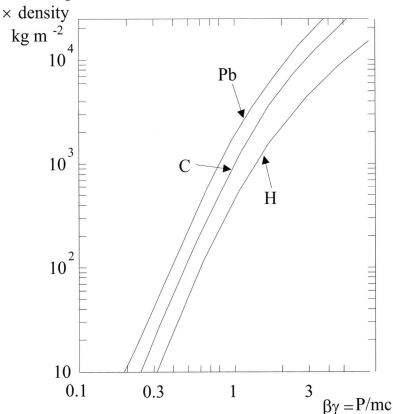

Fig. 3.13 Universal curves for hydrogen, carbon and lead showing proton range times density as a function of the dimensionless $\beta\gamma = P/mc$, for unit charge. [Originally from the Particle Data Group website.]
For particles of mass $m = \mu$, the range scales as m_P/μ. For particles of charge z, the range scales as $1/z^2$.

but give large contributions to the mean scattering. If the beam is along the z-axis, there is scattering in both the xz- and yz-planes. These two distributions are the same and symmetric. It is usual to refer to this projected distribution and its RMS width, θ_{MS}. This introduces a factor of $\frac{1}{2}$ in the expression for the mean square scattering angle:

$$\theta_{MS}^2 = \frac{1}{2} \int \theta^2 \, N\ell \, \frac{d\sigma}{d\Omega} \, d\Omega \qquad (3.49)$$

where N is the nuclear target density with cross section $d\sigma/d\Omega$ given by equation 3.26.[41]

$$\theta_{MS}^2 = \frac{N\ell \, Z(Z+1)}{2} \left(\frac{z\alpha\hbar}{2P\beta} \right)^2 \int \theta^2 \mathrm{cosec}^4(\theta/2) \, 2\pi \sin\theta \, d\theta \qquad (3.50)$$

We have included the scattering due to the atomic electrons. There are Z of these per atom but their cross section is smaller by factor $1/Z^2$. So we have replaced Z^2 by $Z(Z+1)$.

Unfortunately this integral diverges when evaluated between 0 and π. We need to consider the maximum and minimum scattering angles

[41] For simplicity we have assumed a material composed of a single element with atomic number Z.

for which equation 3.26 applies. The scattering is reduced by electron screening of the nuclear charge at large distance a and by the spread of the nuclear charge due to its finite size at small distance r_N, that is when the momentum transfer is less than h/a or greater than h/r_N.[42] The mean charge radius of the atomic electron distribution is about $a = 0.5 \times 10^{-10} Z^{-1/3}$ m according to the Thomas–Fermi model of the atom. The nuclear radius is $r_N = 1.2 \times 10^{-15} A^{1/3}$ m. So for momentum P we have as the limits of integration a lower limit of $\theta_1 = 2\pi\hbar/(Pa)$ and an upper limit $\theta_2 = 2\pi\hbar/(Pr_N)$. Then the result is

$$\theta_{MS}^2 = \frac{N\ell z^2 Z(Z+1)}{2} \left(\frac{\alpha\hbar}{2P\beta}\right)^2 \int_{\theta_1}^{\theta_2} \theta^2 \mathrm{cosec}^4(\theta/2) 2\pi \sin\theta \, d\theta \quad (3.51)$$

Practical estimation of multiple scattering

Such calculations give fair agreement with accepted values for RMS multiple scattering for all elements from hydrogen to lead within a few per cent. The square root dependence of the mean net angle on the slice thickness ℓ is characteristic of the random walk nature of the scattering process.

The distribution of projected scattering angles is not quite Gaussian. The distribution has a tail at very large angles.[43] The distribution of these unusual single scatters is described by Poisson statistics. However, they are rare in thin samples. From the point of view of the practical applications referenced in this book, the distribution of projected scattering angles may be taken to be a Gaussian distribution with the variance calculated.

In practical applications most materials are composed of more than one element. The scattering may be combined in quadrature, but it is usual to use tabulations of the radiation length X_0 for the composite material concerned. Some figures are given in table 3.1. Thus

$$\theta_{MS} = \frac{13.6}{\beta P} \sqrt{\frac{\ell\rho}{X_0}}, \quad (3.52)$$

where $\ell\rho$ is the thickness multiplied by the density and X_0 is the radiation length of the medium.[44] The X_0 conveniently encapsulates all of the physics of the medium discussed above into a single number. The value 13.6 and the momentum P are in units of MeV/c.

Question 3.7 provides a numerical example of the application of equation 3.52. It concerns the scattering in the carbon ion 'stripper' channel of a radiocarbon accelerator mass spectrometer of the type shown in Fig. 6.12.

3.6 Radiative energy loss by electrons

Radiative Rutherford scattering in the nuclear field is important if the incident charges are electrons. Because of their low mass they accelerate and radiate readily. This bremsstrahlung process may be understood simply in the centre-of-mass frame of the incident electron in terms of Compton scattering of virtual photons. The energy loss due to bremsstrahlung involves the same screened nuclear field as in multiple scattering and pair production. The radiation length of different materials is defined in terms of bremsstrahlung production.

3.6.1 Classical, semi-classical and QED electromagnetism

An incident charge that scatters in the nuclear Coulomb field of a target atom accelerates transversely. If it has low mass μ this acceleration may be large and it is likely to radiate. In classical mechanics the radiated energy is uniquely determined by the acceleration; in quantum mechanics the acceleration only determines the probability of photon emission. This can be calculated by the Planck mechanism of equating the classical radiated energy $I(\omega)\Delta\omega$ in each frequency band $\Delta\omega$ to the product of the probability, $P(\omega)\Delta\omega$, of emitting a photon in that band and the energy per photon $\hbar\omega$. Thus $P(\omega) = I(\omega)/(\hbar\omega)$. This is called the semi-classical approximation. If the particle radiates a significant fraction of its kinetic energy, this approximation fails and a calculation in QED is required. Semi-classical methods are sufficient for most purposes and we present such a picture in the following.

3.6.2 Weissäcker–Williams virtual photon picture

To a stationary observer a passing charged particle gives a pulse of electric field. The pulses are sketched in Fig. 3.14. The longitudinal field forms an odd function of time, rising as the charge approaches and then falling rapidly to zero and reversing sign as the charge passes and retreats away. The transverse field is an even function of time, rising to a maximum as the charge passes the observer and then falling again.

We can analyse this pulse into the frequency components seen by the observer, and interpret the power spectrum in each frequency band as the probability of detecting a photon. Under relativistic conditions these photons behave as if they were free.

Bremsstrahlung in the virtual photon picture

Consider the bremsstrahlung emitted by a high velocity electron passing a lead nucleus which is stationary in the laboratory. What understanding of the process can we derive from the virtual photon picture?

The heavy lead nucleus is not much affected by the passing of a singly charged electron. However, consider the electron's view of the process. It sees a charge of $82e$ passing at the same relativistic speed that the

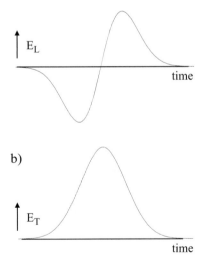

Fig. 3.14 a) The longitudinal electric field pulse, and b) the transverse field pulse, seen as a function of time by an observer close to the trajectory of a passing charge.

electron has in the laboratory. According to the virtual photon picture this rapidly moving charge of $82e$ is seen densely shrouded in photons. The situation is illustrated in cartoon form in Fig. 3.15. Because the charges are point-like over the scale 10^{-15} to 10^{-10} m the Fourier analysis of the field determines that the spectrum of photons is flat over a broad range of frequencies. The amplitude of the field depends on Z and the flux on Z^2. These photons can be Thomson scattered by the electron with[45]

$$\sigma_T = \frac{8\pi}{3}\left(\frac{\alpha\hbar}{\mu c}\right)^2.$$ (3.53)

The angular distribution of scattered photons in the electron rest frame is nearly isotropic. When this distribution is Lorentz-transformed from the electron to the laboratory frame, those photons that were scattered into the forward hemisphere in the electron frame become boosted to small angles $\theta \sim 1/\gamma$ and Doppler blue-shifted.

3.6.3   Radiation length

The dependence of the bremsstrahlung cross section on $1/\mu^2$ and γ means that under normal circumstances such radiation is peculiar to electron beams. The virtual photon flux (and thence the bremsstrahlung) is proportional to Z^2, so that this is largest for high Z materials. Thus a typical X-ray tube consists of an electron beam hitting a heavy metal target such as copper or tungsten. Calculations of flux involve the same type of integral that we encountered in the calculation of multiple scattering. There are the same limits on impact parameter associated with the screening distance due to atomic electrons on the one hand and finite nuclear size (or maximum energy) on the other. The same effect appears in the calculation of pair production by photons incident on a high Z target where the cross section also depends on Z^2.

Because the charges are point-like, the cross section for the electron to lose a fraction η of its energy is independent of η. As a result the bremsstrahlung energy loss process scales, that is the energy loss is proportional to the incident energy. The energy loss is subject to large fluctuations, unlike the recoil and ionisation energy loss.[46] Because of the scaling, when bremsstrahlung dominates over the collisional energy loss, the mean energy E of an incident electron falls exponentially with distance as

$$\frac{\mathrm{d}E}{\mathrm{d}x} = -\frac{E}{X_0}.$$ (3.54)

This is the definition of X_0, the radiation length.

The radiation length X_0 for a material with atomic number Z is given to a good approximation by

$$X_0 = \frac{716.4A}{Z(Z+1)\ln\left(287/\sqrt{Z}\right)}$$ (3.55)

in units of g cm^{-2}. Bremsstrahlung cross sections in mixtures of elements behave additively. In table 3.1 are given values of the radiation length

a)

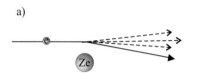

b)

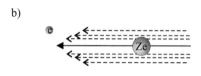

c)

Fig. 3.15 a) Radiation by an electron in a beam passing a nucleus, as seen in the laboratory frame. b) The EM field of the approaching charge Ze as seen by an observer co-travelling with the beam electron. c) The resulting EM field, scattered as a dipole field in the rest frame of the electron.

[46]Together, these two are often referred to as the collisional energy loss as opposed to the radiative energy loss.

and the magnitude of collisional dE/dx in the limit $\beta = 1$ for a wide variety of materials. According to equation 3.54, the radiative energy loss increases linearly with electron energy whereas the collisional energy loss is almost constant. The electron energy at which they are equal can be determined directly from the table. The ratio of these processes in X-ray targets gives an estimate of the efficiency of photon production and also the cooling required. It follows that a high Z target and a high energy electron beam are the most efficient. The energy of the electron beam also determines the photon energy spectrum from the scaling argument.[47]

[47] In addition to the bremsstrahlung spectrum there are also the characteristic X-ray peaks generated by radiative relaxation of inner shell vacancies.

Read more in books

A Century of X-rays and Radioactivity in Medicine, RF Mould, IOP (1993), a book of early pictures and interesting authoritative history

Introduction to Nuclear Physics, DA Greenwood and WN Cottingham, CUP, 2nd edn (2001), a standard book on the interaction of radiation in matter and on nuclear physics

Nuclear and Particle Physics, WSC Williams, OUP (1991), a standard book on the interaction of radiation in matter and on nuclear physics

Quantum Mechanics, F Mandl, Manchester Physics Series, Wiley (1992), a standard book on quantum mechanics including time-dependent perturbation theory

Quantum Physics, S Gasiorowicz, Wiley (1974), a standard book on quantum mechanics including time dependent-perturbation theory

Relativistic Charged Particle Identification by Energy Loss, WWM Allison and JH Cobb, *Annual Reviews of Nuclear and Particle Science,* 30, 253 (1980), further discussion of energy loss

The Physics of Charged Particle Identification: dE/dx, Cherenkov and Transition Radiation, WWM Allison and PRS Wright, *Experimental Techniques in High Energy Nuclear and Particle Physics,* ed. T Ferbel, 2nd edn, World Scientific (1991), further discussion of energy loss

Look on the Web

Look at the book website at
www.physics.ox.ac.uk/users/allison/booksite.htm

Find information on coherent X-ray sources with
FEL and *synchrotron light source*

Get data on the X-ray spectra of elements from the NIST site
NIST X-ray

Look at the Particle Data Group website for information on range–momentum, multiple scattering, radiation length, etc., for charged particles and photons in matter
PDG, particle data group

Questions

3.1 A carbon target of mass m kg is irradiated by a beam of electrons with flux F m^{-2} s^{-1}. A detector of area A at a distance r from the target detects f scattered electrons per second when placed at a polar angle θ with respect to the incident beam. Write down an expression for the differential cross section per atom at this angle?

3.2 Calculate the mass of the target constituent indicated by the position of the peak in the spectrum shown in Fig. 3.5.

3.3 Show that in Compton scattering the change in wavelength of the radiation scattered through an angle θ is given by

$$\Delta\lambda = \frac{2\pi\hbar}{m_e c^2}(1 - \cos\theta).$$

3.4 Prove the relativistic result given by equation 3.9.

3.5 An electromagnetic wave with electric field $E_0 \exp(-i\omega t)$ is incident on an electron. As a result the electron oscillates harmonically with amplitude a.
a) Find an expression for a.
b) Find the incident energy flux.
c) An electron that oscillates in this way radiates an electromagnetic wave with a transverse electric field which at large r is given by

$$E(r,\theta) = \frac{ea\omega^2 \sin\theta}{4\pi\varepsilon_0 c^2 r}.$$

Calculate the scattered power per unit solid angle.
d) By integrating the scattered power over angle derive the Thomson cross section, equation 3.11. Calculate its numerical value.

3.6 A beam of deuterons has a range of 0.05 m in a carbon block. Using the information in Fig. 3.13 and the data in table 3.1, find the momentum and kinetic energy of the deuteron beam. A beam of protons has the same rate of energy loss on entering the carbon block. Estimate its range.

3.7 A beam of triply ionised carbon atoms is accelerated through a potential 3 MV before passing through 10^{-2} g cm^{-2} of argon. Estimate the resulting RMS multiple scattering angle. [Ignore processes other than elastic scattering.]

3.8 An electron beam of kinetic energy 3 MeV enters a copper target of thickness 1.0 mm. Using the data given in table 3.1, estimate the radiative and non-radiative energy loss. Comment on the effect of any assumption that you make.

3.9 Particles of charge ze and velocity βc are incident on a material containing stationary charged constituents of mass M and charge Ze. In a single elastic collision the incident particle loses energy ΔE and transfers momentum Δp to the struck constituent. Treating the constituent as unbound and assuming that it recoils non-relativistically, show that

$$\frac{(\Delta p)^2}{2M\Delta E} = 1.$$

According to a simple classical representation of Rutherford scattering, in a collision with impact parameter b the momentum transfer is given by

$$\Delta p = \frac{2zZe^2}{4\pi\epsilon_0\beta cb}.$$

Use this relation to show that the cross section for a momentum transfer greater than ΔP is

$$\sigma = \frac{z^2 Z^2 e^4}{4\pi\epsilon_0^2 \beta^2 c^2 (\Delta P)^2}.$$

Hence show in this simple model that for a medium with a density of N such constituents per unit volume the mean rate of energy loss per unit path length is

$$\frac{z^2 Z^2 e^4}{4\pi\epsilon_0^2 M\beta^2 c^2} N \ln\left(\frac{b_{\max}}{b_{\min}}\right).$$

Explain the significance of $b_{\max}$ and $b_{\min}$.

Hence comment on the relative importance of collisions with target nuclei and electrons in determining the scattering and energy loss of incident charged articles.

Explain why charged particle beams are used for radiotherapy, but not for diagnostic imaging.

Mechanical waves and properties of matter

4

in which is derived the fundamental physics of sound. This provides an explanation of many applications and an introduction to chapter 9.

4.1 Stress, strain and waves in homogeneous materials

The internal forces in a material (stress) are determined by the relative displacement of its elements (strain). Two types of displacement give internal forces, expansion and distortion. Fluids are distinguished from solids by the fact that internal forces are not generated by distortion. For small changes forces and displacements are related linearly (Hooke's law). This law may fail due to non-linearity and inelasticity. In an ideal fluid expansion and stress are related by the bulk modulus. Waves with longitudinal displacement and isotropic pressure may propagate. In a solid both stress and strain are described by second rank tensors. Each tensor may be separated into irreducible parts, a scalar, a vector and a 5-component symmetric traceless second-rank tensor. In an elastic linear isotropic solid the stress is related to the strain by two moduli. There are three independent wave equations, two with transverse displacements and one with longitudinal displacement.

4.1.1 Relative displacements and internal forces

Every material has an equation of state that describes how it responds to different externally imposed conditions. Under given thermodynamic conditions each individual element of that material acts on its neighbouring elements as their displacements change.[1] These actions and reactions are the internal forces.

To keep track of what happens to each element we choose a reference state. Each part of the material is then labelled by its position (x, y, z) in that reference, and its displacement vector $\boldsymbol{u} = (u_x, u_y, u_z)$ is the shift between its actual position and its reference position.[2] There are internal forces in the reference state. The relations of interest are between the changes in internal forces and the displacements, all relative to the reference state. Looking at the displacement of an elemental volume, it may undergo a combination of the following:

4.1	Stress, strain and waves in homogeneous materials	85
4.2	Reflection and transmission of waves in bounded media	99
4.3	Surface waves and normal modes	111
4.4	Structured media	120
	Read more in books	129
	Look on the Web	129
	Questions	130

[1] We start by thinking of simple isothermal conditions, correcting this later where appropriate.

[2] Here we look at materials macroscopically and treat $\boldsymbol{u}$ as a smooth and continuous function of position, ignoring the fact that at a microscopic level constituents may be moving about, as molecules in a gas, for example.

◇ displacement as a whole;

◇ rotation as a whole;

◇ expansion, so that its volume changes;

◇ distortion, such that it is squeezed in one direction and stretched in another.

Some types of displacement field, $u(r)$, within a material are resisted by internal forces and some are not. Displacement as a whole and rotation as a whole do not change the relative positions of neighbouring parts within a material and therefore are not associated with changes of internal forces. In a solid, expansion and distortion do generate changes in local separations and so are generally accompanied by changes to internal forces. In a fluid, only changes in expansion generate changes in internal forces.[3]

[3]Viscous forces oppose the rate of change of distortion, but there are no internal forces associated with changes in static distortion.

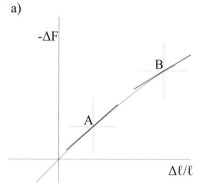

a)

b)

Fig. 4.1 Illustrations of change in force against relative displacement. a) An elastic force, linear to point A but non-linear at B relative to A. Redefining the reference state at B, displacements near B are seen to be linear. b) An inelastic force which is not uniquely determined by the displacement.

[4]In thermodynamic language, the internal energy is not a function of state.

Linearity and elasticity

The archetype of the relation between displacement and the change in force, Hooke's law, assumes that these are small. For such displacement of the end of a spring of length ℓ there is a linear change in the force F,

$$-\Delta F = F_0 - F = K\frac{\Delta\ell}{\ell}, \qquad (4.1)$$

where F_0 is the spring force in the reference state. The coefficient K is known as the modulus and the force is always in the opposite direction to the displacement. This is required for mechanical stability.

Real springs do not quite obey Hooke's law. Figure 4.1 illustrates the two ways in which the law can fail. The relation between extension and change in force may be non-linear, and it may be inelastic.

The plot of force against extension is typically straight only locally. It may be linear near point A and also near point B. But relative to A as reference point, extensions near B are not linear. Mathematically, changing the reference point is doing no more than expanding the description of the problem in a Taylor's series about a closer point at which the series converges more rapidly, requiring no more than a linear term. So questions of non-linearity are intimately connected with the choice of reference state.

Figure 4.1b illustrates what happens when the force is inelastic. If a material is inelastic the relation between force change and relative displacement is not unique. Net work is done when an inelastic material is deformed and then returned to the original state of deformation. The force is not conservative.[4] Everyday experience shows that such materials get hot when repeatedly deformed in this way. This shows explicitly that mechanical energy is not conserved, and that from a thermodynamic point of view such changes are irreversible. Then the relation between force change and relative displacement is typically described by a hysteresis loop.

Elasticity and linearity are essentially different. A deformation may be elastic or inelastic, and linear or non-linear.

Stress and strain in a material

In a material an internal force is expressed as a force per unit area and known as a component of stress. The generalisation of the ratio $\Delta\ell/\ell$, displacement over distance, is known as a component of strain. The modulus for expansion is known as the bulk modulus, K_B. The modulus for distortion is known as the shear modulus, μ. The shear modulus of a fluid is zero.[5] When subjected to distortion, fluids simply flow and relax. There are shear displacements, but no shear forces. Liquids and gases may be described to a fair approximation as ideal fluids. We discuss fluids first. This is followed by a discussion of solids for which a more general approach is required.

[5] We ignore the effect of viscosity for the time being.

4.1.2   Elastic fluids

Stress and strain in a fluid

A fluid reacts in the simplest way to stress, deforming or flowing freely such that only isotropic stress is maintained. The stress T has three components, in the x-direction on the plane $x = 0$, in the y-direction on the plane $y = 0$, in the z-direction on the plane $z = 0$. Each is equal to $-P$, the pressure.[6] Any change in stress is similarly constrained with all three diagonal components of ΔT equal to $-\Delta P$.

A fluid can be subjected to many different elements of strain, ε.[7] But only the expansion in x, in y and in z contribute to the volumetric expansion,

[6] The use of the letter T for the stress denotes the fact that it is conventionally signed as a tensile stress, opposite in sign to pressure.

$$\varepsilon = \frac{\Delta V}{V_0} = \frac{\partial u_x}{\partial x} + \frac{\partial u_y}{\partial y} + \frac{\partial u_z}{\partial z}. \tag{4.2}$$

This volumetric strain is related to the stress by the K_B,

[7] When we study solids we shall need to worry more about the details of strain components. Here we only need the components that contribute to a change of volume.

$$\Delta T = K_B \Delta V/V_0. \tag{4.3}$$

Thence

$$K_B = -V_0 \frac{\partial P}{\partial V}. \tag{4.4}$$

The equation of state of the fluid relates the pressure to the volume at any given value of temperature, so that K_B may be found.

Isothermal bulk modulus for a perfect gas

The equation of state of a perfect gas is

$$PV = nRT, \tag{4.5}$$

where P is the pressure, V the volume, T the temperature,[8] n the number of moles and R the gas constant. In the reference state the volume is V_0 and the stress is $T_0 = -P_0$. This suggests that $K_B = P_0$.

[8] We use T to denote temperature as well as stress. In the following the meaning will usually be stress unless noted otherwise.

Thermodynamic conditions on expansion or deformation

An important question is which thermodynamic parameters are to be held constant in the partial differentiation involved in the definition of the modulus. We have made the mistake, implicitly, of considering changes[9] at constant temperature when differentiating equation 4.5. In general, no heat is supplied when materials are deformed or expanded. So the process takes place at constant entropy, not at constant temperature. Therefore K_B should be defined at constant entropy, and likewise for other moduli.

A constant entropy change in a perfect gas obeys the relation $PV^\gamma =$ constant where γ is the ratio of the specific heats, C_P/C_V. So for a perfect gas,

$$K_B = \gamma P_0, \tag{4.6}$$

instead of P_0. The value of γ depends on n, the number of degrees of freedom of the gas, $\gamma = (n+2)/n$. For a monatomic gas there are only the three translational motions and the expected value is $\gamma = 5/3 = 1.667$. For a diatomic gas rotation about two axes leads to the expectation $7/5 = 1.400$, unless the temperature is high enough for longitudinal vibrational excitation to occur. For polyatomic gases n is 6 or more and $\gamma \leq 1.333$. Observed values, table 4.1, conform to these expectations.

4.1.3 Longitudinal waves in fluids

The stress change and the strain are also related by Newton's second law. The forces described by the change in stress generate accelerations which are related to the second time derivative of the strain. These result in wave motion.

In the case of a plane wave moving along the z-axis, partial derivatives with respect to x or y are zero. The only nonzero element of volumetric strain is $\partial u_z/\partial z$. Therefore the displacement is along z. We track the displacement of a point defined by its reference position z in the material. At time t the displacement in the z-direction is $u_z(z,t)$. At the same time a neighbouring point with reference $z+dz$ will be displaced by a distance $u_z + du_z/dz \times dz$ as illustrated in Fig. 4.2. The subsequent behaviour of the medium is determined by Newton's second law applied in the z-direction on the element dz. If the reference density is ρ_0 and an element of the slice has unit area in the xy-plane, then the mass of the element is $\rho_0 dz$, a constant. The force in the negative z-direction on the element through the boundary at z is $\Delta T(z)$ and the force in the positive z-direction through the boundary at $z+dz$ is $\Delta T(z) + \partial \Delta T(z)/\partial z\, dz$.[10] The mass times the acceleration is then equal to the net force in the positive z-direction acting *on* the slice

$$\rho_0 dz \frac{\partial^2 u_z}{\partial t^2} = \frac{\partial \Delta T}{\partial z} dz. \tag{4.7}$$

Substituting

$$\Delta T = K_B \frac{\partial u_z}{\partial z}. \tag{4.8}$$

[9]We are in good company for Newton also made this mistake, as pointed out by Laplace.

Table 4.1 Values of γ, the ratio C_P/C_V, for some gases.

Argon	1.667
Neon	1.642
Mercury at 550 K	1.666
Hydrogen	1.407
Air	1.402
Carbon monoxide	1.297
Steam at 373 K	1.334
Carbon dioxide	1.300
Methane	1.313
Ethane	1.22
Propane	1.130
Carbon tetrachloride	1.130

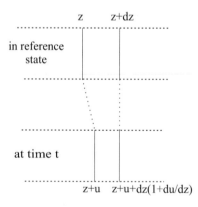

Fig. 4.2 The position of the boundaries of a slice, defined as being at z and $z+dz$ in the reference state, as seen at time t.

[10]Recall that T is positive for a tensile stress.

The final differential equation is[11]

$$\frac{\partial^2 u_z}{\partial t^2} = \left(\frac{K_B}{\rho_0}\right)\frac{\partial^2 u_z}{\partial z^2}. \tag{4.9}$$

Comparing this equation with the standard form of the one dimensional linear wave equation

$$\frac{\partial^2 \phi}{\partial t^2} = c^2\frac{\partial^2 \phi}{\partial z^2}, \tag{4.10}$$

it is evident that $u_z(z,t)$ obeys a wave equation with wave velocity

$$c = \sqrt{\frac{K_B}{\rho_0}}. \tag{4.11}$$

Note that this is the velocity of the waves as they propagate in the z-direction and is a fixed property of the medium independent of wave amplitude. It should not be confused with the instantaneous velocity of elements of the medium $v_z = \partial u_z/\partial t$, which is proportional to the amplitude and oscillates with time. This is called the particle velocity. For this longitudinal wave it is also in the z-direction.

D'Alembert solutions in one dimension

Since the relation between stress and strain does not depend explicitly on time and Newton's second law is symmetric in time, for every solution to the wave equation in an unbounded material that involves t, there is another in which the sign of t is reversed. These pairs of solutions in $\pm t$ are called progressive or travelling wave solutions.

Such pairs of solutions exist for all elastic media, even when the waves are not linear. However, only in linear media does the principle of superposition apply, such that progressive wave solutions may be freely added together. We return to consider the non-linear case in section 9.4.

In the linear case these pairs of solutions may be found by the d'Alembert method. We change variables in the wave equation 4.10 from z and t to $\xi = z - ct$ and $\zeta = z + ct$. Using the chain rule for partial differentiation

$$\frac{\partial}{\partial t} = \frac{\partial \xi}{\partial t}\frac{\partial}{\partial \xi} + \frac{\partial \zeta}{\partial t}\frac{\partial}{\partial \zeta} \tag{4.12}$$

and a similar expression for the $\partial/\partial z$ operator, the wave equation reduces to

$$\frac{\partial^2 u}{\partial \xi \partial \zeta} = 0. \tag{4.13}$$

The general solution is found by integration to be

$$u = f(z - ct) + g(z + ct), \tag{4.14}$$

where the arbitrary functions f and g arise from the two constants of integration for which ζ and ξ are, in turn, held constant.

[11] Note that there are also changes in transverse stress because the pressure is isotropic. However, these are in equilibrium and so cause no motion.

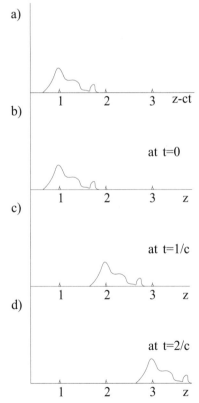

Fig. 4.3 A progressive wave $f(z - ct)$ travelling in the positive z-direction: a) plotted as a function of $z - ct$, b) plotted at $t = 0$, c) plotted at $t = 1/c$, d) plotted at $t = 2/c$.

To understand these solutions, consider first $f(z - ct)$. Such a wave is illustrated in Fig. 4.3. The peak at $z - ct = 1$ appears at $z = 1$ when $t = 0$, $z = 2$ when $t = 1/c$, $z = 3$ when $t = 2/c$, and so on. The profile of the wave moves in the positive z-direction at wave velocity c. These are the properties that we expect of an ideal travelling wave. It is ideal in the sense that it suffers neither distortion nor attenuation as it progresses. A wave of the form $g(z + ct)$ behaves similarly but with the sign of time t reversed. It therefore describes an unattenuated distortion-free wave that moves in the negative z direction.

Wave impedance of fluids

To complete the solution for a travelling wave we need to determine the remaining variables. For a longitudinal wave in the z-direction the particle velocity v_z is the time derivative of u_z

$$v_z(z, t) = \frac{\partial u_z}{\partial t} = -cf'(z - ct) \tag{4.15}$$

[12]The impedance is often written as the ratio $Z = P/v$ for a positive going wave. Since the excess pressure $P = -\Delta T$, the signs are consistent.

where f' is the derivative of f with respect to its argument, $\xi = z - ct$. The acceleration is $c^2 f''(z - ct)$. Equation 4.8 gives

$$\Delta T = K_{\mathrm{B}} f'(z - ct). \tag{4.16}$$

This shows that in a travelling wave ΔT and v_z are proportional, and

$$\Delta T = \mp Z v_z, \tag{4.17}$$

where the positive and negative signs apply for negative and positive going waves respectively. The impedance of the fluid is

$$Z = \rho_0 c = \sqrt{\rho_0 K_{\mathrm{B}}}. \tag{4.18}$$

Table 4.2 Values of density (ρ in kg m^{-3}), wave velocity (c_L and c_S m s^{-1}) for longitudinal and transversely polarised waves for various materials at temperature $T = 293$ K (except at 273 K for He, ice and water).

Material	ρ	c_L	c_S
Be	1848	12800	8880
Al	2700	6420	3040
Steel	7870	5940	3220
Lead	11350	1960	690
Sandstone	2323	2920	1840
SiO$_2$	2200	5968	3764
Ice	910	3760	2000
Nylon	1140	2620	1070
Polythene	930	1950	540
Muscle	1040	1567	var.
Bone	1940	3550	1970
Blood	1060	1547	–
Hg	13600	1450	–
Water	1000	1401	–
Air	1.29	344	–
He	0.17	965	–

The ratio of stress to particle velocity changes sign according to the direction in which the progressive waves are moving. For a superposition of waves in both directions the ratio $\Delta T / v$ varies in space and time and may have any value.[12]

Values for moduli and impedances can be deduced from the table of densities and velocities for longitudinal and shear waves for a wide variety of materials given in table 4.2. Notice that gases have both low density and low elastic modulus compared with solids. The low sound velocity indicates that the low elastic modulus is the dominant effect.[13]

Energy density

The energy density relative to the reference state in a linear medium can be expressed as the sum of the local kinetic and potential energy densities. The kinetic energy density is

[13]In fact, as discussed below, the interpretation for waves in solids is not so straightforward. Specifically the modulus for longitudinal waves differs from the bulk modulus.

$$U_{\mathrm{KE}}(x, t) = \frac{1}{2} \rho_0 v^2. \tag{4.19}$$

The potential energy density can be expressed in a number of equivalent ways using the relations $\Delta T = K\varepsilon$ and $\varepsilon = \Delta V/V_0$:

$$U_{\mathrm{PE}}(x,t) = \frac{1}{2}K\varepsilon^2 = \frac{1}{2}\Delta T\varepsilon. \tag{4.20}$$

For a progressive wave $\Delta T = \pm Zv$. Therefore the kinetic energy and potential energy densities are equal everywhere. In the absence of dispersion, the energy flux N in a progressive wave is the sum of kinetic and potential energy densities multiplied by the wave velocity,

$$N = \rho_0 v^2 c = Zv^2 = \Delta Tv = (\Delta T)^2/Z. \tag{4.21}$$

These are instantaneous values in SI units of W m^{-2}. For the mean flux the RMS value of v or ΔT should be used. Note that the transverse forces present in a longitudinal compression wave are in equilibrium and do no work. It is the longitudinal force that is relevant.

Dispersion for progressive waves

So far we have implicitly assumed that the bulk modulus K_{B} is constant. At low frequencies this is a good approximation. At higher frequencies it is not true and consequently the values of c and Z are functions of frequency. Then the different Fourier components in a signal travel at different speeds.

The result is that the signal profile changes as it propagates. A Fourier component (frequency ω and wavenumber k) has phase velocity $c = \omega/k$. In general a study of the physical mechanics of the wave motion gives a relation between ω and k from which c may be determined. Such an equation is called a dispersion relation.

The envelope of a pulse composed of frequencies between ω and $\omega+\mathrm{d}\omega$ and wavenumber between k and $k+\mathrm{d}k$ moves with velocity $\mathrm{d}\omega/\mathrm{d}k$, the slope of the dispersion curve, plotted as ω against k. This is called the group velocity. If the group velocity itself varies with frequency within the signal bandwidth then the pulse will distort and spread out in time, even if there is no actual absorption. If the dispersion relation has complex parts then there is energy absorption. The reason for this may be understood by considering a progressive harmonic wave in the form $\exp\left[\mathrm{i}(kx - \omega t)\right]$. Any imaginary part of k (or ω) implies an exponentially damped progression in x (or t). This describes a wave that is being attenuated.

Absorption of harmonic linear waves may be described by considering complex moduli. An imaginary part of K_{B} describes a phase difference between stress and strain. Of course this is just a formalism. We shall need to dig deeper to understand what is happening in absorptive processes.

We shall see later in section 4.4.2 that the absorption of sound waves is associated with a delay in the strain relative to the causative stress. Every effect must follow its cause so that such a delay may not be negative without violating causality. There is a theorem called the Kramers–Kronig relation based on this causality principle. It has widespread application in electromagnetism and electronics. It links the real and imaginary parts of the complex modulus of the material. We do not pursue this matter here beyond noting that dispersion and absorption are linked and considering a qualitative example.

It is observed that the velocity of sound in diatomic and polyatomic gases, $\sqrt{\gamma P_0/\rho}$, depends on frequency. This is understood in terms of a frequency dependence of γ. At low frequency there is time to establish thermal equilibrium between translational and other degrees of freedom during the cycle of the wave. At high frequency this is not so. The effective number of degrees of freedom, and therefore γ, depends on frequency. Further, any energy that flows into rotational modes is transferred under thermodynamically irreversible conditions, and so absorption occurs. In this example it is seen how one mechanism can be responsible for both dispersion and absorption. In fact this link is general.

4.1.4   Stress and strain in solids †

Stress tensor

The internal forces inside a material at a point x, y, z are defined relative to the three planes, the yz-plane, the zx-plane and the xy-plane. The material on one side of each of these imaginary planes acts on the material on the other side with a force (or rather a force per unit area) which is balanced by an equal and opposite reaction on the other side. For each plane the force per unit area may have components in the x-direction, in the y-direction and in the z-direction, making nine components in all, as shown in Fig. 4.4. Each of these components may be a function of (x, y, z). The nine numbers at any point form a complete description of the mechanical internal forces there. Together they form the second rank stress tensor at that point, T_{ij} where the subscripts i, j run from 1 to 3, denoting x to z

$$T = \begin{bmatrix} T_{xx} & T_{xy} & T_{xz} \\ T_{yx} & T_{yy} & T_{yz} \\ T_{zx} & T_{zy} & T_{zz} \end{bmatrix}. \tag{4.22}$$

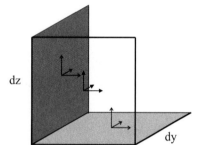

dz

(x,y,z)   dx   dy

Fig. 4.4 Three elemental orthogonal planes in a material at point (x, y, z) upon each of which three forces per unit area may act.

Reduction of a second rank tensor

The concept of a tensor is an extension of the concept of a vector. A vector is identified by its transformation properties when referred to new axes. A physical vector, such as position, has three components and may be described as a first rank tensor. Under changes of axes the three components are transformed amongst themselves. A second rank tensor transforms under rotation like the outer product of two vectors.

The nine elements of a second rank tensor T referred to one set of axes may be written in terms of the elements of T' in a different primed set of axes as

$$T_{ij} = \sum_{l=1}^{3}\sum_{k=1}^{3} c_{il} T'_{lk} c_{jk} \tag{4.23}$$

where the c_{il} are the cosines of the angles between directions i and primed directions l. However, there are subsets of the nine components whose transformation is simpler and which only transform amongst themselves: a singlet, a triplet and a subset of five.[14] These subsets cannot be simplified further and are termed irreducible tensors. Thus we write symbolically

$$9 = 1 \oplus 3 \oplus 5.$$

The scalar is the trace of the tensor (4.22), $T_{xx} + T_{yy} + T_{zz}$. The three antisymmetric elements of T, the combinations

$$T_{yz} - T_{zy},\ T_{zx} - T_{xz},\ T_{xy} - T_{yx},$$

transform as a vector. The remaining five elements of the original nine form an irreducible symmetric traceless second-rank tensor, and transform amongst themselves with equation 4.23.

Not only mathematically but physically, the three subsets have their own identity. In the case of stress there is the isotropic tension which is just the bulk pressure with opposite sign. Then there are the parts of the stress tensor that describe body torque. This subset is a vector, or first rank tensor. The remaining five elements form the irreducible stress tensor.

In the following, in order to simplify the discussion we minimise the use of tensor algebra as far as possible by considering isotropic media.[15]

Strain tensor

The displacement vector field $u = (u_x, u_y, u_z)$ which is a measure of how far the point at (x, y, z) has moved generates nine first order stretches which may oppose the elements of the stress change tensor in a generalisation of Hooke's law for small displacements. These general stretches are given locally by the nine gradients of u

$$\mathcal{E} = \begin{bmatrix} \partial u_x/\partial x & \partial u_x/\partial y & \partial u_x/\partial z \\ \partial u_y/\partial x & \partial u_y/\partial y & \partial u_y/\partial z \\ \partial u_z/\partial x & \partial u_z/\partial y & \partial u_z/\partial z \end{bmatrix}.$$

This is the strain tensor, defined relative to the reference state. Each element is a ratio of distances and is therefore physically dimensionless. As for the stress tensor we need to separate this into its irreducible parts.

To help understand tensor reduction, consider first a two dimensional strain tensor. Figure 4.5 illustrates the ideas in pictorial terms. In two dimensions the 2×2 strain tensor may be reduced to a compression, a rotation and a two-component shear. In component terms this reduction may be illustrated as

$$\begin{bmatrix} r & s \\ t & u \end{bmatrix} = \begin{bmatrix} a & 0 \\ 0 & a \end{bmatrix} + \begin{bmatrix} 0 & \theta \\ -\theta & 0 \end{bmatrix} + \begin{bmatrix} b & c \\ c & -b \end{bmatrix}. \tag{4.24}$$

In tensor terms we write

$$4 = 1 \oplus 1 \oplus 2.$$

[14] In terms of the outer product of two vectors these subsets may be visualised as the scalar product, the vector cross product and the five remaining elements of the outer product.

[15] The books by Nye and by Fano and Racah may be consulted to understand how less symmetric materials may be handled.

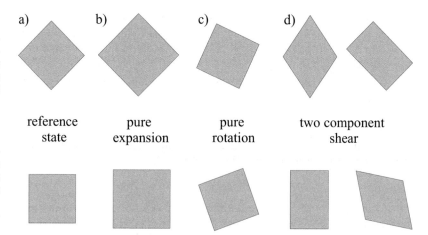

Fig. 4.5 Illustrations of the effect of ir-reducible strain elements in two dimensions using a geometrical figure. In the first row the elements are applied to the shape of a square at 45°. In the second they are applied to a square at 0°.
a) The un-strained shape.
b) A pure expansion.
c) A pure rotation.
d) Two independent elements of symmetric volume-conserving distortion (shear). Referring to equation 4.24, the first has d non-zero, the second has c non-zero.
For clarity, these strains are shown exaggerated.

In three dimensions the whole strain tensor has nine elements and the reduction follows that of the stress tensor. The trace $\varepsilon_{11} + \varepsilon_{22} + \varepsilon_{33}$ is the scalar, the isotropic volumetric expansion. The subset of three antisymmetric elements describes pure rotations with angles

$$\theta_1 = \frac{\varepsilon_{23} - \varepsilon_{32}}{2}, \; \theta_2 = \frac{\varepsilon_{31} - \varepsilon_{13}}{2}, \; \theta_3 = \frac{\varepsilon_{12} - \varepsilon_{21}}{2}.$$

[16]Note that all these strain elements are small.

about the three cartesian axes.[16] The remaining five elements form the irreducible symmetric second rank compressionless strain tensor.

Relation between strain and stress in an isotropic solid

In the reference state the strain is zero by definition, but there is a certain stress $T = T_0$. Changes in stress $\Delta T = T - T_0$ generate non-zero strain $\mathcal{E}$. An elastic material is one in which the ΔT and $\mathcal{E}$ are uniquely related. If, in addition, the relationship is linear we can write

$$\Delta T = K\mathcal{E}, \tag{4.25}$$

where K is the general elastic modulus. Since ΔT and $\mathcal{E}$ are second rank tensors, K is a fourth rank tensor whose elements are a property of the material.

We restrict our attention to materials that are sufficiently isotropic that K is independent of direction. But we should note that even cubic crystals with maximal internal symmetry are not isotropic in this sense, and their stiffness tensor depends on the orientation of the crystal axes. Here we consider only materials that are truly isotropic and non-chiral with no defined crystal axes. They should be thought of as having a random polycrystalline internal structure. Then in the most general case, the three irreducible tensors of ΔT and $\mathcal{E}$ are separately related to one another. The modulus for the antisymmetric parts is zero because a rotation generates no internal forces. That leaves just two numbers. Put most simply, these describe the processes of compression and distortion. The compression modulus relates the scalar ΔT and scalar $\mathcal{E}$, and the shear modulus relates the elements of the irreducible second rank ΔT and the irreducible second rank $\mathcal{E}$. For historical reasons these are written as $3\lambda + 2\mu$ and 2μ, respectively. λ and μ are the Lamé constants. Any more complex relation between stress and strain in a linear elastic medium would not be compatible with the assumed isotropy of the material.[17]

In Fig. 4.6 we show how the stress associated with a single diagonal element of strain is composed. The strain is analysed into its irreducible parts, each of which is multiplied by the appropriate modulus. The stress is found by recomposing the resulting elements.

[17]This is a simple application of the idea behind the Wigner–Eckart theorem (see the book by Fano and Racah). This theorem may be used to describe relations between stress and strain in anisotropic materials of a particular symmetry class using further irreducible elements of the rank-4 stiffness tensor.

$$\text{strain} = \begin{bmatrix} 0 & 0 & 0 \\ 0 & 0 & 0 \\ 0 & 0 & 1 \end{bmatrix} \Rightarrow \begin{bmatrix} -\tfrac{1}{3} & 0 & 0 \\ 0 & -\tfrac{1}{3} & 0 \\ 0 & 0 & \tfrac{2}{3} \end{bmatrix} \oplus \begin{bmatrix} \tfrac{1}{3} & 0 & 0 \\ 0 & \tfrac{1}{3} & 0 \\ 0 & 0 & \tfrac{1}{3} \end{bmatrix}$$

$$\text{modulus} \quad \Downarrow \quad\quad \otimes 2\mu \Downarrow \quad\quad \otimes (3\lambda + 2\mu)\Downarrow$$

$$\text{stress} = \begin{bmatrix} \lambda & 0 & 0 \\ 0 & \lambda & 0 \\ 0 & 0 & \lambda+2\mu \end{bmatrix} \Leftarrow \begin{bmatrix} -\tfrac{2}{3}\mu & 0 & 0 \\ 0 & -\tfrac{2}{3}\mu & 0 \\ 0 & 0 & \tfrac{4}{3}\mu \end{bmatrix} \oplus \begin{bmatrix} \lambda+\tfrac{2}{3}\mu & 0 & 0 \\ 0 & \lambda+\tfrac{2}{3}\mu & 0 \\ 0 & 0 & \lambda+\tfrac{2}{3}\mu \end{bmatrix}$$

Fig. 4.6 The calculation of the stress in a solid associated with a unit diagonal element of strain. In the first row the strain is decomposed into its irreducible parts, a symmetric traceless strain and a uniform expansion. In the second row each is related by its modulus to the corresponding stress. The net stress is given by the addition of these.

This procedure is repeated for a single off-diagonal element of strain in Fig. 4.7.

$$\text{strain} = \begin{bmatrix} 0 & 0 & 0 \\ 0 & 0 & 1 \\ 0 & 0 & 0 \end{bmatrix} \Rightarrow \begin{bmatrix} 0 & 0 & 0 \\ 0 & 0 & \tfrac{1}{2} \\ 0 & \tfrac{1}{2} & 0 \end{bmatrix} \oplus \begin{bmatrix} 0 & 0 & 0 \\ 0 & 0 & \tfrac{1}{2} \\ 0 & -\tfrac{1}{2} & 0 \end{bmatrix}$$

$$\text{modulus} \quad\quad \otimes 2\mu \Downarrow \quad\quad\quad \otimes 0 \Downarrow$$

$$\text{shear stress} = \begin{bmatrix} 0 & 0 & 0 \\ 0 & 0 & \mu \\ 0 & \mu & 0 \end{bmatrix} \oplus \quad 0$$

Fig. 4.7 The calculation of the stress in a solid associated with a unit off-diagonal element of strain. The first row shows its irreducible parts, a symmetric traceless part and an antisymmetric part. In the second row the symmetric part is related by its modulus to a stress (third row). The antisymmetric part of the strain has modulus zero, because a pure rotation generates no internal stress.

Uniform stress change and the bulk modulus

A material subjected to a uniform stress change,

$$\Delta T_{11} = \Delta T_{22} = \Delta T_{33} = -\Delta P,$$

experiences a strain which is also rank 0,

$$\frac{\Delta V}{V} = \frac{\partial u_x}{\partial x} + \frac{\partial u_y}{\partial y} + \frac{\partial u_z}{\partial z}.$$

Under these conditions only, the bulk modulus may be defined as $K_B = \lambda + \frac{2}{3}\mu$. In a fluid $\mu = 0$ and the restriction to uniform compression goes away.

4.1.5   Polarisation of waves in solids †

Longitudinal and transverse polarisations

As in a fluid, the stress change and the strain in a solid are also related by Newton's second law, and this results in wave motion. For a plane wave moving along the z-axis all partial derivatives with respect to x or y are zero. So the only non-zero elements of strain are

$$\mathcal{E} = \begin{bmatrix} 0 & 0 & \partial u_x/\partial z \\ 0 & 0 & \partial u_y/\partial z \\ 0 & 0 & \partial u_z/\partial z \end{bmatrix}.$$

These three elements are associated with three independent polarisations: a transverse shear wave polarised along x, a transverse shear wave polarised along y, and a longitudinal compressive wave with motion along z. We can write down the stress tensor for each using the decompositions shown in Figs 4.6 and 4.7.

◇ In a longitudinal wave the only strain component is ε_{zz}, and this is accompanied by a longitudinal stress change,

$$\Delta T_{zz} = (\lambda + 2\mu)\varepsilon_{zz} \qquad (4.26)$$

but also by stress changes in both orthogonal directions,

$$\Delta T_{xx} = \Delta T_{yy} = \lambda\varepsilon_{zz}. \qquad (4.27)$$

◇ In one transverse wave the only strain component is the off-diagonal element, $\varepsilon_{xz} = \partial u_x/\partial z$. The shear stress change is then in the same xz-plane

$$\Delta T_{xz} = \Delta T_{zx} = \mu\varepsilon_{xz}. \qquad (4.28)$$

◇ The other transverse wave involves the other independent off-diagonal element $\varepsilon_{yz} = \partial u_y/\partial z$ with stress change components

$$\Delta T_{yz} = \Delta T_{zy} = \mu\varepsilon_{yz}. \qquad (4.29)$$

Longitudinal and transverse waves

For a wave with longitudinal polarisation the longitudinal strain is driven by the longitudinal force described with the modulus $\lambda + 2\mu$. The velocity of longitudinal waves is therefore

$$c_L = \sqrt{\frac{\lambda + 2\mu}{\rho_0}}. \qquad (4.30)$$

The transverse forces present in the wave are in equilibrium and cause no oscillation. This does not mean that they do not matter. Where the wave meets a boundary at oblique incidence, the transverse forces do not balance and have to be taken into account. Indeed this is true even in a fluid where there are also transverse forces present in a longitudinal wave.

For each of the two independent waves with transverse polarisation the transverse strain is driven by the transverse force described with the modulus μ. The velocity of transverse waves is therefore

$$c_S = \sqrt{\frac{\mu}{\rho_0}}. \tag{4.31}$$

The longitudinal force present in the wave is in equilibrium and causes no oscillation. Again, where the wave meets a boundary, this force has to be taken into account in the boundary conditions. In a fluid $\mu = 0$ and there are no transverse waves.[18]

Table 4.2 shows that c_S is always significantly smaller than c_L. Since λ and μ are both positive, c_L must be at least $\sqrt{2}$ times c_S. In a solid the longitudinal impedance $Z_L = \rho c_L$ and the transverse or shear impedance $Z_S = \rho c_S$ can be defined also. However, the relation between stress and velocity in a solid is more involved than in a fluid.

Stress and velocity for waves travelling in the z-direction

For a longitudinal wave the particle velocity v_z is the time derivative of u_z

$$v_z(z,t) = \frac{\partial u_z}{\partial t} = -c_L f'(z - c_L t) \tag{4.32}$$

where f' is the derivative of $f(\xi)$ with respect to $\xi = z - c_L t$. The corresponding stress change is

$$\Delta T_L = \begin{bmatrix} \lambda & 0 & 0 \\ 0 & \lambda & 0 \\ 0 & 0 & \lambda + 2\mu \end{bmatrix} \times f'(z - c_L t). \tag{4.33}$$

For a wave transversely polarised along y, we can write

$$\Delta T_{S1} = \begin{bmatrix} 0 & 0 & 0 \\ 0 & 0 & \mu \\ 0 & \mu & 0 \end{bmatrix} \times f'(z - c_S t). \tag{4.34}$$

For a wave transversely polarised along x we have

$$\Delta T_{S2} = \begin{bmatrix} 0 & 0 & \mu \\ 0 & 0 & 0 \\ \mu & 0 & 0 \end{bmatrix} \times f'(z - c_S t). \tag{4.35}$$

In a solid the kinetic energy density is $\frac{1}{2}\rho v^2$ for each polarisation, as in a fluid.[19]

[18] In a viscous fluid the dissipative resistance to shear motion can be described for harmonic motion by an imaginary value of μ. Such motion is oscillatory in time but decays exponential in space.

[19] The potential energy density, equal to the kinetic energy density, is one half the contraction of the stress and strain tensors, $\Sigma_{i,j} \frac{1}{2} \varepsilon_{ij} \Delta T_{ij}$, but with only one non-vanishing element of ε for each polarisation this is simple.

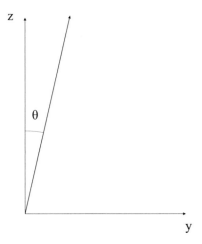

Fig. 4.8 A wave travels at an angle θ to the z-axis in the yz-plane.

Stress and strain in a wave travelling at an angle

Because stress and strain are tensors we need to derive their components for a wave travelling at an angle to the z-axis. In the following we label the transverse wave with displacement in the x-direction with the subscript $S2$. The orthogonal one is labelled $S1$.

When a plane wave with one of these polarisations travels at an angle θ in the yz-plane instead of along the z-axis the displacement for each polarisation u is rotated through θ according to the transformation for vectors (Fig. 4.8). Thus for the wave $f(\xi)$ where $\xi = z\cos\theta + y\sin\theta - ct$, the components of the displacement vector u is one of the following vectors, depending on the polarisation,

$$u_L = \begin{bmatrix} 0 & \sin\theta & \cos\theta \end{bmatrix} f(\xi),$$

$$u_{S1} = \begin{bmatrix} 0 & \cos\theta & -\sin\theta \end{bmatrix} f(\xi), \qquad (4.36)$$

$$u_{S2} = \begin{bmatrix} 1 & 0 & 0 \end{bmatrix} f(\xi).$$

The tensor ΔT for each polarisation transforms under rotations according to equation 4.23 with the following result:

$$\Delta T_L = \begin{bmatrix} 1 & 0 & 0 \\ 0 & 1 & 0 \\ 0 & 0 & 1 \end{bmatrix} \times \lambda f'(\xi)$$

$$+ \begin{bmatrix} 0 & 0 & 0 \\ 0 & 1 - \cos 2\theta & -\sin 2\theta \\ 0 & -\sin 2\theta & 1 + \cos 2\theta \end{bmatrix} \times \mu f'(\xi), \qquad (4.37)$$

$$\Delta T_{S1} = \begin{bmatrix} 0 & 0 & 0 \\ 0 & -\sin 2\theta & \cos 2\theta \\ 0 & \cos 2\theta & \sin 2\theta \end{bmatrix} \times \mu f'(\xi),$$

$$\Delta T_{S2} = \begin{bmatrix} 0 & -\sin\theta & \cos\theta \\ -\sin\theta & 0 & 0 \\ \cos\theta & 0 & 0 \end{bmatrix} \times \mu f'(\xi).$$

We shall need these results when we consider a wave incident on a plane boundary. The boundary conditions for force concern the three components of the vector force transmitted through unit area of the boundary. For a boundary in the plane $z = $ constant, for example, these are given by just picking off the third column of the stress matrix for the polarisation concerned,

$$\boldsymbol{F}_{zL} = \begin{bmatrix} 0 & 0 & 1 \end{bmatrix} \lambda f'(\xi) + \begin{bmatrix} 0 & -\sin 2\theta & 1 + \cos 2\theta \end{bmatrix} \mu f'(\xi),$$

$$\boldsymbol{F}_{zS1} = \begin{bmatrix} 0 & \cos 2\theta & \sin 2\theta \end{bmatrix} \mu f'(\xi), \qquad (4.38)$$

$$\boldsymbol{F}_{zS2} = \begin{bmatrix} \cos\theta & 0 & 0 \end{bmatrix} \mu f'(\xi).$$

4.2   Reflection and transmission of waves in bounded media

Reflection and transmission amplitudes for waves at normal incidence on a boundary are determined for each polarisation. At oblique incidence wave solutions in the two media must satisfy continuity of displacement and force on that boundary along all three axes. These six conditions apply over the whole surface and for all time. They are sufficient to determine the frequency and directions of the six waves that may be generated by a single incident wave. Parity considerations determine that no more than four outgoing waves are generated in any one case. The detailed conditions themselves determine the various reflection and transmission coefficients. In each case the component of energy flux normal to the boundary is found to satisfy the constraint that the energy carried in by the incident wave is equal to the sum of the energies carried away by the generated waves. If the velocity of the generated wave exceeds the velocity of the incident wave, the generalised law of reflection or transmission can give a generated wave which is evanescent perpendicular to the boundary. Reflection and transmission coefficients are calculated for interfaces of water, air and rock. Typically large changes in polarisation are found and evanescent waves that travel parallel to the surface may be generated.

4.2.1   Reflection and transmission at normal incidence

At normal incidence the reflected and transmitted waves are also normal to the boundary. As a result incident waves of any given polarisation generate reflected and transmitted waves of the same polarisation, and there is no mixing of modes. The boundary conditions[20] require continuity of velocity and force. These conditions are sufficient to determine that the amplitude reflection and transmission coefficients are

[20]These will be examined more carefully later when we consider oblique incidence.

$$\frac{u_r}{u_i} = \frac{Z_1 - Z_2}{Z_1 + Z_2}$$
$$\frac{u_t}{u_i} = \frac{2Z_1}{Z_1 + Z_2},$$

(4.39)

where Z_1 and Z_2 are the impedances of the first and second materials for the polarisation concerned. The demonstration of this result and the fact that it conserves energy flux is left to question 4.2.

These results cannot be extended simply to other angles of incidence. When all the force components in a given polarisation are taken into account, the different polarisation modes become coupled. At a fluid–fluid interface where no transverse waves are expected, there are difficulties in satisfying boundary conditions for an incident longitudinal wave at oblique incidence.

4.2.2 Relative directions of waves at boundaries †

Need to match waves at material boundaries

Consider two half-spaces, $z < 0$ occupied by one homogeneous material, and $z > 0$ occupied by a second homogeneous material. At the boundary there is a change in the density (from ρ_1 to ρ_2) and elastic moduli, and thence in the wave velocity and impedance.[21] When we derived the wave equation we wrote down the gradient of the stress change ΔT in equation 4.7. In stepping from there to equation 4.9, we assumed that the modulus K was not dependent on z. This is untrue on the boundary, and consequently the wave equation is invalid there. That is the reason we have to solve for the waves in each half-space separately, and then impose the constraints at the boundary on those solutions, instead of solving the problem all at once. In each material progressive waves with various directions of wave motion may propagate with various polarisations, longitudinal and transverse, some approaching the surface and some moving away. To determine the relation between the waves on either side of the boundary we need to consider the problem in three dimensions. In the first step we shall prove that two dimensions are sufficient. This will then make drawing diagrams and visualising the waves easier.

We consider the case where one single plane wave moves towards the boundary with a specific polarisation, a single frequency ω and wavenumber $\boldsymbol{k}$.[22] This incident wave generates a number of waves in both materials which move away from the boundary. We have already chosen the z-axis as normal to the boundary. We choose the y-axis such that the incident $\boldsymbol{k}$-vector of the incident wave lies in the yz-plane. Thus $\boldsymbol{k} = (0, k_y, k_z)$ with $|k| = 2\pi/\lambda$, and $\omega = |k|c$. As we have seen, the wave velocity c depends only on the choice of longitudinal or transverse polarisation and the properties of the medium, assumed linear and isotropic. The polarisation indicates the displacement and particle velocity vectors, see equation 4.36. The stress is described by a tensor, but it is the components of the force vector per unit area of the surface $z = 0$ which we shall need, see equation 4.38.

Boundary conditions for displacement and force

The displacement of material perpendicular to the boundary must be continuous across the boundary, otherwise either a gap would open up or the two materials would coexist in the same space. If the former is conceivable under shock conditions, the latter is not.

This condition must be satisfied, not just at one point on the surface and one time, but over the whole surface and for all time. Continuity of the particle velocity perpendicular to the boundary is then also assured.

A discontinuity in either of the two components of particle velocity parallel to the boundary between two materials is imaginable, but a finite viscous friction between the materials would eliminate this. So all three components of displacement at the boundary must be continuous

[21]We use subscripts 1 and 2 to denote variables in the first and second medium respectively. Similarly we use subscripts i, r and t to denote incident, reflected and transmitted wave variables.

[22]The wavenumber vector points normal to the plane wavefront with magnitude $2\pi/\lambda$ where λ is the wavelength.

at the boundary.[23]

Equation 4.38 describes the components of force in the x-, y- and z-directions per unit area of the boundary $z = 0$ due to an incident wave at angle θ. The z-component must be continuous to avoid a finite force acting on the zero mass of a infinitesimal surface layer spanning the boundary. Similarly any discontinuity in the force acting parallel to the boundary would produce unacceptable angular acceleration.

In summary, when all of the incident, reflected and transmitted waves are taken into account, the three components of displacement and three components of force per unit area of the boundary must be continuous across the boundary. These conditions apply, not just at one point in x, y and time t, but for all x, y and t together. As we now show, this requirement alone is sufficient to determine the relative directions of each of the wavevectors in terms of the incident wavevector and to reduce the 3-D problem to 2-D. The detail of the six constraints in force and displacement then determine the amplitudes of the outgoing waves relative to the incident one.

Footprint of all waves on the boundary plane

Each harmonic plane wave of amplitude A, wavenumber (k_x, k_y, k_z) and angular frequency ω has the form[24]

$$A \exp\left[\mathrm{i}(k_x x + k_y y + k_z z - \omega t)\right]. \tag{4.40}$$

In the plane $z = 0$ the incident wave looks like

$$A \exp\left[\mathrm{i}(k_y y - \omega t)\right] \tag{4.41}$$

since $k_x = 0$ by choice of axes. In this plane the various reflected and transmitted waves could be of the form

$$A' \exp\left[\mathrm{i}(k'_x x + k'_y y - \omega' t)\right]. \tag{4.42}$$

In general the matching conditions for all x, y and t together would represent an infinite number of conditions to be satisfied. This is clearly not possible with a finite number of wave amplitudes, A', unless the dependence of every wave in each material on the plane $z = 0$ has exactly the same dependence on x, y and t as the incident wave. Each wave has to match the same footprint in x, y and t.

Then these phase factors are identical for all waves and simply cancel when applying the boundary conditions. The consequence is that the values of k_x, k_y and ω must be the same for every wave, incident, reflected and transmitted, in the two materials.[25] The conditions then reduce to relations between the different overall amplitudes A and A'.

So the frequency of each generated wave is the same as the incident wave. Then the value of k_x is the same for each wave. However, the axes were chosen such that $k_x = 0$ for the incident wave. It follows that $k_x = 0$ for all the other waves also. This means that all the wavevectors and the surface normal are perpendicular to the x-axis and lie in the yz-plane. This is called the plane of incidence. It makes it much easier to draw diagrams of the various wavevectors $\boldsymbol{k}$ for they are all coplanar.

[23]We argue that the parallel velocity is continuous across the boundary over the whole boundary for all time. This is equivalent to continuity of displacement since the possibility of a constant parallel displacement of one material relative to the other is trivial.

[24]Later we shall need to include the effect of the polarisation, that is that A is a vector. However, that does not affect the initial stages of the discussion.

[25]Note that this matching does not extend to k_z.

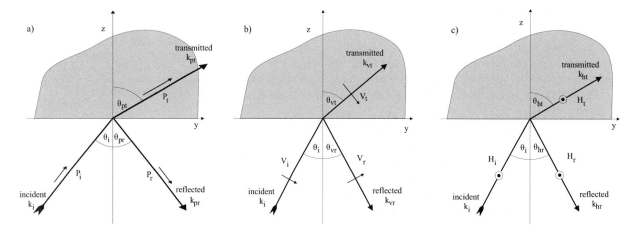

Fig. 4.9 Diagrams of wavevectors for each of the three polarisations of plane waves incident, reflected and transmitted at a plane boundary. The small arrows show the direction of the displacement vectors in each case: a) for compression waves (P), b) for shear waves (SV), c) for shear waves (SH). These diagrams simply describe the variables for the three polarisations. As discussed later, polarisations are changed on reflection and transmission. Note that reflected angles here are defined in the conventional way, such that reflected wavevectors are at angles $\pi - \theta_r$ to the z-axis.

Polarisation of waves

Each wave, incident, reflected and transmitted, can have one of three polarisations. A longitudinal or 'compression' wave is called a P-wave in geophysics. The two transverse shear waves are distinguished. A wave polarised with displacement in the plane of incidence is called an SV wave.[26] A wave polarised perpendicular to the plane of incidence is called an SH wave. Wavevector diagrams for each of the polarisations are drawn in the plane of incidence in Figs. 4.9a–c. These simply describe the variables. All three possible polarisations for the reflected and transmitted waves have to be considered in principle for each polarisation of the incident wave. Each generated wave has its own amplitude and value for k_z, but its values for ω, k_x and k_y are fixed, being the same as the incident wave.

The value of k_z is determined for each wave as follows. The phase velocity is given by $c = \sqrt{K/\rho}$ with $K = \lambda + 2\mu$ for longitudinal waves, and $K = \mu$ for shear waves. For each wave $c = \omega/k$, so that

$$\omega^2 = \frac{K}{\rho}(k_y^2 + k_z^2). \tag{4.43}$$

This fixes the value of k_z, except for its sign, for each wave,

$$k_z = \pm\sqrt{\frac{\rho\omega^2}{K} - k_y^2}. \tag{4.44}$$

The sign of k_z is determined by whether the wave concerned is moving towards the surface or away. Thus incident and transmitted waves have a positive sign, and reflected waves have a negative sign.

[26] In geophysics the plane of incidence is typically vertical, hence the choice of labels, V for vertical and H for horizontal.

General laws of reflection and transmission

From the geometry of the k-vector and the relation between ω, k and c, the angle of the wave vector with respect to the normal is given by

$$\sin \theta = \frac{k_y}{|k|} = c\frac{k_y}{\omega}. \tag{4.45}$$

In particular the incident wave with phase velocity c_i has an angle given by

$$\sin \theta_i = \sqrt{\frac{K_i}{\rho_1}} \frac{k_y}{\omega}. \tag{4.46}$$

Similar expressions for θ_r and θ_t, the angle of each reflected and transmitted wave, can differ only by virtue of their phase velocities:

$$\sin \theta_r = c_r\frac{k_y}{\omega} \tag{4.47}$$

$$\sin \theta_t = c_t\frac{k_y}{\omega}. \tag{4.48}$$

The general law of reflection is therefore

$$\frac{\sin \theta_i}{\sin \theta_r} = \frac{c_i}{c_r} = \sqrt{\frac{K_i}{K_r}}. \tag{4.49}$$

This includes the case of a reflected wave with a different polarisation to the incident wave. If the polarisation of the reflected wave is the same as the incident wave we get the familiar law of reflection,

$$\sin \theta_i = \sin \theta_r. \tag{4.50}$$

Applying the same argument to the transmitted wavevector, the general law of transmission is

$$\frac{\sin \theta_i}{\sin \theta_t} = \frac{c_i}{c_t} = \sqrt{\frac{K_i}{K_t}\frac{\rho_2}{\rho_1}}. \tag{4.51}$$

As we shall see below, an incident P wave gives rise both to a reflected P wave and to a reflected SV wave, and the same is true of an incident SV wave. Similarly there may be more than one transmitted wave, each with its own amplitude and polarisation, travelling at its own angle given by the generalised law of transmission, equation 4.51. Snell's law, where there is no change in polarisation on transmission, is a particular case.

4.2.3   Relative amplitudes of waves at boundaries †

Possible polarisation changes on reflection and transmission

In the following we consider three cases separately: an incident SH wave, an incident P wave and an incident SV wave. In each case the wave amplitude is described by its displacement with a positive amplitude, meaning that there is a positive component of the displacement along the positive x-axis (for SH waves) or positive y-axis (for SV or P waves).

There is an important difference between SH waves on the one hand, and SV and P waves on the other. They have different parities and therefore do not mix. We explore this statement briefly. The x-axis is the normal to the plane of incidence and, as such, is defined by a vector cross product. This therefore changes sign if the waves are described in a left-handed coordinate system instead of a right-handed one. The displacement amplitude of SH waves is measured along this axis. The amplitudes of SV and P waves lie in the plane of incidence and have a sign defined by the direction of the y-axis, the direction of the incident k on the xy-plane. This is independent of the choice of left- or right-handed axes. Therefore the relative sign of SH and SV/P waves depends on whether we use left- or right-handed axes. If they were to mix on reflection or transmission at the boundary, this relative sign would have observable consequences. Since the physics of sound does not depend on the handedness of the axes, this may not happen.

The conclusion is that the problem simplifies. An incident SH wave generates only reflected and transmitted SH waves. An incident SV or P wave generates a superposition of reflected and transmitted SV and P waves, but no SH waves.

Calculations with geophysical examples

The reflection and transmission amplitude coefficients are calculated below for a solid–solid boundary. As examples, numerical values of these coefficients for the Earth's mantle–crust boundary (table 4.3) are plotted as a function of angle of incidence for an incident SH wave, P wave and SV wave in Figs 4.12 and 4.14. For a solid–fluid interface we have used rock–water and rock–air boundaries as examples, and the coefficients are plotted in Figs 4.15–4.16.

We discuss the simpler incident SH wave first, followed by the incident P wave and then the SV wave. We use a notation in which the subscripts S and P refer to shear and compression waves respectively. The symbol used for the wave displacement amplitude, H, P or V denotes the polarisation. The angle θ is subscripted h, p or v to indicate the polarisation of the wave to which it refers. Calculations of energy flux may be used as a check that the boundary conditions have been satisfied correctly. If fluxes are calculated per unit area normal to the boundary, the incident flux towards the boundary should equal the sum of all of the outgoing fluxes away from it.

Table 4.3 Material properties for the Earth's mantle–crust interface used in the calculation of Figs 4.12 and 4.14.

	Units	First material	Second material
c_P	m s^{-1}	8000	2920
c_S	m s^{-1}	4300	1840
ρ	kg m^{-3}	4000	2700

Incident shear wave polarised perpendicular to the plane of incidence

The only components of displacement or force in the x-direction (perpendicular to the plane of incidence) arise from SH waves, as shown in equations 4.36 and 4.38. Continuity of these two quantities across the boundary gives the two constraints that determine the two unknowns, the reflection amplitude coefficient H_r/H_i and the transmission amplitude coefficient H_t/H_i for the two outgoing SH waves.

The displacement amplitudes indicated in Fig. 4.10 are to be understood as positive along the x-axis standing out of the paper. Then the constraint of continuity of displacement is

$$H_i + H_r = H_t \tag{4.52}$$

and the constraint of continuity of force is

$$H_i \cos \theta_{hi} k_i \mu_1 - H_r \cos \theta_{hr} k_r \mu_1 = H_t \cos \theta_{ht} k_t \mu_2. \tag{4.53}$$

The negative sign for the reflected ray comes from the fact that the reflected wave is travelling in a direction $\theta = \pi - \theta_{hr}$. The magnitude of each k is ω/c_S where c_S is the wave velocity for shear waves in the material. The solutions for the amplitude transmission and reflection coefficients are

$$\frac{H_t}{H_i} = \frac{2Z_{1S} \cos \theta_{hi}}{Z_{1S} \cos \theta_{hi} + Z_{2S} \cos \theta_{ht}} \tag{4.54}$$

$$\frac{H_r}{H_i} = \frac{Z_{1S} \cos \theta_{hi} - Z_{2S} \cos \theta_{ht}}{Z_{1S} \cos \theta_{hi} + Z_{2S} \cos \theta_{ht}}, \tag{4.55}$$

where $Z_{1S} = \rho_1 c_{1S}$ and $Z_{2S} = \rho_2 c_{2S}$. These solutions satisfy the conservation of energy flux, as shown in Fig. 4.12b.

These coefficients are plotted in Figs 4.12a and 4.12b. They show that for the mantle–crust boundary at most angles the amplitude of the transmitted wave in the lighter and less rigid second medium is larger than in the incident amplitude. However, the associated energy in the lower density of the second medium is seen to be smaller. This is a general feature. For example, the displacement amplitude of earthquakes is typically largest in regions of low density and rigidity, such as landfill sites. The large displacement does not carry large energy, although this may be where the greatest damage occurs.

Incident compression wave

For an incident P wave there are both reflected and transmitted SV waves as well as reflected and transmitted P waves, as shown in Fig. 4.11. To match the unknown amplitudes of the four outgoing waves there are four constraints coming from the y- and z-components of displacement and force. The four constraint equations are linear in the amplitudes of the four generated waves, and can be written as the matrix equation

$$M \times C_P = I_P, \tag{4.56}$$

where the vector of the four amplitudes is

$$C_P = \begin{bmatrix} P_r/P_i \\ V_r/P_i \\ P_t/P_i \\ V_t/P_i \end{bmatrix}, \tag{4.57}$$

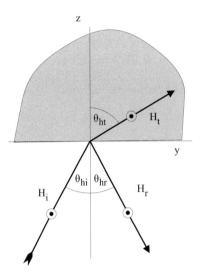

Fig. 4.10 A diagram for wave amplitudes generated by an incident shear wave polarised normal to the plane of incidence (SH).

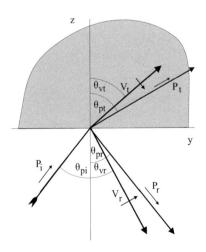

Fig. 4.11 A diagram for wave amplitudes generated by an incident compression wave (P).

Incident SH wave on mantle-crust boundary

a) Relative displacement

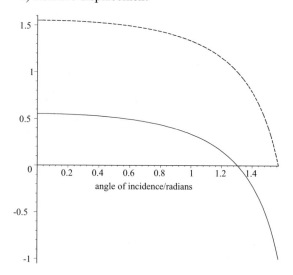

b) Relative normal energy flux

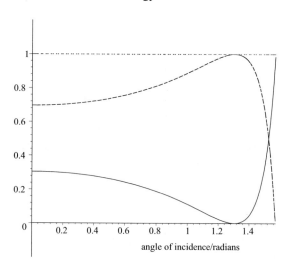

Incident P wave on mantle-crust boundary

c) Relative displacement

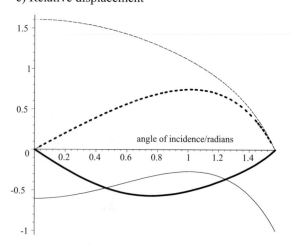

d) Relative normal energy flux

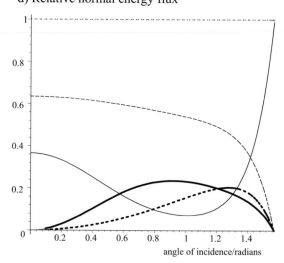

Fig. 4.12 Reflection and transmission coefficients as a function of the angle of incidence for the mantle–crust interface (table 4.3). a) and b) for incident SH wave: full curve = reflected SH, dash curve = transmitted SH, dotted line = calculated energy sum. c) and d) for an incident P wave: full thin curve = reflected P, full heavy curve = reflected SV, thin dashed curve = transmitted P, heavy dashed curve = transmitted SV, dotted line = calculated energy sum.

the vector of displacements and forces due to the incident wave is

$$I_P = \begin{bmatrix} \sin\theta_i \\ \cos\theta_i \\ Z_{1S}(c_{1S}/c_{1P})\sin 2\theta_i \\ Z_{1P} - Z_{1S}(c_{1S}/c_{1P})(1-\cos 2\theta_i) \end{bmatrix} \qquad (4.58)$$

and the matrix $M =$

$$\begin{bmatrix} -\sin\theta_{pr} & \cos\theta_{vr} & \sin\theta_{pt} & \cos\theta_{vt} \\ \cos\theta_{pr} & \sin\theta_{vr} & \cos\theta_{pt} & -\sin\theta_{vt} \\ Z_{1S}(c_{1S}/c_{1P})\sin 2\theta_{pr} & -Z_{1S}\cos 2\theta_{vr} & Z_{2S}(c_{2S}/c_{2P})\sin 2\theta_{pt} & Z_{2S}\cos 2\theta_{vt} \\ -Z_{1P}+Z_{1S}(c_{1S}/c_{1P})(1-\cos 2\theta_{pr}) & -Z_{1S}\sin 2\theta_{vr} & Z_{2P}-Z_{2S}(c_{2S}/c_{2P})(1-\cos 2\theta_{pt}) & -Z_{2S}\sin 2\theta_{vt} \end{bmatrix}$$

$$(4.59)$$

By inverting the matrix the amplitude coefficients due to the incident P wave may be found,

$$C_P = M^{-1}I_P. \qquad (4.60)$$

The amplitude ratios and the corresponding energy fluxes are shown in Figs 4.12c and 4.12d for the mantle–crust boundary. While at normal incidence there is no change in polarisation,[27] at other angles there are significant amplitudes of shear as well as compressive waves. Because in this case the phase velocities of these generated secondary waves are lower than the incident wave, their angles are real and less than the angle of incidence, whatever it is.

[27] Such a change is forbidden by parity conservation.

Incident shear wave polarised in the plane of incidence

A similar analysis can be made for waves generated by an SV wave incident on the boundary with wavevectors as shown in Fig. 4.13. The same matrix (equation 4.59) links the generated wave amplitudes to the forces and displacements of the incident wave. The amplitudes

$$C_V = \begin{bmatrix} P_r/V_i \\ V_r/V_i \\ P_t/V_i \\ V_t/V_i \end{bmatrix} \qquad (4.61)$$

are given by

$$C_V = M^{-1}I_V, \qquad (4.62)$$

where

$$I_V = \begin{bmatrix} \cos\theta_i \\ -\sin\theta_i \\ Z_{1S}\cos 2\theta_i \\ -Z_{1S}\sin 2\theta_i \end{bmatrix}. \qquad (4.63)$$

As an example, the amplitudes are plotted in Fig. 4.14 for the same mantle–crust interface (table 4.3).

These show new features which do not arise for incident SH or P waves on this boundary. For angles of incidence greater than 0.56 all four

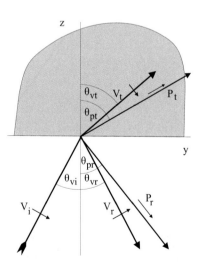

Fig. 4.13 A diagram for wave amplitudes generated by an incident shear wave polarised in the plane of incidence (SV).

Incident SV wave on mantle-crust boundary

a) Real part of relative displacement amplitude

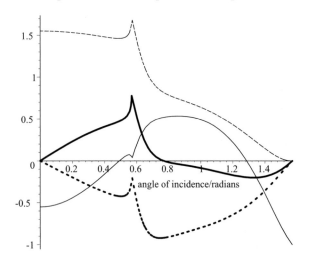

b) Imaginary part of relative displacement amplitude

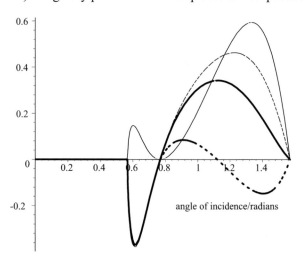

c) Squared cosine of wavevector angle

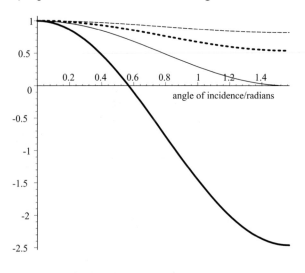

d) Relative normal energy flux

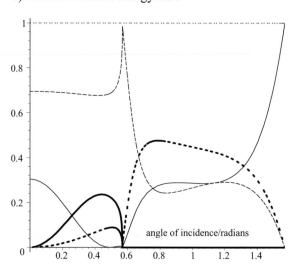

Fig. 4.14 Reflected and transmitted waves generated by an incident SV wave as a function of the angle of incidence for the solid–solid interface described in table 4.3.
Full thin curve = reflected SV, full heavy curve = reflected P, thin dashed curve = transmitted SV, heavy dashed curve = transmitted P, dotted line = calculated energy sum.
Note the absence of a normal energy flux for the reflected P wave if the angle of incidence is greater than 0.56.

generated waves have a non-trivial phase angle relative to the incident wave. Figures 4.14a and 4.14b show the imaginary as well as the real parts. For angles above 0.56 the reflected P wave has a negative value of k_z^2 (Fig. 4.14c). In this region there is zero normal energy flow associated with the reflected P wave (Fig. 4.14d), although the calculated energy flux for the remaining three waves still adds up to the incoming flux.

Evanescent waves

If k_z^2 is negative, then k_z is pure imaginary. The phase factor $\exp[ik_z z]$ for the reflected P wave normally represents a harmonic dependence. However, when k_z is imaginary this describes a real exponential of the form $\exp -z/z_0$ where the range is given by $z_0 = 1/\sqrt{-k_z^2}$. Such an evanescent wave is familiar from other branches of physics, such as the total internal reflection of light and the exponential behaviour of quantum wavefunctions in regions where the classical kinetic energy would be negative. These have important applications in the physics of fibre optics and field effect transistors as well as describing the phenomenon of alpha radioactive decay.

As given by equations 4.49 and 4.51 whenever an incident wave generates an outgoing wave with an increased wave velocity, the value of $\sin\theta$ for that wave is greater than for the incident wave. So for some range of the angle of incidence, the sine of the angle of reflection or transmission exceeds unity. Then its cosine is pure imaginary. Such a wave decays in z but oscillates in time and in x or y in the normal way. It propagates along the boundary with a finite amplitude but carries no energy away from the surface. To match the boundary conditions the other waves acquire phases and display threshold effects as shown in Fig. 4.14. The threshold is at an angle with $\sin\theta_i = c_1/c_2$. There may be up to three such thresholds, one for each generated wave whose phase velocity exceeds the phase velocity of the incident wave. At the threshold angle the exponential range z_0 of the evanescent wave starts from infinity and falls as the angle of incidence is increased further.

Solid–liquid interface

Table 4.4 gives constants for a rock–water interface, such as might apply at the bottom of the oceans. As the second material is a fluid the impedance $Z_{2S} = 0$ and there can be no propagation of energy by shear wave in the second medium. Equations 4.54 and 4.55 show that an incident SH wave is totally reflected at all angles. Because the surface is unrestrained in shear, there is an antinode there. There is a transmitted SH wave. It has a displacement amplitude twice that of the incident wave but transmits no force or energy because the modulus is zero.[28]

For an incident SV wave[29] the reflection and transmission coefficients are plotted out in Fig. 4.15. The amplitude and energy of the reflected P wave is large away from normal incidence, showing again that polarisation changes are first order effects, not just corrections. The large transmitted SV wave amplitude has the same interpretation as for the

Table 4.4 Material properties for the rock–water boundary used in the calculation of Fig. 4.15.

	Units	First material	Second material
c_P	m s^{-1}	2920	1407
c_S	m s^{-1}	1840	0
ρ	kg m^{-3}	2700	1000

[28]Because this amplitude does not depend on the angle of incidence the plot is featureless and is not shown.

[29]An incident P wave generates no evanescent wave as no wave has a phase velocity faster than the incident wave. The reflection and transmission coefficients are all real and unremarkable. They are not plotted here.

Incident SV wave on rock-water boundary

a) Real part of relative displacement amplitude

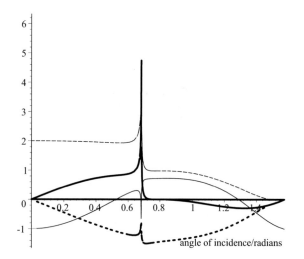

b) Imaginary part of relative displacement amplitude

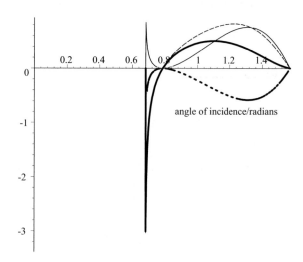

c) Squared cosine of wavevector angle

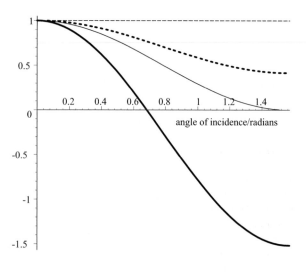

d) Relative normal energy flux

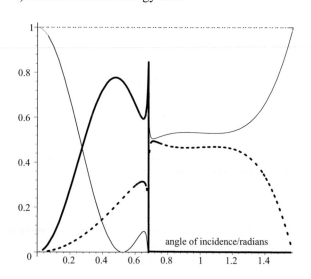

Fig. 4.15 Reflected and transmitted waves generated by an incident SV wave as a function of the angle of incidence for the rock–water interface described in table 4.4.
Full thin curve = reflected SV, full heavy curve = reflected P, dashed thin curve = transmitted SV, dashed heavy curve = transmitted P, dotted line = calculated energy sum.
Note the absence of a normal energy flux for the reflected P wave for an angle of incidence of greater than 0.68.

transmitted SH wave, amplitude but no velocity or impedance and therefore no energy flux. The reflected P wave is evanescent for angles of
incidence above $\sin\theta_i = c_{1S}/c_{1P} = 0.682$. In this range only the transmitted P wave and the reflected SV wave can carry energy, although
the other two waves have amplitude. The transmitted SV wave has zero
modulus, and the reflected P wave is an evanescent wave. Neither carry
energy.

Solid–gas interface

Table 4.5 describes a rock–air interface, such as might apply at the
Earth's surface. Air not only is a fluid but also has a very low impedance
for transmitted P waves. For incident SH waves the situation is the same
as it was for the rock–water interface. For incident SV waves or P waves
there is a near antinode at the surface and the large transmitted P wave
amplitude carries negligible energy. Figure 4.16 shows curves for an
incident SV wave. The energy is divided between the reflected P wave
and SV wave, again with a large change in polarisation. With no energy
in the transmitted P wave, above the critical angle for the reflected P
wave all the energy is reflected in the SV polarisation.

Table 4.5 Material properties for the rock–air boundary used in the calculation of Fig. 4.16.

	Units	First material	Second material
c_P	m s^{-1}	2920	330
c_S	m s^{-1}	1840	0
ρ	kg m^{-3}	2700	1.2

4.3 Surface waves and normal modes

*We consider surface waves on the plane boundary between two homoge-
neous materials and the three dimensional oscillation modes of bodies as
a whole.*

*Under certain circumstances combinations of evanescent waves alone
can satisfy the boundary conditions. These combinations are known as
surface waves although they penetrate the bulk material to an important
extent. At a solid–fluid interface a superposition of evanescent longitudinal and evanescent transverse waves can form a surface wave. This is
a Rayleigh wave. Such waves have important applications in geophysics,
geological prospecting and modern electronics. A different surface wave
may propagate along a fluid–fluid interface. Gravity or surface tension
allows a single evanescent longitudinal wave to satisfy the boundary conditions. Such waves describe a range of phenomena from short ripples to
long ocean tidal waves. Compared with free bulk waves they have low wave
velocity. Their properties are of considerable environmental importance.*

*In a medium bounded on all sides wave solutions form a set of discrete
modes with well-defined frequencies. Such modes are realised in the tune
of musical instruments and in the bodily vibrations of the Earth and the
Sun.*

Incident SV wave on rock-air boundary

a) Real part of relative displacement amplitude

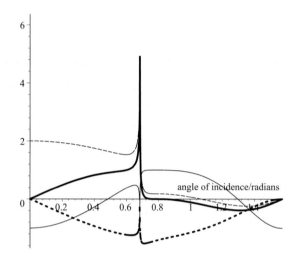

b) Imaginary part of relative displacement amplitude

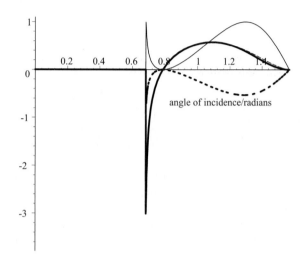

c) Squared cosine of wavevector angle

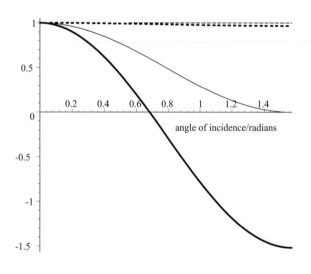

d) Relative normal energy flux

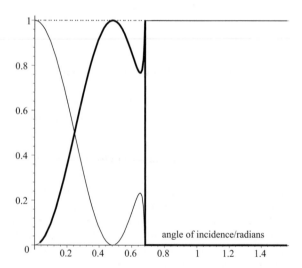

Fig. 4.16 Reflected and transmitted waves generated by an incident SV wave as a function of the angle of incidence for the rock–air interface described in table 4.5.

Full thin curve = reflected SV, full heavy curve = reflected P, dashed thin curve = transmitted SV, dashed heavy curve = transmitted P, dotted line = calculated energy sum.

Note that if the angle of incidence is greater than 0.68 only the reflected SV wave has significant energy flux.

4.3.1   General surface waves

The evanescent waves discussed so far are generated by the incidence of a bulk S or P wave on a surface. This happens whenever the angle of incidence is large enough to satisfy equation 4.49 or 4.51 with $\sin\theta_r > 1$ or $\sin\theta_t > 1$. In table 4.6 we list all the possible combinations of polarisation that can generate such evanescent waves. These evanescent waves only persist as long as the incident bulk S or P waves are present. In that sense they are not free.

In addition there exist surface wave solutions which are freely propagating and which, though confined to the surface and the region close to it, do not require incident free bulk waves. Such surface waves may be formed by particular superpositions of evanescent S and P waves for which the boundary conditions can be matched. These are called Rayleigh waves and can be excited at a solid–light-fluid interface.[30]

A quite different surface wave can be excited at a fluid–fluid interface where a restoring force enables the boundary conditions to be satisfied. This force may be due either to gravity and a density difference between the fluids or to the effect of surface tension.

Table 4.6 The complete list of possible combinations of incident wave (i-wave) and evanescent wave (e-wave) polarisations, transmitted or reflected, with the necessary condition on the phase velocities in the first and second media.

Polarisation		necessary
i-wave	e-wave	condition
P	P transmitted	$c_{1P} < c_{2P}$
SV	P transmitted	$c_{1S} < c_{2P}$
SV	SV transmitted	$c_{1S} < c_{2S}$
SH	SH transmitted	$c_{1S} < c_{2S}$
P	SV transmitted	$c_{1P} < c_{2S}$
SV	P reflected	$c_{1S} < c_{1P}$

[30]The light fluid implies that the solid has an essentially free surface. Strictly this couples to a P wave in the fluid which if sufficiently light carries away no energy.

4.3.2   Rayleigh waves on free solid surfaces

A Rayleigh wave is a particular linear combination of an evanescent SV wave and an evanescent P wave on the solid side of a free unloaded surface. Consider a P wave with amplitude u_P and wavevector $\mathbf{k}_P$, and a S wave with amplitude u_S and wavevector $\mathbf{k}_S$. We choose axes with the interface being the plane $z = 0$ and $\mathbf{k}_P$ lying in the yz-plane, as before. To satisfy the boundary conditions for all (x, y, t), the 'footprint' of the SV wave and the P wave on the boundary must match as discussed in section 23. Therefore the frequencies of the two waves are the same, their k-vectors lie in the yz-plane and they have the same value $k_y = k_{Py} = k_{Sy}$. The magnitudes of the $\mathbf{k}$-vectors are $k_P = \omega/c_P$ and $k_S = \omega/c_S$. So far all quantities are real. Defining angles θ_P and θ_S for the direction of the $\mathbf{k}$-vectors relative to the interface normal, we deduce

$$k_y = \frac{\omega}{c_P}\sin\theta_P = \frac{\omega}{c_S}\sin\theta_S. \tag{4.64}$$

Both of these sines are greater than 1 in a Rayleigh wave, and the corresponding cosines pure imaginary. Because the two waves are on the same side of the boundary, the signs of the two imaginary cosines are the same.

The constraints for a free surface are that the components of stress through the surface in the z- and y-directions for the two waves should add to zero. The forces per unit area, $\mathbf{F}_{zL}$ and $\mathbf{F}_{zS1}$, may be read off from equation 4.38 in terms of the moduli and angles. Defining the velocity $\omega/k_y = c_R$, these two constraints may be reduced to the equations

$$\left(2 - \frac{c_R^2}{c_S^2}\right)^4 = 16\left(1 - \frac{c_R^2}{c_S^2}\right)\left(1 - \frac{c_R^2}{c_P^2}\right) \tag{4.65}$$

Fig. 4.17 a) The ratio of the velocity of Rayleigh waves to free shear waves plotted as a function of the ratio of the velocities of free shear and longitudinal waves. b) The depth penetration of Rayleigh waves in terms of the free shear wavelength/2π plotted as a function of the same ratio of the velocities of free shear and longitudinal waves. The upper curve is the penetration of the evanescent shear wave component. The lower curve is the penetration of the evanescent longitudinal wave component.

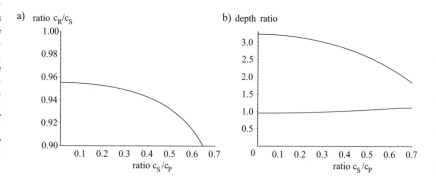

$$\frac{u_P}{u_S} = -\mathrm{i}\left(\frac{2c_S^2 - c_R^2}{2c_P\sqrt{c_P^2 - c_R^2}}\right). \tag{4.66}$$

The velocity c_R is the phase velocity of both evanescent waves along the surface. These relations do not depend on ω, so that Rayleigh waves are not dispersive, unless the free longitudinal and shear waves themselves are.

Figure 4.17a shows a plot of the velocity of the ratio of the velocity of Rayleigh waves to free shear waves as a function of the ratio of the velocity of free shear waves to longitudinal waves as given by equation 4.65. The velocity of Rayleigh waves is seen to be 0.92–0.95 of the free shear velocity, more or less independent of the velocity of the longitudinal velocity.

Equation 4.66 gives the relative amplitude of the two evanescent components at the surface. They are $\pi/2$ out of phase, and so combine to give an elliptical motion. The depth penetration of the two constituent evanescent waves is different. It may be calculated in each case as the reciprocal of k_z, determined by Pythagoras's theorem from $k = \omega c$ and k_y. The shear wave component is only just evanescent and so has a relatively deep penetration. The longitudinal component is more strongly evanescent, and consequently has a short range. Both ranges are plotted in Fig. 4.17b. The consequence is that near the surface there is a significant component of strain parallel to the surface and the motion is elliptical. Further from the surface the motion is more closely that of a simple shear wave with motion perpendicular to the surface.

In geophysics Rayleigh waves may be distinguished from bulk waves by their slow speed and by their characteristic polarisation, involving an elliptical motion, both longitudinal and transverse.[31] They are generated by earthquakes within a few wavelengths of the surface. Typically earthquakes involve sudden shear failure of structures over the area of a fault. Geological seismometers are also sensitive to the signals emitted by underground nuclear weapon tests. These involve spherically symmetric point-like P wave emission but no SH or SV waves. Such events do not create Rayleigh waves. By measuring the polarisation of seis-

[31]They are seen to be slow because they follow the longer route along the Earth's surface while the bulk waves follow a more direct route.

mic signals and their relative timings, earthquakes and explosions from nuclear weapon testing may be distinguished.

Rayleigh waves are of particular importance in modern seismology because their propagation is sensitive to density and modulus changes close to the surface. As the shear modulus of any fluid vanishes, oil or gas bearing strata show up clearly. The change of shear modulus in regions of loose sand or landfill where they adjoin rigid compacted strata can also be imaged.

In our analysis so far we have seen that evanescent waves (including Rayleigh waves) carry no energy compared with bulk waves. This is not the whole story. There is a dimensional point here. Rayleigh waves are two dimensional while bulk waves are three dimensional. Bulk waves carry an energy flux *per square metre* of wavefront, while a Rayleigh wave carries energy *per linear metre* in the surface. Indeed there is a real energy flux parallel to the surface but only within the depth penetration of the wave. The energy from a localised source of such a two dimensional wave spreads out with a circular wavefront and an intensity falling as $1/r$. This is in contrast to bulk waves that spread out from a source with an intensity falling as $1/r^2$. Therefore signals carried by surface waves, guided by the boundary, propagate effectively over longer distances.

Variants on Rayleigh waves as described above occur for thin sheets. The single evanescent wave in each polarisation is replaced by sinh and cosh waves built from the evanescent behaviour at the two surfaces. Such waves are dispersive. Further modifications also arise if the elastic properties of the lighter medium are taken into account.

Surface waves of a different type may occur if the medium is layered or graded in elasticity or density. Such layers may act as a waveguide thereby trapping the energy in two dimensions. In a geological context such waves are known as Love waves.

Surface acoustic waves

On a different scale Rayleigh waves may be excited in the surface of thin wafers that form important components of modern electronics. A shear wave is launched on a thin piezoelectric wafer by an array of alternating electrodes laid down on the surface as illustrated in Fig. 4.18. In this context thin-film Rayleigh waves are described as surface acoustic waves (SAW). In electronic engineering they are the basis of most high frequency filters and delay lines in modern communication systems. It has been estimated that in 2005 annual production exceeded 10^9 such devices worldwide.

4.3.3 Waves at fluid–fluid interfaces

Shear motion in fluids

At solid–fluid interfaces the transmitted shear wave carries no energy as shown in Figs 4.15 and 4.16. However, the displacement amplitude of the shear wave in the fluid required by the boundary conditions is

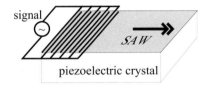

Fig. 4.18 An electrode array deposited on the surface of a piezoelectric crystal to transmit a surface acoustic wave (SAW). It may also act as a receiver, transforming a SAW into a high frequency electrical signal.

[32]The real part of the shear modulus of fluids is zero. The only shear force generated by shear motion in a fluid is viscous damping. This is proportional to the shear velocity, not the shear displacement to which it is $\pi/2$ out of phase, and is therefore described by a small imaginary value of μ proportional to ω. This small imaginary value of μ gives small complex values for the impedance and the wave velocity for shear waves.

[33]These observations have an obvious connection with the use of ultrasound as a cleaning agent and emulsifying technique.

very large.[32] The structural integrity of the surface depends essentially on the rigidity of the solid and the resulting reflected SV wave.

The surface at a fluid–fluid interface does not have the structural integrity of a solid on either side. When a P wave is incident at an oblique angle interesting problems arise. In the absence of any other forces the boundary conditions cannot be met and the molecules of the two fluids would simply mix and the surface dissolve.[33] However, there are two other weaker forces, gravity and surface tension, that can maintain the surface under these conditions. The most familiar manifestation is the 'water wave'. Analogous waves exist between all fluids including between layers of different density in the atmosphere and between layers of water of different temperature or salinity. However, these forces are weak, the velocities are slow and there is generally a large displacement amplitude at the interface.

Gravity and surface tension waves

Water waves travel so slowly compared with any free bulk P wave that water behaves as if it were incompressible. In other words, the associated evanescent P wave that underlies a surface wave with real k in y has a $\boldsymbol{k}$-vector with $k^2 \approx 0$, and thence $k_z^2 = -k_y^2$ and $k_z = \pm i k_y$. The general P wave solution is then a superposition of the solutions with these two signs, each of which has $\boldsymbol{k}$ parallel to the displacement for all space and time,

$$\psi_y = A \exp[i(k_y y - \omega t)] \exp(-k_y z) + B \exp[i(k_y y - \omega t)] \exp(+k_y z)$$
$$\psi_z = iA \exp[i(k_y y - \omega t)] \exp(-k_y z) - iB \exp[i(k_y y - \omega t)] \exp(+k_y z).$$
$$(4.67)$$

Each describes circular motion in the yz-plane, one decreasing exponentially with depth z and the other increasing. From this point we drop the subscript y in k_y, and describe the wave as a surface wave of wavenumber k. Note that the same value of k also describes the exponential depth behaviour.

If the depth of water is z_0 the bottom forms a rigid boundary where there is no vertical motion. This fixes $B = A \exp(-2kz_0)$. Thence

$$\psi_y = A \exp[i(ky - \omega t)] \times [\exp(-kz) + \exp(kz - 2kz_0)] \quad (4.68)$$
$$\psi_z = iA \exp[i(ky - \omega t)] \times [\exp(-kz) - \exp(kz - 2kz_0)].$$

At the surface $z = 0$ the difference between the zero pressure above the surface and the P wave pressure below the surface is balanced by the hydrostatic weight of the surface wave amplitude. Figure 4.19 illustrates how the horizontal pressure gradient depends on the wave height h,

$$\frac{\partial P}{\partial y} = \rho g \frac{\partial h}{\partial y}. \tag{4.69}$$

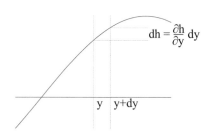

Fig. 4.19 The horizontal pressure gradient near the surface of a water wave is due to the gradient of the wave height, multiplied by density and acceleration due to gravity g.

We substitute $h = \psi_z$ at $z = 0$ and equate this horizontal pressure gradient to the product of density and acceleration in y,

$$-\rho \omega^2 \psi_y = i \rho g k \psi_z. \tag{4.70}$$

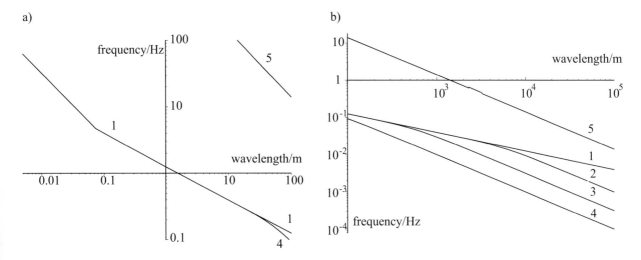

Fig. 4.20 The dispersion relation for water waves on Earth shown as plots of frequency against wavelength using $T = 0.072$ N m^{-2}. Curve 1, for 'deep' waves; curves 2–4 show the effects of a depth of 1000 m, 100 m and 10 m depth respectively. Curve 5 shows the relation for free P waves, the velocity of sound in water. Plots a) and b) cover different logarithmic wavelength scales.

This is satisfied provided

$$\omega^2 = \left(gk + \frac{T}{\rho}k^3\right)\left(\frac{1 - \exp(-2kz_0)}{1 + \exp(-2kz_0)}\right). \qquad (4.71)$$

To complete this dispersion relation we have added the term that describes surface tension T. This gives a force proportional to surface curvature Tk^2 where T is the surface tension in N m^{-1}. This term is much smaller than the effect of gravity for water waves on Earth for wavelengths greater than 2 cm.

Properties of gravity and surface tension waves

For gravity waves in the deep water limit, $z_0 \gg \lambda$, the dispersion equation 4.71 reduces to

$$\omega^2 = gk. \qquad (4.72)$$

The phase velocity is then

$$c_{\text{gravity}} = \sqrt{\frac{g\lambda}{2\pi}} \qquad (4.73)$$

which is dispersive. Long wavelength waves travel much faster than short ones. The group velocity is half the phase velocity. This dispersion is a dominant effect in naval architecture as it determines the maximum speed of a non-planing ship's hull. As a boat moves forward it displaces the water in front and, if it is not to find itself trying to climb its own bow wave, relies on this displaced water flowing round to the stern fast enough to fill the volume left by the retreating hull, hopefully nearly reversibly. This requires water motion with wavelength not longer than the waterline length of the boat, from which the limiting boat speed may be deduced.[34]

In shallow water gravity waves travel more slowly than in deep water. If z_0 is of order λ or less, the dispersion equation 4.71 gives a phase velocity

$$c_g = \sqrt{g z_0}. \tag{4.74}$$

This is not dispersive.[35]

In Fig. 4.20 this dependence of frequency on wavelength is plotted out for water waves in these different ranges. The relation for bulk P waves is shown on the same plot. The contribution of surface tension to water waves is evident from the change in slope for very short wavelengths. At long wavelengths for which the depth is significant, the slope for gravity waves steepens to a gradient of -1. There is no depth and wavelength for which the velocity of gravity waves approaches the velocity of bulk P waves c_p, even in the deepest ocean on Earth.

In shallow water there is a node in the vertical motion at the bottom $z = z_0$, as required by the boundary conditions. The horizontal motion at the bottom is non-zero, as shown by setting $z = z_0$ in equation 4.68. This relative shear motion is peculiar to a fluid, and is why gravity waves of this type only occur in fluids. If the water is deep the exponential factor ensures that this shear motion is small. However, where long wavelength waves enter shallower water, the scouring action of this shear motion on a rough seabed has significant consequences for shifting sands and coastal erosion. It is also a significant damping mechanism for the wave itself.

Gravity waves may be excited from the top fluid–fluid surface or from the bottom fluid–solid interface. Since the horizontal particle velocity at the surface is in the direction of the $\boldsymbol{k}$-vector on the peak of the wave and in the reverse direction in the trough, a horizontal wind is effective in exciting a gravity wave moving in the same direction. Excitation from the bottom is inhibited by the exponential factor. Only wavelengths of the same order as the depth of the water or longer can generate surface waves efficiently. This also means that such generated waves are shallow and non-dispersive, until they move into significantly deeper water at least. This has important consequences for tsunami generation. Tsunami waves are not dispersive and have a wavelength spectrum equal to the depth of water or longer. Thus a tsunami generated in a depth of 5000 m of water consists of a wave of length 5000 m or longer. Shorter wavelengths are attenuated by the exponential factor in equation 4.68.

When such a wave reaches shallower water the wave velocity drops. As the velocity falls the wave amplitude (and particle velocity) increases as $1/\sqrt{c_{\text{gravity}}}$ to conserve the energy flux. The amplitude therefore increases as $z_0^{-1/4}$, although there is still no dispersion. If the depth decreases and the wave height increases further, eventually the wave height becomes a significant fraction of the depth. At this point the wave starts to propagate non-linearly with the peak of the wave moving forward faster than the trough. Eventually it breaks as discussed briefly in section 9.4.

On the other hand, if the tsunami wave propagates into deeper water where the velocity is dispersive, the pulse spreads out. Non-linear effects would then be reduced, if and when the wave subsequently reached shallow water again.

Prospectively, the most dangerous tsunamis are associated with earthquakes in shallow water, for these have no filtering of short wavelengths by the exponential depth factor.

4.3.4   Normal mode oscillations

Waves in bounded volumes

If a space is enclosed by boundaries, there are too many conditions and the propagation of waves in that space is overdetermined. This means that waves may propagate only at certain frequencies. This happens to the sound waves in a musical instrument, or the vibrations on a finite string. The surfaces may be constrained or free. If they are constrained, there is a node at the surface. If they are free, there is an antinode. The allowed states are defined in either case. Another possibility is that the space is joined to itself, so that it has periodic boundary conditions. The important examples are waves on a cylinder or circle, and also on a sphere. A string attached at both ends, a drum held at its periphery, or a cavity with well-defined surfaces are familiar eigenvalue problems of this type. Other examples are the tune of musical instruments and the bound state energies of atoms and molecules.

In a cartesian geometry for a volume of dimensions ℓ_x by ℓ_y by ℓ_z the wave equation determines that

$$\omega^2/c^2 = k_x^2 + k_y^2 + k_z^2 = 4\pi^2 \left(\frac{1}{\lambda_x^2} + \frac{1}{\lambda_y^2} + \frac{1}{\lambda_z^2} \right), \qquad (4.75)$$

where the boundary conditions require that ℓ_x matches an integer (or half integer) number of wavelengths λ_x, with similar relations in y and z. This defines a spectrum of allowable values of ω. Such solutions are called normal modes. In spherical and other geometries the calculation of the spectrum of modes becomes more involved, although symmetry may be used to classify and understand the results. The mathematics is analogous, whether the waves are vibrational, electromagnetic or quantum mechanical.

Boundary conditions are not exact. For example, the point at the end of string of a musical instrument is not quite a node. It is a high impedance point where the transverse tension is high and the displacement low. This small displacement is used to generate the sound via the sound box and causes damping of the resonance giving it a finite width γ_r. In general a frequency spectrum of modes consists of sequence of peaks each with its particular width. The width for a given mode may be described by the dimensionless ratio $Q = \omega/\gamma_r$.

Vibrational modes of the Earth and Sun

Normal modes are important if the wavelengths of interest become a significant fraction of the size of the region. On the other hand, at short wavelengths and high frequencies the resonances become sufficiently close that they may be treated as a continuum. The solid Earth has a spectrum of vibrational resonances. The lowest is at 468 μHz. Many such modes may be distinguished with Q between 350 and 900. Their identification is limited only when their spacing in frequency becomes smaller than their widths.[36] Some of these resonances are excited by large earthquakes and their subsequent oscillation has been followed for significant times. Comparison of experimental data with the results of finite-element calculations of the frequency, width and symmetry of such modes can be made to constrain our understanding of the elastic structure of the Earth's interior.

These are free oscillations. In addition there are data on forced oscillations driven by lunar and solar tidal forces. The excitation of such off-resonance steady-state oscillations of the solid Earth have an amplitude of a some centimetres.[37]

Similar observations and calculations have been made for a large number of normal modes of oscillation of the Sun. They confirm that our understanding of the density and elasticity structure of the Sun's interior is good enough to give agreement between calculated and observed frequencies of better than 1%.

Normal modes of piezoelectric crystals

If a small piezoelectric crystal, typically a quartz crystal, is set up to resonate, the oscillator has a very high Q because of the low coupling with its environment. Such quartz oscillators have generated a revolution in cheap and accurate time-keeping that is now largely taken for granted.

[36]The Earth's oceans are too heavily damped and obstructed by land masses to show any such worldwide resonance structure. Isolated lakes and seas, such as the Lake of Geneva, do show resonances, called seiches.

[37]Famously, when account was taken of the effect on the geometry of the LEP electron–positron accelerator ring at CERN of this tidal dilation of the surface of the Earth, the machine energy calibration was improved by an order of magnitude with considerable benefit to our knowledge of fundamental physics.

4.4   Structured media

The mechanical properties of materials derive from the potential between atoms and molecules, in the simplest model between nearest neighbours. This potential is hard and repulsive at short distance and softly attractive at long distance. This asymmetry is not important for small changes relative to a reference state for which materials behave linearly. The discreteness of the atomic chain is responsible for breaking the dispersion curve of oscillations into energy bands. The usual low frequency band is the acoustic band. In addition there may be higher frequency optical bands. The heat capacity of a material body is largely associated with the thermal excitation of these same oscillation modes. The absorption of sound takes the form of the scattering and degradation of these waves. Any delay in the strain response to an applied stress in a periodic wave absorbs energy. Such delay can come from other energy modes such as molecular rotation whose speed of response to changes in temperature may be slow compared with the period of the oscillation. The asymmetry of the potential gives rise naturally to non-linear behaviour at large amplitude, and this is responsible for further absorption processes.

4.4.1 Interatomic potential wells

Solids and liquids

As discussed in section 4.1.2 the strain response of gases to changes in pressure (or negative stress) may be described by the isentropic behaviour of a perfect gas with modulus γP_0. However, there is no general prescription for solids and liquids. Compared with gases they have a much lower compressibility, or a much higher bulk modulus.

Solids have limited tensile strength, which means that they can resist positive tension up to the point where they fracture. Liquids and gases have no tensile strength at all. Any attempt to apply a negative pressure to a liquid causes cavitation, involving an irreversible separation of liquid and vacuum (or vapour).

What is happening within liquids and solids and how may their response to stress be described? There is a mechanical answer based on the forces between atoms and molecules, and then there are the thermodynamic modifications that enter when thermal flows of energy are taken into account. In the following we ignore the thermodynamics in the first instance. Also, when looking at macroscopic properties in terms of microscopic mechanisms, we ignore the fact that each atom or molecule has a number of neighbours in three dimensions, not just two as in a one dimensional chain model. It will be sufficient for our purposes to consider qualitatively the material properties that would derive from the interaction between nearest neighbours on a one dimensional chain. On occasion this simple model will be significantly in error. Usually it will be illustrative and useful in teaching us how to think.

We assume that the microscopic force acting between nearest neighbours is a function only of their relative position. Such forces are elastic and can be described by a potential. These forces are repulsive at small separation and repel atoms by the following mechanism. Because of the so-called Pauli exclusion principle[38] all electrons must have wavefunctions different from one another. Hence, if matter with its electrons is made to occupy a smaller region, these electrons must occupy ever higher kinetic energy states. The need to provide such extra energy on compression gives rise to a pressure called electron degeneracy. Without electron degeneracy pressure all matter as we know it would collapse under the attraction of Coulomb (and gravitational) forces.

At large separations neighbouring atoms and molecules are attracted in several different ways. We consider just two, the Coulomb force re-

[38] Historically this was introduced as a principle but today that description is inappropriate. In quantum mechanics it follows from the indistinguishability of identical particles, and the observation that electrons are fermions rather than bosons.

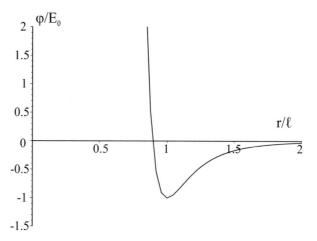

Fig. 4.21 The Lennard–Jones interatomic potential described in the text, shown relative to its equilibrium depth E_0 and equilibrium separation ℓ.

sponsible for the ionic solid, and the van der Waals force responsible for most other solids and liquids. An ionic solid is composed of ions, that is atoms or molecules that are charged. Between these positive and negative ions there is a Coulomb potential with the basic $1/r$ dependence. This is strong and long range so that forces between next-to-nearest neighbours as well as nearest neighbours are important.

By contrast, in a van der Waals solid the force is weak and shorter range, but acts between neutral atoms and molecules as well. It arises in the following way.[39] If, by fluctuation, a small electric dipole moment were to occur on one molecule, it would create a field proportional to $1/r^3$ in a second neighbouring molecule. This field would induce a dipole moment in the second molecule whose field at the first molecule would then be proportional to $1/r^6$. The energy of the initial dipole in this field makes the initial fluctuation energetically favourable. The consequence is that such fluctuations occur and there is a weak attractive term in the potential energy varying as $1/r^6$. Figure 4.21 shows such a potential, the Lennard–Jones potential,

$$\phi(r) = E_0 \left[\left(\frac{\ell}{r} \right)^{12} - 2 \left(\frac{\ell}{r} \right)^6 \right]. \qquad (4.76)$$

The short range repulsion is parametrised by the $1/r^{12}$ term. The depth of the potential is E_0 at the equilibrium distance of ℓ.

The value of the binding energy E_0 is of order 0.1 eV so that a typical van der Waals solid or liquid vaporises at a temperature not far removed from ambient. An ionic solid has a much deeper well and such materials do not vaporise at normal temperatures. For a metal the shape of the potential well is similar to the Lennard–Jones potential although the binding is tighter because, qualitatively, electric dipole moments are easily induced in conductors and so the effect of the dipole–dipole attraction is enhanced.

The position of the potential minimum determines the separation of atoms at low temperature. At higher temperature, T, classically the

[39]This is a reasonably convincing classical justification for a second order perturbation effect in quantum mechanics.

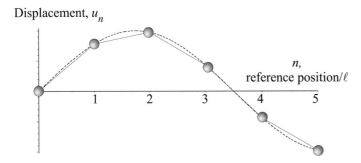

Displacement, u_n

n,
reference position/ℓ

Fig. 4.22 A plot of displacement against reference position for a one dimensional chain of identical masses. The reference position is given as a multiple of the reference separation ℓ. In the model forces act between nearest neighbours. A possible solution is that at any time the displacements follow a sine wave as shown. However, the lines between the masses have no physical significance and the wave exists only at the reference position of the masses.

atoms bounce back and forth inside the well with energy of order $k_{\mathrm{B}}T$. In quantum mechanics they occupy states of energy E within the well with probabilities related to the Boltzmann factor, $\exp(-E/k_{\mathrm{B}}T)$. Effectively the differences between the quantum and classical descriptions are small. In the following we use a classical description.

The asymmetry of the potential is such that the mean separation increases with increased temperature, an effect familiar as the thermal expansion of solids and liquids. This is a direct result of the non-linearity of the force, or equivalently the non-parabolic shape of the potential. In the following we neglect the effect of finite temperature in the first instance.

Motion of masses with nearest-neighbour potential

Such a constituent potential establishes the linear terms that determine the mechanical properties for small amplitude distortions and the non-linear terms as well. It also determines the periodic structure that gives rise to the band structure at short wavelengths which distinguishes a constituent model from a continuum model.[40]

The motion of two point masses, initially separated in x, can be described by six coordinates. If they do not interact, each can move independently in x, y and z. If they interact, there are still six modes and these may be described as the relative motion and the combined motion of the centre of mass in each direction, x, y and z. By extension, the normal modes of N identical masses along x-may be described by their N different amplitudes, and the same for their transverse positions in y and z.

We consider a chain of atoms extended in x, for simplicity looking only at motion in the x-direction. Figure 4.22 shows such a line for which the reference separation of the masses is ℓ and the displacement of mass n relative to its reference position is u_n where n is its integer label. To simplify the diagram the displacements are shown transversely, although in this case they are actually longitudinal. According to this model the force on mass n due to mass $n+1$ depends only on the strain in their relative positions $\varepsilon_{n+} = (u_{n+1} - u_n)/\ell$ and may be written in terms of

[40] For a non-trivial band structure there needs to be more than one kind of mass, but this is a modest elaboration of the chain.

the derivative of a potential function $\phi(\varepsilon)$,

$$F_{n+} = \frac{\mathrm{d}\phi(\varepsilon)}{\mathrm{d}\varepsilon}\,\varepsilon_{n+}. \tag{4.77}$$

The force due to the mass $n-1$ is of opposite sign. They only fail to cancel exactly because the strain is different there. The difference in strain on the two sides is $(u_{n+1} + u_{n-1} - 2u_n)/\ell$. Therefore the net force on n from its two neighbours is

$$F_n = \frac{(u_{n+1} + u_{n-1} - 2u_n)}{\ell}\frac{\mathrm{d}^2\phi(\varepsilon)}{\mathrm{d}\varepsilon^2}. \tag{4.78}$$

For small strain ε relative to the reference state we may assume that $\mathrm{d}^2\phi(\varepsilon)/\mathrm{d}\varepsilon^2$ is a constant[41] evaluated at $\varepsilon = 0$. We call this ϕ_ℓ''. Then the motion of mass n is found by applying Newton's law,

[41] In section 9.4 we shall consider what happens when the strain is large enough that this is not true.

$$m\frac{\partial^2 u_n}{\partial t^2} = \frac{(u_{n+1} + u_{n-1} - 2u_n)}{\ell}\phi_\ell''. \tag{4.79}$$

Chain dynamics in the linear approximation

Equation 4.79 is linear in the displacements u. We try a solution of the form

$$u_n(t) = A\exp\left[\mathrm{i}(nk\ell - \omega t)\right]. \tag{4.80}$$

Displacements for neighbouring masses that follow such a spatial dependence are sketched in Fig. 4.22. Substituting this into equation 4.79 we get

$$m\omega^2 = \frac{2(1 - \cos k\ell)}{\ell}\phi_\ell''. \tag{4.81}$$

This is the dispersion relation, the relation between k and ω, that must be satisfied for equation 4.80 to be a possible solution. We recall that it assumes that ϕ_ℓ'' is a constant.

For long wavelengths (small $k\ell$) the equation becomes $m\omega^2 = k^2\ell\phi_\ell''$, and in terms of the reference density per unit length ρ_ℓ,

$$\rho_\ell\omega^2 = \frac{k^2}{\ell^2}\phi_\ell''. \tag{4.82}$$

Comparing this with the macroscopic dispersion relation in the form $\rho\omega^2 = k^2 K$, we see that the macroscopic elastic modulus at the reference point $\varepsilon = 0$ is given by

$$K = \phi_\ell''/\ell^2. \tag{4.83}$$

For shorter wavelengths we use the full equation 4.81. This relation between ω and k is sketched out as curve 1 in Fig. 4.23. As k reaches $\pm\pi/\ell$, the curve indicates a maximum value of ω. Beyond these values of k the curve turns down again. The ratio ω/k represents the phase velocity of these sound waves, and the slope $\mathrm{d}\omega/\mathrm{d}k$ is the group velocity. At long wavelengths compared with the atomic spacing, the curve is a straight line and the velocity of sound is constant. At shorter wavelengths the phase velocity (ω/k) and the group velocity ($\mathrm{d}\omega/\mathrm{d}k$) both

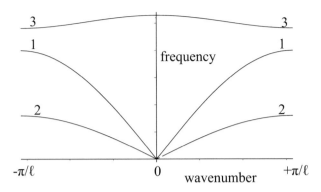

Fig. 4.23 Illustrative plots of dispersion curves, ω against k. Curve 1 for longitudinal acoustic modes; curve 2 for transverse or shear modes; curve 3 for optical modes.

start to fall. When $k = \pm\pi/\ell$, alternate atoms are π out of phase in their relative motion, and curve 1 is flat. The group velocity is zero.

What does this mean? The amplitude of any wave in a discrete structure only has meaning at the position of the elements. When the motions of neighbouring atoms are exactly π out of phase, there is no physical difference between a wave travelling to the left and a wave travelling to the right, or a standing wave. Such a solution cannot transmit energy. That is why the group velocity is zero. This is the maximum frequency that is possible in this 'band'. Higher frequencies incident on the material are attenuated. This means that the atoms are simply too heavy and the restoring forces too loose to respond fast enough.

These principles apply to shear waves just as to longitudinal waves. The dispersion of shear waves, illustrated as curve 2 in Fig. 4.23, is similar although the frequency is lower because the restoring force is weaker. The shear wave frequency band is therefore narrower than the compression frequency band.

If the single atoms in this simple model are replaced by bound pairs of atoms[42] with their own mutual but stronger interaction, their internal oscillation will be influenced by the phase of neighbouring pairs. The result is a spread of normal mode frequencies. The phase between neighbouring pairs can only vary between $-\pi$ and π. This phase advance is described by the wavenumber k, and the speed at which the phase moves is ω/k as before. An illustrative example is shown as curve 3 in Fig. 4.23. Such modes may have much higher frequency than the modes which we have discussed previously, because the higher restoring force within the bound pairs is excited. They are called optical modes and often have infrared frequencies. Consequently they have strong coupling to electromagnetic radiation in the infrared. This is a picture of how each single resonant frequency characteristic of an isolated individual atom or molecule is spread into a band by mutual interaction with neighbours when it forms part of a condensed material of similar atoms or molecules.

[42]In condensed matter terms this is described as having more than one atom per unit cell.

Thermal excitation of chains

In the absence of a source of sound the atoms of a material are not stationary. The thermal behaviour may be described in terms of the mean energy of each vibrational mode. This energy depends only on the temperature T, the degeneracy n and the excitation energy of the mode $\hbar\omega$. The mean energy is given by the same Planck distribution that describes the mean energy of electromagnetic radiation modes of a body or cavity,

$$E = \frac{n\hbar\omega}{\exp(\hbar\omega/k_{\mathrm{B}}T) - 1}. \tag{4.84}$$

It follows that there are thermally induced sound waves criss-crossing all materials at finite temperature. In so far as the behaviour is linear, these waves do not interact with other sound waves, such as the ones which carry the information in which we may be interested.

However, materials are not exactly linear. The principle of superposition is an approximation and waves do scatter off one another as a result of anharmonic terms in the potential. Then in a further approximation the frequency of normal modes may be shifted slightly if other modes are excited. This is most simply described by a temperature dependence of ϕ'' associated with a change in the reference state. These non-linear effects are discussed further in chapter 9.

4.4.2   Linear absorption

Sound and thermal energy

Here we discuss linear absorption processes both macroscopically and microscopically. Later in section 9.4 we discuss non-linear processes that lead to absorption. A macroscopic description of linear absorption is phenomenological, but does not explain fully what is happening. A microscopic explanation discusses different mechanisms.

A wave may suffer absorption or scattering. By absorption it is understood that the energy is changed into another form. By scattering it is understood that the wave changes the direction in which it travels, possibly with some frequency change also. In the case of sound waves at the microscopic level there is a difficulty in making such a clear distinction. When sound waves are absorbed the energy becomes thermal. Thermal energy in a material is itself largely composed of sound waves, although there is a small contribution from the thermal excitation of electromagnetic waves also.[43] So at a microscopic level at least, the processes of absorption and scattering are not distinguishable in principle.

In practice sound coherently scattered from large objects is described as scattered. Energy incoherently scattered from small objects or other thermal waves may be shared out and thermalised rather more quickly. This may therefore be counted as 'absorbed' though still vibrational sound energy.

[43]Such electromagnetic waves constitute the black body radiation which persists even in an empty cavity. These waves have the same spatial modes as sound but each mode has a frequency which is higher by the ratio of the phase velocities, which is typically 10^6. In each case the net energy in the mode is given by equation 4.84. For wavelengths greater than 100 μm the contributions of EM and acoustic waves are similar. However, EM waves make a completely negligible contribution to the integrated thermal energy of any massive body.

Absorption through delayed strain

An absorptive medium is inelastic. In thermodynamic terms this implies that changes are irreversible and that the internal energy is not a function of state. In a macroscopic description the components of stress of the material are not uniquely described by the strain and the material shows hysteresis. For example, it is observed that many materials take time to respond when deformed by a sudden change in stress. Such a delay causes absorption.

Consider first the changes of stress and strain in an elastic material. Figure 4.24a shows a plot of stress change ΔT against strain ε. As a progressive wave passes a small element of material, the stress and strain in that element run back and forth along this curve. This is a straight line of slope equal to the modulus if the material is elastic and linear. The work done per unit volume is the area under an element of this curve, $\Delta T \, d\varepsilon$. If the curve is single valued, this integrates to zero over a pulse.

If the contour is a loop as in Fig. 4.24b, there would be net work done every cycle. This is what happens in a dynamic situation when the strain lags the stress. This is an example of hysteresis.

As a simple model we consider a harmonic wave, which at the point z is given by $\varepsilon = \alpha \exp(i\omega t)$ and $\Delta T = \beta \exp(i\omega t)$. If the stress wave and the strain wave are in phase, the contour encloses no area and no energy is lost, Fig. 4.24a. However, if the strain lags the stress in phase the contour will be an ellipse. Mathematically this may be described as a phase difference between β and α, so that β/α is a complex number. This ratio is the elasticity modulus concerned. Thus absorption in a linear model may be described in terms of a complex elastic modulus. Thus the complex phase of the mechanical modulus applicable to Fig. 4.24b is 0.1 radians. The energy lost per unit volume per cycle is given by the area enclosed by the ΔT–ε contour.

This is a linear picture of energy absorption in a medium showing delayed response. The rate of loss of energy is proportional to the rate at which the material is taken round the loop. So for a given wave velocity and hysteresis loop, the rate of energy loss of the wave is linearly proportional to the frequency. A complex value of the modulus K means that the phase velocity $c = \sqrt{K/\rho}$ and the impedance $Z = \sqrt{K\rho}$ are also complex numbers. The wave whose amplitude in space is of the form $\exp(i\omega x/c)$ therefore includes an attenuation factor that describes the energy loss of the wave in a consistent way.

Figure 4.25 shows that sound attenuation in air is quite well described by a linear frequency dependence. However, the observation that damp air has a much higher attenuation indicates that we need to understand the cause of the delay and why it might be greater, for instance, in damp than in dry air.

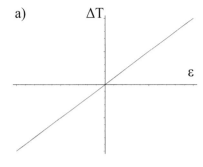

a)

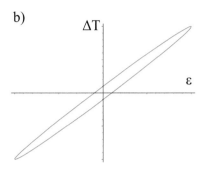

b)

Fig. 4.24 Illustrative plots of stress against strain: a) for stress and strain in phase, b) for strain lagging stress by 0.1 radians.

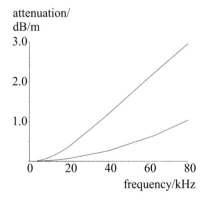

Fig. 4.25 Frequency dependence of the attenuation of sound in air at 293 K, dry air (lower curve) and 60% humidity (upper curve).

Microscopic mechanisms of delayed strain

When we considered the motion of a small volume of material to a change in stress in equation 4.7, we omitted the possibility of velocity-dependent forces. We assumed that the material was homogeneous. We ignored the effects of any time required to reach thermodynamic equilibrium and the possibility of other thermal effects.

Velocity-dependent forces that act in material are described as viscosity. They contribute imaginary parts to both λ and μ. Viscosity arises from momentum diffusion by molecules within the wave. For example, shear wave viscosity arises when molecules transmit momentum forwards and backwards to neighbouring volumes with a different mean transverse velocity. Similar momentum transfer in the longitudinal direction for longitudinal waves also gives rise to viscosity.

The exchange of thermal energy within the wave is also important in the attenuation of sound. Acoustic waves are isentropic not isothermal, and the various mechanical moduli concerned are defined at constant entropy. As a result there are temperature differences within the wave, and the exchange of energy by thermal conduction between the peaks and troughs with different temperatures is a significant dissipative process. This thermal conduction process is distinct from the viscous mechanism discussed above.

A related point concerns the internal energy of non-translational degrees of freedom, such as the rotation or internal vibration of molecules. If acoustic waves had been isothermal, this energy would not have been involved. However, because the temperature oscillates up and down in an acoustic wave, energy flows in and out of these otherwise translationally irrelevant modes. This flow takes time because the coupling is weak. The time required for the translational energy to reach thermal equilibrium with rotational modes is often significant compared with the period of the acoustic waves. Such thermodynamic changes taking place under non-equilibrium conditions cannot be reversible. The delayed exchange of this energy is a major cause of absorption in diatomic and polyatomic gases, and in more complex solid and liquid materials also. This mechanism with the extra degrees of freedom of the water molecule is responsible for the extra attenuation of damp compared with dry air shown in Fig. 4.25.

Sound attenuation by scattering and non-linearity

So far materials have been assumed homogeneous. Inhomogeneities cause attenuation through scattering and changes in polarisation. Scattering from an inhomogeneity of size a depends critically on the ratio a/λ. If this is not too small, scattering is determined by the detailed boundary conditions at the inhomogeneity. If it is small, scattering is more isotropic and the scattering cross section depends on λ^{-4}, so that the scattering (and consequently the beam absorption) is much larger at short wavelengths. This is known as Rayleigh scattering.[44] This is of great technical importance in the use of ultrasound for industrial quality

[44]Rayleigh scattering is unconnected with Rayleigh waves. Lord Rayleigh, like Gauss, belongs to that small band of physicists whose contributions are so extensive that it is sometimes not easy to distinguish which eponymous effect is being referenced.

control, fish location and medical imaging. It is discussed in more detail in chapter 9.

Other mechanisms of energy absorption involve the non-linearity of the medium. In particular there is the scattering of sound waves by other acoustic waves, including thermally generated waves. This does not happen in a linear medium.

Read more in books

Waves, Berkeley Physics Course Vol. 3, FS Crawford, McGraw–Hill (1965), a student text on waves

The Physics of Vibrations and Waves, HJ Pain, Wiley (1999), a student text on waves

Properties of Matter, Manchester Physics Series, BH Flowers and E Mendoza, Wiley (1970), a student text on condensed matter

Introduction to Solid State Physics, C Kittel, Wiley (1971), a student text on condensed matter

Physical Properties of Crystals, JF Nye, Oxford (1957), a classic text on tensors

Irreducible Tensorial Sets, U Fano and G Racah, Academic Press (1959), a classic text on tensors

Fundamentals of Acoustics, LE Kinsler, AR Frey, AB Coppens, JV Sanders, Wiley, 3rd edn (1982), a standard textbook on sound

The Theory of Sound, Vols 1 and 2, JWS Rayleigh, reprinted Dover (1945), Rayleigh's original texts on sound

The Solid Earth, CMR Fowler, Cambridge (1990), a standard textbook on geology and geophysics

Look on the Web

Look at the book website at
   www.physics.ox.ac.uk/users/allison/booksite.htm

Discover how far geophysicists have progressed with the difficulties of recording, analysing and understanding seismic waves with different polarisations
   Geophysics SV SH P waves

Find out about the current state of knowledge of the Earth's structure
   USGS Earth structure

Study more about Rayleigh waves, high frequency surface waves and their applications
   Rayleigh waves or *surface acoustic waves*

Get data on the normal vibration modes of the Earth and Sun
   Earth normal modes and *helioseismology*

Questions

4.1 A material is subject to a general strain,

$$\begin{bmatrix} a & b & c \\ d & e & f \\ g & h & i \end{bmatrix}.$$

Separate this into the components of compression strain, shear strain and rotation.

The material has Lamé constants, λ and μ. Determine the related stress tensor.

4.2 Prove that at normal incidence the reflection and transmission coefficients are given by equation 4.39 for any polarisation.

Show that with these coefficients the energy flux is conserved.

4.3 A planet of uniform composition has a radius of 5000 km. The velocity of compression waves is 3.0 km s^{-1} and of shear waves is 1.9 km s^{-1}. The Rayleigh wave generated by a quake arrives 500 s after the compression wave. When did the shear wave arrive and how far away was the quake?

4.4 A primitive musical instrument consists of an open square box section of side a. A piston closes off one end to make a channel of length ℓ. What are the normal modes of this cavity?

At first sight the sides of width a do not appear to influence the displacement of this longitudinal wave. Why do they need to be there?

4.5 Prove that for a gravity wave of height h and wave number k travelling in fluid of depth z_0 the displacement amplitude of the horizontal scouring motion at the bottom is

$$h/\sinh(kz_0).$$

Hence that, if the depth is small compared with the wavelength, the scouring velocity has amplitude

$$h\sqrt{g/z_0}.$$

Put in numbers for $h = 1$ m and depth $z_0 = 3$ m. What happens as z_0 gets even smaller?

Information and data analysis

in which the physics and mathematics of information and data analysis are discussed

5.1 Conservation of information
131

5.2 Linear transformations 135

5.3 Analysis of data using models
143

Read more in books 155
Look on the Web 156
Questions 156

5.1 Conservation of information

Information may be quantified in terms of the required number of binary bits, and is equivalent to negative entropy. The second law of thermodynamics requires that information may not increase in an isolated system. Imaging and other data are collected for a purpose, that is to answer specific questions. The processing of data involves the transformation of information available from collected data and previous knowledge. This transformation is intended to separate information related to the purpose (the signal) from the rest (the noise). The separation is uncertain and the calculated signal depends on the estimated noise and the previous knowledge. Statistical errors derive from the effects of random noise and may be reduced by taking further data. Systematic errors derive from uncertainties in the previous information and their effect on the signal. These are not reduced just by taking more data.

Information and thermodynamics

Information can be quantified by counting the number N of independent binary bits required to store it. This definition of information makes no reference to any value judgement. The same quantity of information may be interesting, unintelligible, or dominated by noise. The definition is, however, objective and of practical importance. The requirement that the state of the binary bits be independent is significant. For example, if a record of information is copied, the quantity of information is not thereby increased. If the vector of N bits is linearly transformed by multiplying by a unitary $N \times N$ matrix, the quantity of information is also unchanged. If the matrix is not unitary, the independence of the binary bits is reduced and information reduced.

In statistical mechanics a system whose macroscopic state is equally consistent with any of W different microscopic quantum states has an entropy S given by

$$S = k_{\mathrm{B}} \log_e W, \tag{5.1}$$

where k_{B} is Boltzmann's constant. To specify the microstate of the system unambiguously the number of binary bits needed would be

$$N = \log_2 W. \tag{5.2}$$

Therefore such a macrostate may be said to lack N binary bits for its complete specification. This leads to the identification of information with negative entropy.

The second law of thermodynamics can be summarised in the form, 'the entropy of an isolated system cannot decrease'. It follows that information cannot increase in an isolated system. This strict law is consistent with common sense. Information can be conserved, lost or destroyed but it cannot appear spontaneously in isolation. So an isolated system is one cut off, in respect of both thermal and information contact.[1]

Separating signal information from noise information

The law of information conservation has important consequences for any kind of computation including image processing. The only information available is the raw data and previous knowledge. The latter may include previously acquired data, such as a calibration, or preconceptions that may be encoded into the processing. There is no other information available and there is no image processing strategy or algorithm so clever that it can itself add to the information content. More informative output must be related to more input information in the form of either further data or 'better' preconceptions.[2]

If the objective of processing is not to increase information, what is it for? The information has been gathered for some end purpose, and the task of the computing and image processing is to separate this information according to whether it is useful for this purpose or not. This is illustrated in the diagram, Fig. 5.1. The useful information is referred to as the signal. That which is not useful is called noise and is to be discarded. While the result of one experiment is known precisely, how much of that result is noise and how much signal is not known exactly. Consequently one experiment gives information about the signal with an uncertainty described by its error. If more knowledge is supposed about the distinction between signal and noise, this error may be reduced, but it will be conditional on that knowledge.

Consider an example. Suppose that it is found that the number of cancers of a particular type that occurred in a population of 10^6 living near a reactor in the year before an accident is 1356, and that the number occurring in the year after the accident is 1372. These numbers are given and so certain, but the separation into those related to the accident (the signal) and those related to background noise is not certain at all. The background noise is random and varies from year to year with standard deviation $\sqrt{\text{mean}}$, and this applies to both the year before and the year after. We can improve our knowledge of the mean background rate by averaging over, say, 20 years preceding the accident. Suppose that this mean is 1348 with a rather small uncertainty. Then the number

[1]The difference between thermal and information contact relates to the difference between contact which involves thermal equilibrium and where it does not. This difference is not important here as both kinds of contact are excluded.

[2]As must always be the case, any such output is conditional on the truth of the inputs. So risky assumptions carry a condition, a health warning, with their outputs.

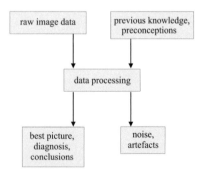

Fig. 5.1 An information flow diagram describing the input and output of image or data processing.

of background cancers in the year in question that contribute to 1372 is $1348 \pm \sqrt{1348}$. Because the separation of signal from noise involves this large statistical error, the number of cancers that can be attributed to the accident is consistent with zero at the one standard deviation level. If the difference had been larger any conclusion would still have been conditional on implicit assumptions. For example, if the background count had risen because of an increase in smoking-induced cancers caused by socio-economic collapse triggered by the accident, the information alone could not be used to relate cancer to radiation exposure.

If the number of possible outcomes of an experiment[3] is W, then, when that experiment is done and yields one result, by equation 5.2 that experiment conveys information $N = \log_2(W)$. This allows that some of this information may be uninteresting, that is to say noise. If the results of experiments were truly continuous variables then, by the above definition, they would yield an infinite amount of information, almost all of which would be noise. This is not helpful. In practice, either by recording results with limited precision, or by explicitly binning readings in discrete bins, we may reduce the number of possible outcomes of the experiment W to a finite, if large, number. When precision is discarded in this way, care is taken to ensure that signal information is not lost, either by truncating significant figures too hard, or by digitising or binning too coarsely. This might, for example, lead to a decision to digitise the square or the logarithm of an analogue signal. Anyway, the recording of the experiment or image in terms of limited information and discrete variables is the first stage in the separation of signal and noise.

[3] For simplicity we assume that the a priori probabilities of all such outcomes are equal.

Reproducible and non-reproducible noise

We distinguish two types of noise, random noise and artefactual noise, by their reproducibility. Random noise changes if the experiment is repeated, and this is the way in which its separation from the signal may be improved. Systematic effects and artefactual noise are reproducible and causal. They differ from the signal only in that they are not wanted in the image or in answer to the question. Only previous knowledge can separate these from the signal. For example, unwanted reflections and distortions may be subtracted from an image if previous knowledge is available. However, if random noise is present as well, there may be uncertainty in such a separation. The error on the corrected signal would then be increased by the artefact. In practice therefore information may be correlated between types of noise as well as with signal.

In the absence of noise the result of the experiment is a certain signal. In the presence of noise the result of the experiment is a certain outcome. But the outcome for the signal alone is uncertain because it is not resolved from the noise. This uncertainty is the error. Signal errors associated with random noise are called statistical errors; those associated with reproducible noise are called systematic errors. Statistical errors can be reduced by taking further data, at a cost in experimental time and inconvenience. Because systematic errors relate to previous knowledge and not to the experiment at hand, they may only be reduced with additional knowledge. If that comes from other observations such as data runs with different conditions, data from other modalities, calibration or registration information, then those may be repeated and the previous information increased.[4]

Noise filtering by transformation and modelling

Separation of the signal from the noise is only possible in so far as the signal and noise are different. Information about this difference comes from previous knowledge, and without this no progress is possible. There are two approaches, one based on transformation and the other on modelling.

If the previous knowledge gives information about the characteristics of the signal, the data are transformed into variables in which these characteristics are most clear. The noise, too, will have characteristics. We consider two simple examples of such previous information.

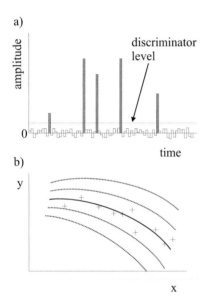

    ◇ If it is known that the signals are episodic pulses in time and that the noise is not, a discriminator may be applied to the data as a time sequence, as illustrated in Fig. 5.2a. Then all elements in time below that level may be reset to zero which erases the noise signals. Such a filter makes two kinds of mistake. It discards some real signals that are very low, and it retains the contribution of the noise within the time window of a signal pulse. Its success depends on the difference between the signal and noise, and on our knowledge of that difference.

    ◇ Perhaps it is known that the signals are waves with quite well-defined frequency. In that case the data should be transformed to a frequency spectrum. A discriminator is then applied to this spectrum in the same way. The spectrum cleaned of low amplitude noise in this way may be reconstituted in time by inverting the transformation.[5]

Fig. 5.2 a) An illustration of signal and noise contributions in time, distinguished by a discriminator in the light of previous knowledge.
b) Measurements of y for different x which are fitted by a model, given by previous knowledge, that generates a family of curves for different values of the physically interesting parameter α.

In any case the success of the filter is only as good as the previous knowledge.

Similar considerations apply when the previous information comes in the form of a model. A simple example is illustrated in Fig. 5.2b, where the data for a number of measurements of y for particular values of x are plotted. The experiment was carried out to measure a parameter α and a model $y = f(\alpha, x)$ relates this to y and x. The model, itself previous knowledge, represents a family of curves in the plot, each for a different value of α. On the assumption that the effect of the noise is random, sometimes increasing a reading y and sometimes decreasing it, the best choice of α would be the curve that is closest to the points with some above and some below, although we need to specify exactly what is meant by 'best'. All differences between the best model curve and the measured points are attributed to noise and discarded. This type of filtering is called model fitting. Its results are conditional on the validity of the model.

Having filtered as much noise from the data as possible we look at the second computation task, that is the reconstructing and plotting of the 'best' image of the object. In the case of a fitted model, the reconstruction is the model with the best values of parameters α and their errors. However, the task may be complicated by artefacts.

In some cases the collected image data are not sufficiently complete for a reconstruction in object space to be possible. The classic example is the structure analysis of molecules by X-ray diffraction. The phases of the scattered X-ray diffraction peaks are not recorded. If they were, crystal structure analysis would be very straightforward. Instead, inspired guesswork or further information from other sources is needed to carry out the reconstruction.[6] Another example is imaging with ultrasound scattering. Each resolvable element of the image may contain a number of individual scattering centres. Although the net phase of each resolvable element is recorded, the relative phases of the individual contributing scattering centres cannot be determined from the recorded data. As a result images are spoiled with a noise field known as speckle, as discussed further in chapter 9. This appears random but is in fact reproducible. It is an artefact of the missing phase information. Other examples of artefacts generated by missing information are important in MRI. These are discussed later in this chapter and in chapter 7.

[6]The analysis may then be a matter of some drama, as recounted in the story of the double helix of DNA.

5.2 Linear transformations

Fourier transformation analyses time-sequenced data in terms of harmonic signals. Ideally, such a transformation may be inverted to reconstitute the time signal. Application of the mathematics to real data is hampered by incompleteness, in both range and continuous detail. Further assumptions are required if Fourier analysis is to be used, and these assumptions generate artefacts, ranging from loss of resolution to the arbitrary superposition in the reconstructed image space of signals from different regions of the object space. Similar problems haunt all transformation methods. Wavelet transforms are suited to the analysis of signals characterised by localisation in both frequency and time.

5.2.1 Fourier transforms

One dimensional continuous Fourier transforms

Suppose data are recorded as a continuous time sequence $f(t)$. As these are actual readings, the values are real numbers. It is often the case that these data are to be interpreted in terms of a superposition of a number of frequencies ω.[7] Thus in general we can express

$$f(t) = \frac{1}{\sqrt{2\pi}} \int_{-\infty}^{+\infty} g(\omega) \exp(-i\omega t) d\omega, \qquad (5.3)$$

where $g(\omega)$, the amplitude of the angular frequency ω, is known as the Fourier transform. (The choice of constant outside the integral is explained below.)

The reason for extending the integral to negative frequencies deserves more explanation. What do we mean by negative frequencies? Because $f(t)$ and ω are real, we may equate the right hand side of equation 5.3 to its conjugate

$$\frac{1}{\sqrt{2\pi}} \int_{-\infty}^{+\infty} g^*(\omega) \exp(+i\omega t) d\omega$$

and deduce that $g(-\omega) = g^*(\omega)$. So g for negative ω is determined when g for positive ω is known, and negative frequencies are just a mathematical convenience. The real and imaginary parts of $g(\omega)$ form the coefficients of $\cos(\omega t)$ and $\sin(\omega t)$ respectively in a real expansion using cos and sin. We have chosen this complex exponential representation because it is the most concise.

But how may the $g(\omega)$ be determined? This is done with a standard trick. We multiply both sides of equation 5.3 by $\exp(i\omega t)$ and integrate over all time t,[8]

$$\int_{-\infty}^{+\infty} f(t) \exp(i\omega t) dt = \frac{1}{\sqrt{2\pi}} \int_{-\infty}^{+\infty} g(\omega') \left[\int_{-\infty}^{+\infty} \exp[i(\omega - \omega')t] dt \right] d\omega'. \qquad (5.4)$$

The integral in square brackets is $2\pi\delta(\omega - \omega')$ where δ is the Dirac delta function. The proof of this is left to question 5.2.

The δ-function, $\delta(x - x_0)$, may be described as a spike at $x = x_0$ of infinite height, infinitesimal width but unit area, as illustrated in Fig. 5.3a for $x_0 = 0$. It has the important property for any function $f(x)$,

$$\int f(x)\delta(x - x_0) dx = f(x_0). \qquad (5.5)$$

The integral of $\delta(x)$ is the Heaviside unit stepfunction $H(x)$ shown in Fig. 5.3b.

Returning to equation 5.4 and using equation 5.5 to complete the integration on the right hand side, we get

$$g(\omega) = \frac{1}{\sqrt{2\pi}} \int_{-\infty}^{+\infty} f(t) \exp(i\omega t) dt. \qquad (5.6)$$

The reason for choosing the factor in equation 5.3 is now seen to be to make the pair of equations, 5.3 and 5.6, of the same form except for the change in sign in the phase factor. These represent the Fourier transformation and the inverse Fourier transformation. The general properties of Fourier transforms are treated in standard mathematical texts, and

[7]For example, such a decomposition is applied by the brain to the pressure signals detected by the ear so that music may be heard. In this decomposition linearity is essential so that the signal heard at any time is a simple superposition of each frequency present.

[8]We have replaced the dummy variable in equation 5.3 by ω'.

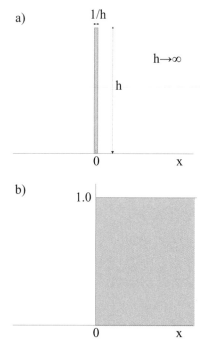

Fig. 5.3 a) The Dirac δ-function, $\delta(x)$, where h is taken to infinity. b) The Heaviside function or unit stepfunction is the integral of the Dirac function, $H(x) = \int_{-\infty}^{x} \delta(x') dx'$.

some are the subject of questions at the end of the chapter. In the following we highlight the physical interpretation as it affects the use of Fourier transforms in imaging. It is important to note that the transformation is linear, possesses an inverse and therefore conserves information.

In many applications the variables are not time and frequency, but their spatial equivalents, distance x and wavenumber $k_x = 2\pi/\lambda_x$. As these transforms are linear, the Fourier transformation may be taken in any cartesian dimensions, separately or simultaneously. Such wavenumbers form a vector $\boldsymbol{k} = (k_x, k_y, k_z)$. Any function of $\boldsymbol{r} = (x, y, z)$ may be expressed as[9]

$$f(\boldsymbol{r}) = \left(\frac{1}{2\pi}\right)^{3/2} \iiint g(\boldsymbol{k}) \exp(i\boldsymbol{k} \cdot \boldsymbol{r}) d^3\boldsymbol{k}, \qquad (5.7)$$

with the three dimensional inverse transform given by

$$g(\boldsymbol{k}) = \left(\frac{1}{2\pi}\right)^{3/2} \iiint f(\boldsymbol{r}) \exp(-i\boldsymbol{k} \cdot \boldsymbol{r}) d^3\boldsymbol{r}, \qquad (5.8)$$

where the integrals are over all three dimensional $\boldsymbol{k}$ and $\boldsymbol{r}$ space respectively.

A wave in space and time, $f(\boldsymbol{r}, t)$, may be expanded by Fourier analysis into a superposition of elemental plane waves with direction and wavenumber $\boldsymbol{k}$ and frequency ω. A plane wave has a constant value of $\boldsymbol{k}$ independent of $\boldsymbol{r}$; its wavelength and direction of travel are independent of position. Loci of constant phase $\phi = \boldsymbol{k} \cdot \boldsymbol{r} - \omega t$ describe wavefronts, families of parallel flat planes. Such an expansion may be written

$$f(\boldsymbol{r}, t) = \left(\frac{1}{2\pi}\right)^2 \iiiint g(\boldsymbol{k}, \omega) \exp[i(\boldsymbol{k} \cdot \boldsymbol{r} - \omega t)] d^3\boldsymbol{k} d\omega. \qquad (5.9)$$

Its transform is then given by

$$g(\boldsymbol{k}, \omega) = \left(\frac{1}{2\pi}\right)^2 \iiiint f(\boldsymbol{r}, t) \exp[-i(\boldsymbol{k} \cdot \boldsymbol{r} - \omega t)] d^3\boldsymbol{r} dt. \qquad (5.10)$$

Convolution theorem

In the application of Fourier techniques to object distributions and related image distributions we need to use the formalism of convolutions.

The convolution of two functions, $h_1(x)$ and $h_2(x)$, is defined by

$$h_{12}(x) = \int_{-\infty}^{\infty} h_1(x') \times h_2(x - x') dx'. \qquad (5.11)$$

An important application is the use of the image of a point source, the point spread function (PSF).[10] If the object distribution is $u(x)$ and the PSF is $w(x)$, then the image $v(x)$ is the convolution of the two,

$$v(x) = \int_{-\infty}^{\infty} u(x') w(x - x') dx'. \qquad (5.12)$$

[9]In anticipation of the analysis of plane waves in the following paragraph we change the sign of the i in this spatial transform. This will ensure that a wave with positive $\boldsymbol{k}$ and ω travels forward in $\boldsymbol{r}$ with time, t.

[10]In the following we describe the 1-D problem. This may be extended simply to 2-D and 3-D problems.

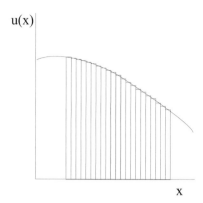

u(x)

x

Fig. 5.4 The object distribution $u(x)$ can be represented in terms of a sequence of bar elements of width d. As d tends to zero, the representation becomes a sequence of δ-functions, each multiplied by its own $u(x)$.

To prove this result consider first a unit point object at $x = x_0$ with $u(x) = \delta(x - x_0)$. By definition the image distribution is the PSF itself in this case, and integration over the δ-function confirms that the image is $v(x) = w(x - x_0)$. Next we verify equation 5.12 for a general extended object $u(x)$. We can represent u by a superposition of such δ-functions side by side, as illustrated in Fig. 5.4. Formally

$$u(x) = \int u(x')\delta(x - x')\mathrm{d}x'. \tag{5.13}$$

Since the relation between u and v is linear, the corresponding image is the superposition of the image of each δ-function. Equation 5.12 is therefore true for any object function $u(x)$. The image of any object is the convolution of the object and the PSF.

The Fourier transform (FT) of any image is therefore the transform of this convolution. The transform of a convolution of two functions is equal to the product of their transforms. The proof is left to question 5.3. So we deduce the relation,

$$\mathrm{FT}_{\mathrm{image}}(\omega) = \mathrm{FT}_{\mathrm{object}}(\omega) \times \mathrm{FT}_{\mathrm{PSF}}(\omega). \tag{5.14}$$

This is an important result. The FT of the image is determined by the data. The PSF is known from previous knowledge. So the true FT of the object may be determined from the image FT using equation 5.14 for any range of ω for which the FT of the PSF is non-zero. For values of ω for which the PSF is small, the image FT is multiplied by a large factor to determine the object FT. In practice noise in the data limits the efficacy of this. Indeed to reduce noise the contribution at frequencies for which the PSF is small may be suppressed altogether. So for the reconstruction there is a choice between fine detail but amplified noise, and coarse resolution with reduced noise. A balance can be struck. This process is called filtering.

Shortcomings in the practical application of Fourier analysis

The formalism of Fourier transformations is symmetric, instructive and elegant. This is deceptive! The apparent symmetry between the two canonical variables, say ω and t, is broken. In terms of one of them, say t, the measurements are made, the values $f(t)$. These measurements are real, not complex, and there are a finite number of them, that is to say the measurements are samples of f at certain values of t only. The resulting $g(\omega)$ is complex in general, and is not sampled. So in practical terms the symmetry is removed.[11]

The problem is that finite data, whether recorded in ω or t, cannot provide the infinite number of values required to carry out the integration prescribed by the transformation. Whichever is recorded, both the total range of measurements and the step between successive measurements are finite. Data values at intermediate points simply do not exist, and the same is true for values outside the range.[12] These limitations have major effects on the application of Fourier techniques. To implement the Fourier transformation at all, arbitrary decisions have to be

[11] If the measurements are samples as a function of ω, the roles are reversed but in no case is the symmetry maintained.

[12] It is not the case that we just need to reconstruct the missing values from those that we have. Any such reconstruction depends on the model or prescription used. There is no way to check with the data whether the prescription used is correct, or even just sensible. It is better to say that the data are undefined except at the points where they are explicitly measured. Then we may appreciate fully the artefacts generated by using different prescriptions to force the application of a Fourier decomposition.

made about the non-existent data, and the effect of these decisions is to create artefacts, the artificial differences between reconstruction and true object.

Formally we may say that the measurements in t are the product of the continuous function of possible measurements $f(t)$ and a sampling function $S(t)$. Thus the actual measurements are the non-zero values of

$$F(t) = S(t) \times f(t). \tag{5.15}$$

The sampling function has value 1 for values of t for which a measurement is made, and 0 otherwise. This provides a symmetry with equation 5.14. However, this relation is singular, and is not much help. It does not tell us the values of $f(t)$ for which no measurement is available.

The problem of missing information is the same whichever of the two conjugate variables, ω or t, is measured. Random noise enters symmetrically between them. Noise power is constant with time. So the squared noise amplitude increases linearly with the length of the measurement period, Δt. This is a simple example of the random walk.[13] The squared noise amplitude is also linearly dependent on bandwidth, $\Delta \omega$, in the same way.

Depending on the way in which the incomplete Fourier information is circumvented, the artefacts generated are different. Thus understanding artefacts in images is largely a matter of understanding the effect of these circumventions. It is usual to simplify the discussion by distinguishing the range of sampling from the intersampling spacing, assumed uniform within that range. Clearly similar artefacts would be generated by more general sampling schemes.

In Fig. 5.5 we give a calculated graphical example where the function $f(t)$ is measured at the discrete values $t = -2, -1, 0, 1, 2$ and found to have the values $-4, 3, -2, 3, -4$ respectively. Four different assumptions (in the left-hand column: a, c, e, g), all equally reasonable, are made about the values of the continuous function $f(t)$ that might have been found at other values of t, if these had been measured. The four resulting transforms $g(\omega)$ are very different (in the right-hand column: b, d, f, h). In the following paragraphs we discuss these differences and how they relate to previous knowledge, preconceptions or guesswork.[14]

Circumvention of the limited range of measurement points

If data are only available for a limited range, $\Delta t = 5$ in our example, it is often supposed that at other values of t the data would have been periodic, thus

$$f(t + n\Delta t) = f(t), \tag{5.16}$$

for all integer n, Fig. 5.5a. The FT $g(\omega)$ of such a periodic function is discrete, vanishing for all values of ω except $2\pi m/\Delta t$ for any integer m.[15] This is illustrated in Fig. 5.5b; the FT is non-zero only at discrete values of ω. These spikes are artefacts created by the supposition.

A more elementary assumption is that the function $f(t)$ is zero outside the measured range, Fig. 5.5c. Through the convolution theorem this

[13] A coherent signal amplitude is not a random walk. It integrates linearly over the same time and is seen rising above the noise as Δt increases. The difference is that the signal is coherent with defined phase, while the noise is not.

[14] The measurements are symmetric about $t = 0$ and the prescriptions illustrated maintain this symmetry. Otherwise the FTs would have had an imaginary part. We have avoided this to simplify the example.

[15] See question 5.4.

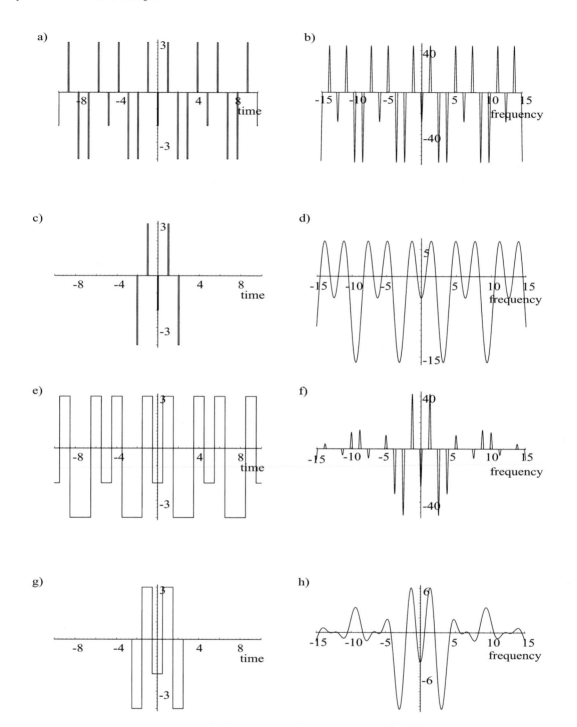

Fig. 5.5 The four distributions $f(t)$ shown in the left-hand column (a,c,e,g) have identical measured values (at times: $t = -2, -1, 0, 1, 2$). For a) the measured values of $f(t)$ are assumed to be part of a infinitely repeated discrete sequence of values. For c) values of $f(t)$ are assumed zero unless measured. Plot e) is like a) except that the sampled readings are assumed to apply over a range in t of $\pm\frac{1}{2}$ units. Plot g) is like c) but with the local sampling of e). The right-hand column shows the corresponding FTs (b,d,f,h).

limited range of t measurements generates a PSF in ω given by

$$\frac{1}{\sqrt{2\pi}} \int_{-\Delta t/2}^{\Delta t/2} \exp(-\mathrm{i}\omega t)\mathrm{d}t = \sqrt{\frac{2}{\pi}} \frac{\sin(\omega \Delta t)}{\Delta t}. \qquad (5.17)$$

This is a sinc function of width $1/\Delta t$. Convoluted with the discrete elements of plot c this gives rise to the rounded FT shown in plot d, in place of the sharp spikes of plot b.

These two strategies for coping with a limited range of data in t are quite different and the results likewise, although the data are the same.

Circumvention of the discreteness of measurement points

Further artefacts are associated with assumptions about how $f(t)$ inter-polates *between* measurements.

In general, if there are p measurements spaced $\Delta t/p$ apart, the sampling function is 1 at values of $t = m\Delta t/p$ for integer m, and 0 otherwise. What happens to $f(t)$ at the values of t between? The data say nothing but some assumption is required if a Fourier transformation is to be carried out at all. One extreme example has already been illustrated in Fig. 5.5a–d where it is assumed that $f(t)$ is zero in between the discrete points. Alternatively, as shown in Fig. 5.5e–h, a rectangular wave interpolation might be chosen. These FTs are no longer periodic and are localised in ω. In fact the smoother the interpolation of $f(t)$, the more compact is the localisation of the FT in ω-space.

Further, any such smoothly varying function might be multiplied by any linear phase factor

$$\exp\left[2\pi n\mathrm{i}\frac{pt}{\Delta t}\right]$$

with arbitrary integer n. Such factors are unity at the t of each data reading, and would therefore be entirely consistent with all recorded readings.[16] The effect of such factors on the FT is a shift in ω-space by an arbitrary multiple of $2\pi p/\Delta t$. The upshot is that the FT in ω-space is inherently ambiguous with cells of ambiguity forming a periodic structure. Often the assumption is made that the contribution from regions with higher values of ω may be neglected on the basis that detail is less important than gross features. However such an assumption is not based on data.

If data are recorded as a function of ω the roles of t and ω are simply reversed. The above phase ambiguity then generates artefacts in t-space. We shall see important examples of this in the context of MRI data in chapter 7. In spite of these difficulties Fourier analysis is widely used especially where high Q oscillations are involved so that signals are localised in frequency.

[16] Artefacts generated by more general sampling schemes may be understood by generalising this phase factor while maintaining the constraint that it passes through +1 wherever a sampling occurs.

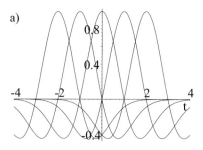

Fig. 5.6 An example of a basic wavelet.

a)

b)

c)

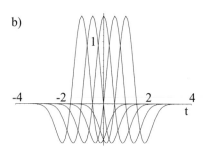

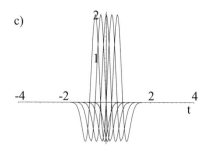

Fig. 5.7 A set of scaled wavelets, in this case generated from the wavelet shown in Fig. 5.6. Only five at each scale have been drawn so as not to confuse the diagram. To span the time interval shown further wavelets are required on either side. So the number of wavelets in b) should be twice the number in a), and in c) four times as many.

5.2.2 Wavelet transforms

Localisation and bandwidth

Data that are localised in time rather than frequency are best analysed in terms of the superposition of the impulse responses following events that occurred at earlier times. These are Laplace transforms which we do not investigate further here. The optimal choice of transformation is based on maximising the separation of signal from noise. With Laplace transforms artefacts are generated by incomplete data, just as with Fourier transforms.

The separation of signal from noise is difficult if signals are not so clearly localised. Thus data composed of finite oscillatory signals need to be analysed in terms of localised harmonic oscillations. The Fourier method is unsuitable; the integration over all time inherent in the usual Fourier prescription drowns the signal in the noise energy present at times when the signal is absent.

An example is a recording of short episodic signals of various frequencies against a background of random noise, such as the ringing of bells of various pitches at different times. The coherence of the ringing of each bell lasts for a time determined by its Q-value. A transform is needed which is sensitive to the short term coherence of the damped resonance of the bell to distinguish signal from noise. Wavelet transforms are matched to this type of problem. Instead of a superposition of perfect sine waves or impulses, a superposition of another finite signal shape is considered. This shape is called a wavelet. In the case of the bells, for example, each bell would emit a damped wave of the same Q-value, but with an appropriate time-delay, amplitude and time-stretch to match the frequency. The set of scaled sine waves used in Fourier analysis is simply an extreme example of a set of such wavelets.

Wavelet analysis is more rigorous and flexible than an attempt at a piecewise Fourier analysis of time segments of data. Actually it covers a range of possible analyses. Here we give just an outline of the ideas, and encourage the reader to learn more from the references given at the end of the chapter and others that may be found on the Web.

A wavelet is a localised waveform, or wave packet, with zero integral. It can be chosen to suit the desired balance between localisation and bandwidth, related to the Q. An example is shown in Fig. 5.6. This particular choice $\psi(t) = (1-t^2)\exp(-t^2/2)$ has a rather large bandwidth (small Q). The choice of a similar form but with more nodes would give a smaller bandwidth (large Q). The function of interest $f(t)$ is expanded in terms of a two dimensional family of functions, $\psi_{jk}(t)$, with the same shape as the basic wavelet

$$f(t) = \sum_j \sum_k c_{jk}\psi_{jk}(t), \qquad (5.18)$$

where each $\psi_{jk}(t)$ is derived from the basic wavelet $\psi(t) = \psi_{00}(t)$ by scaling (j) and shifting (k),

$$\psi_{jk}(t) = 2^{j/2}\psi_{00}(2^j t - k). \qquad (5.19)$$

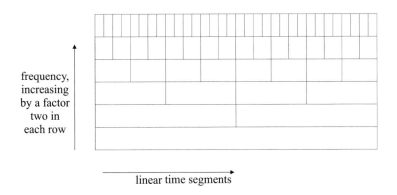

frequency,
increasing
by a factor
two in
each row

linear time segments

Fig. 5.8 A diagram showing the relation between the domains in frequency and time of different members of a wavelet family, such that frequency is proportional to the inverse scaling of the wavelet shape. Each rectangle describes the domain in time and frequency of one wavelet in the expansion.

As an example, Fig. 5.7 shows some members of a wavelet 'family' for shifts $k = -2, -1, 0, 1, 2$ plotted out for scalings $j = 0, 1, 2$ generated from the mother wavelet of Fig. 5.6. These scaled expansion functions have the following properties:

⬦ each new scaling increases the frequency by a factor 2, and halves the wavelet width in t;

⬦ the number of independent shifts in time increases by a factor 2 with each scaling, as illustrated in Fig. 5.8;

⬦ fast algorithms which do not depend on the choice of wavelet are available for implementing the decomposition;

⬦ they are normalised in such a way that their energies are related to c_{jk}^2.

The reconstruction may be filtered of noise by inverting the decomposition omitting those time–frequency elements that show no evidence of signal.

5.3   Analysis of data using models

Data may be analysed by building models based on previous knowledge. Parameters left free in the model may be determined from the data with some error. Three methods are described: least squares, minimum χ^2, and maximum likelihood. In the least squares method the fit is defined by minimising the summed squared differences between the model predictions and the measurements. In the χ^2 method each element of the sum of squares is weighted with its known inverse error squared. This may be generalised to include correlation or off-diagonal effects. Simulated data may be generated according to the model by Monte Carlo, and then compared with real data. We discuss a simple historical example of its use, the analysis of an experiment to scan the Pyramids for hidden chambers.

The maximum likelihood method represents the most complete and flexible use of information. If the data are binned and the errors on those bins are Gaussian, the likelihood method is equivalent to the χ^2 method. The likelihood function enables the errors of potential experiments to be evaluated without the use of Monte Carlo. With the likelihood method, data from different sources can be combined in a rigorous single optimum analysis. This use is important in medical imaging.

5.3.1   General features

Questions and models

At first sight, analysis by transformation and filtering removes noise from an image in a model-independent way, and should therefore be objective. The filter, however, must be designed in the light of a known distinction between signal and noise. The alternative approach is to acknowledge the significance of what is already known and to build a model of how we expect the data to look, and then to compare that with the observed data. Such an analysis attempts to answer two kinds of question. First there is the goodness of fit, 'To what extent does the model give a satisfactory match to the data?' Then there is parameter determination, 'If the model is correct, what are the best values of its unknown parameters?' Of course, if the goodness of fit is unacceptable, the values of the determined parameters are not meaningful.

To minimise dependence on preconceptions the model is given as many free parameters as reasonable.[17] Finding the best values for these parameters is called fitting. With these values is the match to the data satisfactory? This is difficult to answer. The model cannot tell us whether there exists another quite different interpretation or model, of which we are unaware, but which matches the data better.[18]

Noise enters the discussion in two ways. If the model is correct, it should be solely responsible for the observed difference between data and fitted model. Secondly it gives rise to the error on the estimated parameters and their correlations.

The output of these calculations may be a series of simple fitted parameters. Or it may consist of an array of such parameters which themselves constitute the required image. The analysis principles are the same.

Monte Carlo simulation

By combining the prescription of the model with the effect of random noise, artificial experiments may be simulated by Monte Carlo on a computer. The Monte Carlo 'data' generated in this way may be compared with the model, in the same way as real data, to find out how often such comparisons are less favourable than that found for the actual data. Such simulations may also be used to check the variation and value of estimated parameters against their calculated errors. These simulations may be time consuming. Their statistical value improves only as the

[17]Relaxed fits with many free parameters give results with larger errors. Small errors are given by fits tightly constrained by preconceptions. But such apparently accurate results are conditional on the assumptions made.

[18]In an apparently different approach called neural networks, previous data are used as previous knowledge to 'teach' the model, and the software learns dynamically. Nevertheless the overall relation between previous knowledge and data remains.

square root of the number of simulations. However, since they depend on the model, not the data, they may be used with care to optimise the design stage of equipment before any data are available. Blind reliance on Monte Carlo simulation is dangerous since any conclusion is negated if the model is in error.

Noise and errors

An important type of noise derives from the Poisson fluctuations of discrete events whose occurrence is uncorrelated. Other noise such as thermal noise is random but has a well-defined power spectrum whose bandwidth and amplitude need to be understood precisely.

In the model analysis of a set of data, the noise and fitted parameter values become correlated. Thus the observation of extra counts with certain values may be interpreted *either* as an upward chance noise fluctuation *or* as evidence for a stronger signal. A repeat data set would have different noise and different estimated parameters. If the second data set were available the uncertainty of interpretation would be resolved to an extent. Because of this correlation, when the noise information in the given data set is discarded, the information on the estimated parameters suffers an increase in entropy. This negative information takes the form of the calculated statistical errors estimated during the fitting process, usually in the form of an error matrix relating to the uncertainties of all the parameters.

The more free parameters that are included in the model, the looser will be the constraint of the data on the model and the larger will be the estimated errors. Since a loose model may be preferred, this behaviour of calculated errors needs careful interpretation. Should the model be wrong in some respect, not only the parameter values but also their errors are compromised. Often the effect of fitting an inappropriate model is to generate smaller errors in the context of the model whereas a proper overall picture should acknowledge a larger uncertainty.

5.3.2   Least squares and minimum χ^2 methods

Least squares

The least squares method is the simplest way in which data may be fitted to a model. We suppose that we have a set of n data readings of values y, y_i for $i = 1$ to n, forming a set of measurements at values of $x = x_i$. In the absence of noise the model hypothesis is that

$$y = f(x, \alpha, \beta), \qquad (5.20)$$

where α, β are unknown parameters to be determined.[19] The residuals are the n values, $y_i - f(x_i, \alpha, \beta)$, the differences between the measurement for $x = x_i$ and the value calculated by the model. In the least

[19] There may be a whole vector of such quantities. We indicate just two for discussion purposes.

squares method the sum of the squared residuals,

$$S = \sum_{i=1}^{n} (y_i - f(x_i, \alpha, \beta))^2 , \tag{5.21}$$

is formed and minimised with respect to variation of α and β. If this sum is stationary with respect to such variations, the derivatives of the sum with respect to α and β are zero. This leads to the pair of simultaneous equations,

$$0 = \sum_{i=1}^{n} (y_i - f(x_i, \alpha, \beta)) \frac{\partial f(x_i, \alpha, \beta)}{\partial \alpha} \tag{5.22}$$

$$0 = \sum_{i=1}^{n} (y_i - f(x_i, \alpha, \beta)) \frac{\partial f(x_i, \alpha, \beta)}{\partial \beta}.$$

If f is linear in α and β, these equations may be solved for α and β by linear algebra. If f is nearly linear, the solution may be found iteratively. Otherwise, by changing variable or even searching numerically, the values α and β that give the smallest value of S are found. In this simple method there is no gauge of what is an acceptable sum of squares, and therefore no indication of whether the model fit is an adequate description or not.

Minimum χ^2

The minimum χ^2 method is an adaption of least squares in which the measurement error σ_i of each measurement y_i is known. Each element of the sum of squares is now weighted with its inverse error squared,

$$\chi^2 = \sum_{i=1}^{n} \frac{(y_i - f(x_i, \alpha, \beta))^2}{\sigma_i^2} , \tag{5.23}$$

and its contribution to equations 5.22 likewise. The error σ_i is the standard deviation, such that, if the measurement of y_i were repeated a large number of times, the distribution of values would be given by the Gaussian function

$$\frac{1}{\sqrt{2\pi}\sigma_i} \exp\left(\frac{(y_i - \bar{y}_i)^2}{2\sigma_i^2}\right),$$

where $\bar{y}_i$ is the mean. If the measurements are correlated, this distribution may take a two dimensional form using the inverse of σ^2, which is now the error-squared matrix, including its off-diagonal terms. Thus

$$\chi^2 = \sum_{i,j} (y_i - f(x_i, \alpha, \beta)) \left((\sigma^2)^{-1}\right)_{ij} (y_j - f(x_j, \alpha, \beta)). \tag{5.24}$$

The process of minimisation of χ^2 is then similar to the least squares method except that the actual value of χ^2 at the minimum is significant. Its interpretation depends on the number of degrees of freedom ν, defined as the number of measurements (the number of values of i in the summation equation 5.23) less the number of parameters allowed to vary in the minimisation process. The errors on parameters are determined by noting that when χ^2 increases by one, the probability density falls from $\exp(-\chi^2/2)$ to $\exp(-\chi^2/2) \times \exp(-1/2)$, equivalent to one standard deviation. So combinations of parameter values that lie within the surface $\Delta\chi^2 = 1$ are within one standard deviation of the best fit. Acceptable fits, consistent with the errors used, have a distribution of minimum χ^2 with the mean value $\langle\chi^2\rangle = \nu$. The distribution of χ^2 values becomes rather broad for values of ν less than about 8, but at higher ν acceptable fits have $\chi^2 \sim \nu$. The reader is referred to standard texts on statistics for more details.

Sometimes data take the form of counts recorded in a number of bins. The index i is then taken as running over the total of n bins. The model is used to integrate the probability flux within each bin p_i, which may be a function of the parameters α, β, etc., so that the mean number of counts in bin i is expected to be p_i. Then the actual number of counts m_i observed in this bin for an ensemble of experiments should be distributed according to the Poisson distribution,

$$f(m_i) = \frac{p_i^{m_i}}{m_i!}\exp(-p_i), \qquad (5.25)$$

with the fluctuations of the contents of different bins being uncorrelated. The standard deviation is expected to be $\sqrt{p_i}$. If p_i is larger than about 8, the difference between a Poisson and Gaussian distribution becomes small, and the latter may be used. The standard error for the bin content is $\sqrt{p_i}$ in either case and the contribution to χ^2 is $(m_i - p_i)^2/p_i$. However, the χ^2 method does not take into account the skewness of the distribution that applies if Poisson statistics, rather than Gaussian statistics, are required. To this extent χ^2 is not quite right if p_i is less than about 8.

Example: a survey of the Pyramids by cosmic rays

In the 1960s Luis Alvarez speculated how it might be possible to detect the existence of a hidden chamber within the Pyramid of Chephren at Giza in Egypt. As Fig. 5.9 shows, the Pyramid of Cheops, Chephren's father, is well endowed with chambers and these were ransacked in early times. However, the only known chamber in Chephren's Pyramid is the small Belzoni Chamber at ground level near the centre. Alvarez and his Egyptian collaborators built a wide-angle cosmic-ray telescope in the Belzoni Chamber and analysed the cosmic-ray flux as a function of angle in $n = 750$ bins of angle $3° \times 3°$. The telescope was composed of layers of spark chambers and recorded the passage of nearly 1 million tracks.

The way in which the data were analysed is a simple example of early modern data analysis and Fig. 5.10 indicates the stages. The hypothesis being tested is 'Are the data consistent with the absence of chambers?' There are two sources of information that lead eventually down to the

a)

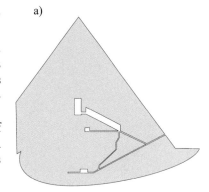

b)

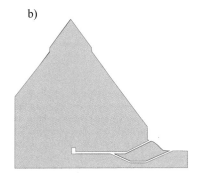

Fig. 5.9 Sections through the Pyramids. a) The Pyramid of Cheops showing the Kings Chamber, the Queens Chamber and the Underground chamber. b) The Pyramid of Chephren showing only the small Belzoni Chamber.

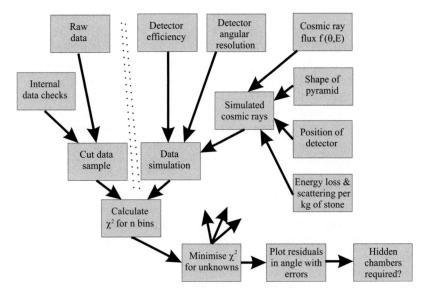

Fig. 5.10 The analysis flow diagram of the search for hidden chambers in the Pyramid

box that asks the question. The first source is the data of recorded tracks from the telescope. This is selected or 'cut' to ensure good data on the basis of internal consistency, records of gas quality and so on. There were 650000 such tracks in the final sample. The result is a series of counts m_i, one for each angular bin i. The second source of information is the previous knowledge. This comprises the expected distribution of cosmic rays in energy and angle, their interaction in stone and the geometry of the Pyramid assuming no hidden chambers. The dotted lines in Fig. 5.10 indicate the separation between the data sample and the simulation. On the right of the diagram is shown the information needed to predict the observed distribution of cosmic-ray angles from: a knowledge of incident cosmic ray flux, the shape of the Pyramid, the detector position, the dE/dx in stone. Thence were constructed the Monte Carlo simulated observations in the detector, $p_i(\alpha, \beta, \ldots)$, for different values of the unknown free parameters $(\alpha, \beta, \ldots)$.[20]

These predictions awee combined with the observed data m_i in the χ^2:

$$\chi^2 = \sum_{\text{bins}} \frac{[m_i - p_i(\alpha, \beta, \ldots)]^2}{p_i}. \tag{5.26}$$

where the p_i in the denominator is the square of the error on the predicted count in each bin. The parameters $\alpha, \beta, \ldots$ are determined by minimising χ^2. If the simulation is acceptable, the mean value of χ^2 should be approximately the number of degrees of freedom.

When carried out the χ^2 minimisation process confirmed the Belzoni Chamber position to be within 2 m of the actual surveyed position.[21] The extra stone cap at the apex of the Pyramid also showed up very clearly in the residuals, as shown in the plots of Fig. 5.11.[22] The final value of χ^2 for the 750 angular bins with useful sensitivity was 1100. In the final step of the analysis, the residuals were plotted as a function of

[20]These parameters included the precise position of the chamber within the Pyramid as discussed below.

[21]This means that when the position of the chamber was left free, as α and β, the minimum χ^2 was found when the chamber position was hypothesised to be in its actual position. This is an example of the way in which confidence can be built in the result of an experiment.

[22]Originally the whole Pyramid was covered with these blocks but in the Middle Ages they were removed except at the top to supply stone to build the city of Cairo.

angle to see whether the discrepancy in χ^2 could not be related to an excess flux in a small angular range, indicating a deficit in the Pyramid rock and a hidden chamber. It was concluded that no chamber in the Chephren Pyramid like those in the Cheops Pyramid exists within 35° of the vertical.

This experiment was carried out 40 years ago. Today a better telescope could be constructed and the statistical analysis could be pursued further. The minimum value of χ^2 was not very good although the discrepancies were not clustered in a particular direction to indicate a chamber. Attempts were made to detect the edges of a hidden chamber by differentiating the flux as a function of angle but these were dominated by noise. If repeated today, emphasis would be placed on increasing the number of counts recorded, improving the angular resolution and using a maximum likelihood analysis instead of χ^2, to avoid the need to define angular bins. Otherwise the principles would be largely the same. Perhaps a hidden chamber lies below the Belzoni Chamber. That would be harder to find!

5.3.3 Maximum likelihood method

The likelihood function

The χ^2 method suffers from two difficulties. Firstly there is the need to bin the data. This is rather arbitrary in execution and discards information within a bin. Secondly, as we already saw, it does not treat correctly bins containing few counts. Both of these shortcomings are handled better in the maximum likelihood method which is used extensively in image analysis.

Consider N measurements of a parameter x. Suppose that we know that the probability distribution[23] of measurements of x would be $f_{\alpha 1}(x)$ if the unknown α were actually to have the value α_1. Since the probability of a measurement having *some* value of x is 100%, the distribution $f_{\alpha 1}(x)$ is normalised to unity. Then the probability of one measurement yielding a value of x between x_1 and $x_1+\mathrm{d}x_1$ is $f_{\alpha 1}(x_1)\mathrm{d}x_1$. And, since separate measurements are independent and uncorrelated, the probability of N measurements with values of x between x_i and $x_i+\mathrm{d}x_i$ (with i running from 1 to N) is

$$P_{\alpha 1} = \prod_{i=1}^{N} f_{\alpha 1}(x_i)\mathrm{d}x_i. \tag{5.27}$$

If on the other hand the true value of α were α_2, the probability $P_{\alpha 2}$ would be different. The ratio of likelihoods is

$$\frac{P_{\alpha 1}}{P_{\alpha 2}} = \prod_{i=1}^{N} \frac{f_{\alpha 1}(x_i)}{f_{\alpha 2}(x_i)}. \tag{5.28}$$

This is as near as we can get to the relative probability of α having the

a) N-S

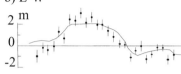

b) E-W

Fig. 5.11 The stone thickness (m) as measured with cosmic rays less the stone thickness calculated for a perfect geometrical pyramid. a) North–south scan through the zenith. b) East–west scan through the zenith. The smooth curves are the simulated corrections for the limestone cap.

[23]This is the distribution of observations in the limit $N \to \infty$.

value α_1 compared with the value α_2. So

$$\mathcal{L}(\alpha) = \prod_{i=1}^{N} f_\alpha(x_i) \tag{5.29}$$

is called the likelihood of α. But it is important to understand that it is not actually what we wanted. It is the likelihood of the data given the value of α, rather than the likelihood of the value of α given the data. So it is sometimes called the inverse probability. The distinction is emphasised by the fact that $\mathcal{L}(\alpha)$ is not normalised.

This is all that is needed. The most likely value of α, labelled α^*, is the one with the highest likelihood $\mathcal{L}$. This is found by setting the derivatives of $\mathcal{L}$ with respect to all parameters α simultaneously to zero. The solution to these simultaneous conditions gives us the best value of the parameters. Previous knowledge enters either through the definition of the problem as described by the likelihood function itself or by an additional factor, or factors, in the likelihood function, as discussed below. The latter method is appropriate if the information is qualified by errors.

The likelihood method may be used under different conditions. Given a set of available experimental data, it can be used to test a model and find the best-fit values of parameters with their errors. In the absence of real data, it can be used to predict the errors of possible experiments by considering the probability function for an ensemble of such experiments.

Determination of parameters by maximum likelihood

General principles and previous knowledge are used to construct the model $f(\alpha, x)$ where x is a quantity we can measure in an experiment and α is a parameter to be determined. Recall that $f(\alpha_0, x)\mathrm{d}x$ is the probability of measuring x with a value between x and $x+\mathrm{d}x$ given that $\alpha = \alpha_0$. The integral or sum of f over x is unity, corresponding to the fact that each measurement yields exactly one value. Real measurements or scans have many measured variables like x, continuous and discrete, and a large number of quantities like α to be determined. In the following, reference to x and α should be understood in each case to describe such long vectors of variables.

In the experiment we make N measurements of x. $\mathcal{L}(\alpha)$ constructed according to equation 5.29 involves the product of a significantly large number of factors, one for each of N measurements. Each is positive and typically small compared with unity. Such a product can lead to a numerical underflow condition, even on a modern computer. It is easier to work with $w(\alpha)$, the natural logarithm of the likelihood function, instead of the likelihood function itself. This converts the awkward product into a sum,

$$w(\alpha) = \log(\mathcal{L}(\alpha)) = \sum_{i=1}^{N} \log(f(\alpha, x_i)). \tag{5.30}$$

The procedures of maximising the likelihood and its logarithm are equivalent. Use of log likelihood also brings further theoretical insights as we shall see. Procedures for maximising functions in many variables are well established. If the log likelihood is quadratic in the parameters α, its derivatives are linear and the solution may be found by matrix inversion.[24]

In the general case we may consider a Taylor expansion of the log likelihood surface in $(\alpha - \alpha^*)$ about its peak value at α^*. This is composed of quadratic and higher terms. With a different choice of variable it may be possible to minimise these higher terms, thereby linearising the determination of the α^*. In the absence of higher terms the confidence levels on the estimation of α become symmetric about α^* and well describable in terms of Gaussian errors.

As a simple example consider observations of an exponential decay process. The data might consist of the observation of a set of decay times t_i with i from 1 to N, confined within an accessible observation window between T_a and T_b. The normalised probability distribution for the observed decay to occur between time t and time $t+dt$ is

$$f(t)dt = \frac{\exp(-t/\tau)}{\tau[\exp(-T_a/\tau) - \exp(-T_b/\tau)]}dt, \qquad (5.31)$$

if the state has a mean lifetime τ. The distribution $f(t)$ is shown in Fig. 5.12 for various values of τ. The log likelihood for the N observations is then

$$w(\tau) = -\frac{\sum_{i=1}^{N} t_i}{\tau} - N \log \tau - N \log[\exp(-T_a/\tau) - \exp(-T_b/\tau)]. \qquad (5.32)$$

The most likely value of τ may be evaluated by setting the derivative of w with respect to τ equal to zero.

Likelihood in the Gaussian approximation

As N becomes larger, the likelihood becomes asymptotically Gaussian,

$$\mathcal{L}(\alpha) = \frac{1}{\sqrt{2\pi}\sigma_\alpha} \exp\left(-\frac{(\alpha - \overline{\alpha})^2}{2\sigma_\alpha^2}\right), \qquad (5.33)$$

where σ_α is the standard error on the reconstructed value of α. This is a general statistical result known as the central limit theorem. In this approximation the logarithm of the likelihood is very simply

$$w(\alpha) = -\frac{(\alpha - \overline{\alpha})^2}{2\sigma_\alpha^2} - \log(\sqrt{2\pi}\sigma_\alpha). \qquad (5.34)$$

Then the procedure of minimising χ^2 is seen to be equivalent to that of maximising likelihood provided σ_α is independent of α.

Differentiating equation 5.34 twice with respect to α gives a relation between $w(\alpha)$ and σ_α, the error on α,

$$\frac{\partial^2 w}{\partial \alpha^2} = -\frac{1}{\sigma_\alpha^2}. \qquad (5.35)$$

[24]This is the Gaussian case, discussed next.

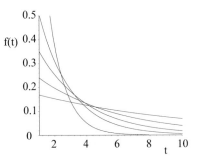

Fig. 5.12 The normalised decay probability functions, equation 5.31, for values of τ equal to 1, 2, 3, 5 and 10, each normalised between $T_a = 1$ and $T_b = 10$.

This result is only correct in the absence of correlations between α and other parameters. The generalisation to a vector of parameters, α_k for $k = 1 \ldots m$, is as follows. The elements of the inverse squared-error matrix H are given by

$$H_{ij} = -\frac{\partial^2 w}{\partial \alpha_i \partial \alpha_j}. \tag{5.36}$$

The complete H-matrix must be inverted to get the matrix of squared errors. The simple errors of interest are the roots of the diagonal terms of H^{-1}. However, the presence of non-trivial correlation terms implies that

$$\sigma_\alpha^2 \neq \left[-\frac{\partial^2 w}{\partial \alpha^2} \right]^{-1}.$$

If the likelihood is not Gaussian, the surface of w is usually plotted and the locus on which w has decreased by $\frac{1}{2}$ relative to its value at α^* interpreted as a standard deviation, and likewise with other confidence-level contours of interest. Such a log likelihood surface is sketched in Fig. 5.13 for a case of two variables. Contours for $n = 1$, 2 and 3 standard deviation are shown. These would be the loci where the log likelihood is lower than the maximum by $n^2/2$. The standard error for each parameter separately is shown by the heavy arrows. If the likelihood surface were Gaussian, the contours would be ellipses and the errors symmetrical. Elliptical contours with rotated axes describe correlations that may be represented by error matrices with off-diagonal terms.

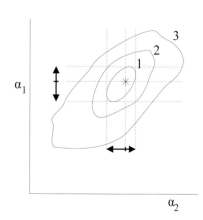

Fig. 5.13 A sketch of a log likelihood surface for two parameters, α_1 and α_2, showing the most probable values marked by a star within contours describing confidence levels.

Combining information from different experiments

Probability distributions that contain uncorrelated information are combined by simply multiplying them together. So if experiment A tells us that the probability of a certain situation is p_A and if a second quite independent experiment B gives the probability p_B, then the probability in the light of both experiments is $p_A \times p_B$. Thus the log likelihood based on two experiments is the sum of the log likelihoods of each of them separately. The inverse squared-error matrices H from each experiment just add for the same reason.

Any other independent information on the relative probability of different values of α can be added to the log likelihood function in the same way.[25] Thus

$$W(\alpha) = w_1(\alpha) + w_2(\alpha) + w_3(\alpha) + \ldots, \tag{5.37}$$

where data from different experiments, different modalities, previous knowledge, etc., each make their own contribution to the combined log likelihood function.

The best value of α is given by the peak of the combined W, and its second derivative gives the error. None of this should surprise. In equation 5.30 the contribution of individual uncorrelated measurements

[25] You may notice that log likelihood, information and entropy are all additive. Boltzmann factors, likelihoods, Gaussian distributions are multiplicative being described by negative exponential quadratic forms. These are closely related. However, this is not an appropriate text in which to mount an expedition into the exciting world of information theory and its recent quantum developments!

within one experiment combined additively to the log likelihood function. It is natural that information from different experiments or scans which are uncorrelated should combine in the same way.

Sometimes the number of readings N is predetermined. However, in other experiments the value of N is itself a measurement. For example, it may represent the number of occurrences in a predetermined time interval. In such cases the value of N contains information.[26] The log likelihood associated purely with the *number* of readings is simply added onto the log likelihood derived from the *values* of those readings.[27] It may be added in this way because there is no correlation between fluctuations in the value of readings and fluctuations in the number of readings. The term to be added comes from the Poisson distribution, equation 5.25. If the expected number is $\mathcal{N}(\alpha)$, the log of the probability of getting N readings is

$$w_{\text{number}}(\alpha, N) = N \log(\mathcal{N}(\alpha)) - \log(N!) - \mathcal{N}(\alpha). \tag{5.38}$$

In the limit of N and $\mathcal{N}$ large, this is given asymptotically by the Gaussian result,

$$w_{\text{number}}(\alpha, N) = -\frac{(N - \mathcal{N}(\alpha))^2}{2\mathcal{N}}, \tag{5.39}$$

where constant terms independent of α have been dropped.

Simulation of experiments by maximum likelihood

Building equipment, optimising conditions and taking multiple data sets are expensive and time consuming. We have already seen that it is possible to simulate experiments by the Monte Carlo technique. However, that brute force method generates data sets of increasingly unmanageable size and comes with its own statistical errors. Such studies can be time consuming and inconvenient. A more elegant simulation of experiments uses the maximum likelihood method. In this method experiments of any statistical precision can be simulated and studied at the same speed.

We consider a very simple example. Suppose that we have a model for our experiment to determine the unknown α from measurements of x between -1 and $+1$. The model is

$$f(\alpha, x) = \frac{1}{2}(1 + \alpha x) \tag{5.40}$$

and this two dimensional surface is shown in Fig. 5.14. The normalisation requires that the integral of the curve of any section at constant α is unity. Let us suppose that the actual value of α is 0.3. Then the actual distribution of x-values would be given, in the limit of large N at least, by the slice through the surface at $\alpha = 0.3$, shown in Fig. 5.15.

We now ask what happens when the experiment is performed without knowing that $\alpha = 0.3$. The likelihood function is used with α undetermined. For each measurement we know that the chance of the value of x being between x and $x + \mathrm{d}x$ is $f(0.3, x)\mathrm{d}x$. To that measurement

[26]The probability function $f(x)$ for each reading must be normalised to unity, as before.

[27]This is usually called the extended maximum likelihood method.

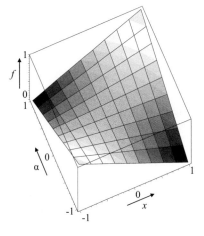

Fig. 5.14 The probability surface, equation 5.40.

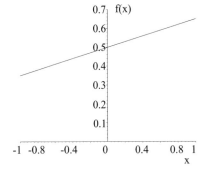

Fig. 5.15 The normalised probability distribution for measurements of x for the experiment in which the actual value of α is 0.3. This is the corresponding section though the surface of Fig. 5.14.

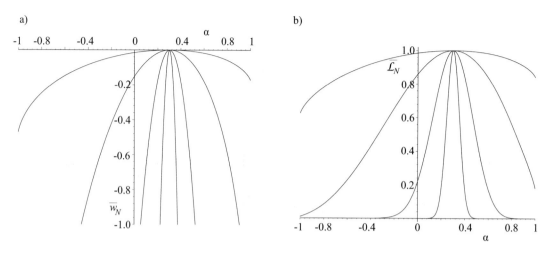

Fig. 5.16 a) The mean log likelihood function $\overline{w}_N$, and b) the average likelihood curves $\overline{\mathcal{L}}_N$, as a function of the hypothesised value of α for experiments where the true value of α is 0.3. The curves describe experiments in which the number of events N is 1,10,100,1000. [To ease comparison the likelihood functions have been shifted to unity at $\alpha=0.3$ and the log likelihoods to zero.]

the program attaches a weight $\log(f(\alpha, x))$ in accordance with the prescription for constructing the log likelihood function. So the *average* log likelihood function that would be found from a *single* measurement is

$$\overline{w}_1(\alpha) = \int_{-1}^{1} \log(f(\alpha, x)) \times f(0.3, x)\mathrm{d}x. \qquad (5.41)$$

As the measurements are uncorrelated the average log likelihood function found for N measurements will be just N times greater (and the likelihood function itself raised to the Nth power),

$$\overline{w}_N(\alpha) = N \int_{-1}^{1} \log(f(\alpha, x)) \times f(0.3, x)\mathrm{d}x. \qquad (5.42)$$

This averaging will be poor when N is small, and this type of analysis will be a poor predictor of an actual experiment, but then such an experiment will convey rather little information anyway. But for larger N, the average will become closer and closer to that found. The shape of the average log likelihood function is shown in Fig. 5.16a for values of N of 1, 10, 100 and 1000. For increased N the log likelihood functions become more closely parabolic. This is equivalent to the central limit theorem which says that as N increases the probability distributions become progressively Gaussian and follow the familiar bell shape. The average error on α that would be found if such an experiment were done, may be read off for different N using the point at which the mean log likelihood has fallen from its peak by 0.5, as in the analysis of an actual experiment. Sometimes it is easier to calculate the second derivative of w, equation 5.35. In Fig. 5.17 we have plotted out the mean error σ_α as a function of the number of measurements N for the case when the

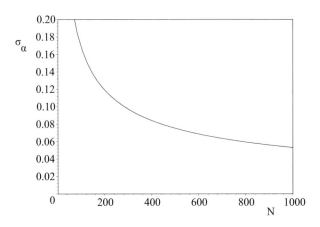

Fig. 5.17 The average error on α that would be found in the experiments discussed in the text, assuming that the true value of α is 0.3, as a function of the number of measurements, N.

true value of α is 0.3. The shapes of the average likelihood functions themselves are shown in Fig. 5.16b.

This method enables us to calculate the average value of any other quantity of interest in the same way. In particular, the average value of the second derivatives of w can be evaluated and then used to derive the elements of the inverse squared-error matrix H following equation 5.36. Thus

$$\overline{\frac{\partial^2 w}{\partial \alpha_i \partial \alpha_j}} = N \int_{-1}^{1} \frac{\partial^2 \log(f(\alpha, x))}{\partial \alpha_i \partial \alpha_j} \times f(\alpha_0, x)\mathrm{d}x, \qquad (5.43)$$

where the known solution has been described by the vector of relations $\alpha = \alpha_0$. The linear dependence of all errors on $1/\sqrt{N}$ is evident. The effect of non-linearities and correlations can all be evaluated cheaply and rapidly in the same way.

It is important to note that these results have been calculated without doing the experiment at all, and without doing any Monte Carlo simulation either. These techniques enable errors to be simulated, experiments to be designed and optimised, and resolution performance to be predicted before the apparatus is built. The only question that maximum likelihood cannot address is the determination of the actual values. To do that, money must be spent and measurements actually made!

Read more in books

The Mathematical Theory of Communication, CE Shannon and W Weaver, University of Illinois Press (1948) is the classical reference for information theory

Science and Information Theory, Leon Brillouin, Academic Press (1962) is another classical reference

Mathematical methods in the physical sciences, Mary L Boas, Wiley, 2nd edn (1983) is a student text covering Fourier analysis

All you ever wanted to know about Mathematics, Vols 1 & 2, L Lyons, Cambridge (1998) is a more basic student reference

Notes on statistics for physicists, J Orear (1958, rev. 1982) is a highly recommended reference on statistics that is classic, readable, concise, but unpublished. It may now be found on the Web, see below

Statistics for nuclear and particle physicists, L Lyons, Cambridge (1986) is a more gentle introduction to statistics

Oxford users' guide to mathematics, ed. E Zeidler, Oxford (2004) includes a section on wavelets

Discovering Alvarez, ed. WP Trower, Chicago (1987) includes a chapter that describes Alvarez's experiment in the Pyramids

Fundamentals of Medical Imaging, P Suetens, Cambridge (2002) discusses the reconstruction of medical images and includes some excellent illustrations and a CD

Look on the Web

Look at the book website at
    www.physics.ox.ac.uk/users/allison/booksite.htm

Read more about information theory by searching on words such as
    information theory, shannon, quantum

Look up a very helpful website that explains more about wavelets
    wavelets clemens

Be sure to read Orear's article on statistics which may be found on a number of websites by searching for
    orear statistics physicists

Questions

5.1 If the mean probability for each of N pixels being black rather than white is f, the mean information content is given by

$$N_{\text{effective}} = N \frac{f \ln f + (1 - f) \ln(1 - f)}{\ln(1/2)}.$$

Consider the implications of this expression for the cases $f = 0.0, 0.5, 1.0$. Hence prove the following statement: 'A digital image with a certain number of pixels or voxels which may be black or white contains maximum information when the illumination is adjusted such that on average the expected fraction of black pixels is 50%.'

5.2 Prove the property of the Dirac δ-function,

$$\int_{-\infty}^{+\infty} \exp[i(\omega - \omega')t]dt = 2\pi\delta(\omega - \omega').$$

5.3 Prove that the transform of the convolution of two functions is equal to the product of their transforms, equation 5.14.

5.4 Consider a periodic function with $f(x) = f(x+nL)$ for all integer values of n. Find its Fourier transform using the convolution theorem.

5.5 A measured *in vivo* NMR spectrum $M(\omega)$ is a superposition of the spectra of 5 metabolites, $A(\omega)$, $B(\omega)$, $C(\omega)$, $D(\omega)$, $E(\omega)$, with a background of the form $X + \omega \times Y$, where X and Y are free parameters. The 5 spectra are known from *in vitro* measurement of each separate metabolite of known concentration. Show how you would determine the concentrations of the 5 metabolites by the least squares method while making allowance for the background. You should assume that each spectrum consists of amplitudes measured at the same set of 100 discrete frequencies.

Analysis and damage by irradiation

in which the detection, health effects, safety, uses and abuses of ionising radiation are discussed.

6.1 Radiation detectors

6.1 Radiation detectors   157

6.2 Analysis methods for elements and isotopes   168

6.3 Radiation exposure of the population at large   179

6.4 Radiation damage to biological tissue   187

6.5 Nuclear energy and applications   192

Read more in books   204

Look on the Web   204

Questions   205

Energy deposited by radiation in matter undergoes prompt changes of form (secondary ionisation electrons, bremsstrahlung and electromagnetic showers) and further changes with some delay (fluorescence radiation and Auger electron emission). In addition to releasing photons and electrons, irradiation also leaves a trail of atomic ions and molecular fragments in excited states. These are highly reactive and are responsible for longer term radiation-induced effects. Charged particle detectors may be based on the use of photographic emulsion, the detection of scintillation light, or the ionisation signal generated in gas-filled or solid state devices. The detection of electromagnetic radiation depends on the photon energy. At the lowest frequency, coherent waves, not photons, are detected. In the mid range individual photons are detected but the signals are amplified externally using photomultipliers, gas or solid state devices. At the highest energies, the self-powered avalanche of an electromagnetic shower generated by the incident photon may be detected by its scintillation light.

6.1.1 Photons and ionisation generated by irradiation

Prompt processes

Much of the energy initially deposited by irradiation is in the form of either excited atoms and molecules, or free electrons and ions. The latter are called primary ionisation. The electrons may have substantial kinetic energy. Historically the most energetic are called δ-rays because they leave their own tracks in detectors.

Other prompt or primary processes include the bremsstrahlung radiation discussed in section 3.6.2 and Cherenkov radiation. In both cases the radiation is coherent and highly directional. Cherenkov radiation is emitted by any charge moving faster than the speed of light in the medium, c/n, where n is the refractive index. Since the speed of the charge is necessarily less than c, such emission is not possible unless n is greater than 1. Figure 6.1 shows the dependence of n on frequency

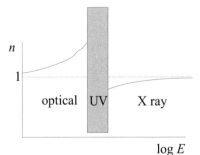

Fig. 6.1 Sketch of the frequency dependence of the refractive index n due to resonances in the ultraviolet region (shaded). In this region radiation is absorbed and n is complex. With the typical resonance shape n may only be real and greater than 1 below resonance in the optical region.

a)   b)

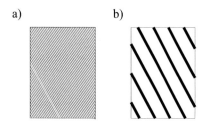

Fig. 6.2 Sketches illustrating how, for a given energy deposition, a) the even distribution due to a large number of relativistic electron tracks (z=1, β=1) contrasts with b) the uneven distribution due to a small number of ion tracks with z^2/β^2 of 10, for example.

a)

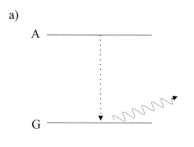

b)

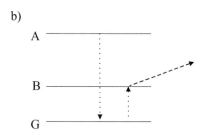

Fig. 6.3 Energy level diagrams showing modes of de-excitation of an atom or molecule in an excited state, A.
a) This may occur on its own by the emission of a fluorescent photon leaving the atom or molecule in its ground state, G.
b) It may also occur by excitation to an ionised state B with emission of a free electron with the remaining kinetic energy. In each case vertical lines denote internal energy changes, the external wavy line is an emitted photon and the dashed line an emitted electron.

[1]We use the terms scintillation and fluorescence interchangeably according to context.

for a typical material. Above the ultraviolet absorption region, n is less than 1. In the ultraviolet region itself, the medium is absorptive and no free Cherenkov radiation propagates. So only in the optical region is Cherenkov emission possible. The flux in this modest bandwidth is small, the spectrum is rather soft and in total accounts for no more than a per cent or so of the total energy loss.

Delayed processes

Many important applications in medical and environmental physics depend on what happens to the deposited energy subsequently and on the timescale on which it evolves. The free electrons do not usually recombine with ions. This is unlikely because they are spatially separated, and recombination requires the involvement of a catalyst if energy and momentum are to be conserved. Generally, primary and secondary ions come to rest within a few tens of microns of the flight path of the original charge. Consequently, except in a gas, the debris or the radiation damage is not evenly spread though the medium but is confined to narrow cylinders along the flight paths of the individual incident charges. This is most pronounced when the density of the elemental energy loss process is high. The linear density of energy loss is given by the Bethe–Bloch formula, equation 3.44, and $dE/dx \sim z^2/\beta^2$. So when the incident particle is highly charged (high z) or non-relativistic (low β), the energy deposition is highly concentrated. Figure 6.2 gives an impression of how, for the same total deposited energy in a material, many singly charged relativistic particles give a much more uniform distribution of irradiation than a few multiply charged non-relativistic ions. The energy deposited along the track is referred to as the linear energy transfer density (LET). It is the dE/dx discussed in chapter 3.

Excited atoms may begin the process of thermalisation by internal transition, either by emitting a photon or by emitting an atomic electron as illustrated in Fig. 6.3. The first process is called fluorescence or scintillation and the second Auger emission.[1] Scintillation emission is isotropic, being incoherent and having no memory of the direction of the incident radiation. However, the emission spectrum has sharp lines characteristic of the atom or molecule. These two de-excitation processes usually compete with one another. A third possibility is that the atom or molecule, instead of de-exciting on its own, has a collision with another atom or molecule. (In condensed matter it may transfer energy to its neighbours by forming a mixed quantum state with them by level broadening, although this is actually equivalent.) No radiation is emitted in such collisions and the energy progresses to thermal equilibrium.

In some regions of the spectrum scintillation photons are not actually observed, simply because they are reabsorbed by the medium within a short distance. Otherwise the radiation can travel away from the point where the initial energy deposition occurred and be detected. In particular scintillation light emitted in a transition to a state that is not the ground state (or within an energy of $k_B T$ of it) will not suffer self-absorption.

Electromagnetic showers

The behaviour of high energy electrons and photons is a special case. Such photons create electron–positron pairs and also recoil Compton electrons. The electrons (and positrons) then create bremsstrahlung. These elemental processes are illustrated in Fig. 6.4. Starting either from an electron or from a photon, each of energy several times the electron rest-mass energy $m_e c^2$, a cascade or shower builds up. As the shower develops the energy of the individual secondary photons eventually falls below the threshold for pair production. Similarly the energy of each electron falls to the point below which non-radiative energy loss dominates over bremsstrahlung. Further statistical multiplication of the shower is then choked off.

Both pair production and bremsstrahlung depend on the radiation length of the medium. In dense high Z materials with short radiation lengths electromagnetic showers are spatially compact. This is ideal for high energy electron and photon detection with containment of the shower and measurement of both the energy and the line of flight of the original electron or photon.

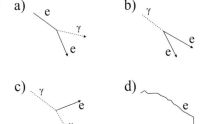

Fig. 6.4 The four processes contributing to the build-up of a shower:
a) emission of radiation by an electron,
b) pair production by a photon,
c) Compton scattering of a photon,
d) multiple scattering of an electron. All processes except c) occur on a scale determined by the radiation length, as discussed in chapter 3.

Chemical radicals

In addition to these free electrons, ionised and excited atoms, irradiation leaves excited molecules and chemical fragments. These highly reactive products are not found in normal materials at room temperature. They may migrate in the gaseous or liquid state and initiate reactions that do not normally occur. In a solid they may remain trapped for a long time, depending on the temperature. Typically such radicals have unpaired electrons and therefore exhibit electron spin resonance with a frequency characteristic of the radical concerned. This offers a method of diagnosing their presence.

One example of a chemical product of radiation is the creation of unstable polymer film. This shrinks rapidly when exposed to heat and is used as a wrapping material in the electrical and food industries. Another example is the generation of a metastable image of a radiation exposure in a photographic emulsion, or analogue photography as we commonly know it.

6.1.2   Task of radiation detection

The basic methods have changed little over the decades. What has changed is the technology with which any method is implemented. Detectors differ in the way in which the incident energy is amplified and recorded. The amplification is important because the basic signals are tiny. A single electronic charge, even after acceleration through a million volts, only has an energy of 1.6×10^{-13} J. Nevertheless individual charged particles can be detected, and their energy, position and angle determined.

The range of energy involved in photon detection is the most extreme. From Fig. 1.3 we see that the energy of a photon of visible light is about 1 eV or 1.6×10^{-19} J. The energy of harder photons is easier to detect. The detection of softer photons with energies falling by nine orders of magnitude below visible involves special consideration.

Detector resolution and efficiency

Some radiation detectors are sensitive to the number of electrons released in the material. Others detect released photons and some detect phonons. In every case the statistics of these information carriers are important. The working assumption is that the average number of carriers released is proportional to the deposited energy for a detector that does not saturate. The number of carriers n will fluctuate as prescribed by Poisson statistics with standard deviation $\sigma_n = \sqrt{\bar{n}}$ where $\bar{n}$ is the mean. These fluctuations will be responsible for a fractional variation in signal size of order $1/\sqrt{\bar{n}}$. In a linear detector this determines the energy resolution. The larger the value of n, the better the energy resolution.

Even for a detector that saturates, $\bar{n}$ determines the detector efficiency in the absence of the effects of other sources of noise. The probability of getting zero carriers is given by $P(0, \bar{n}) = \exp(-\bar{n})$. This is the irreducible inefficiency. So an efficient detector should have a suitably large value of $\bar{n}$. If the energy deposited in the detector is E, then

$$E = W \times \bar{n}. \tag{6.1}$$

The value of W, the average energy deposited per carrier, is an important number for a detector technology. For a given value of E, a lower value of W brings smaller signal fluctuations, higher statistical efficiency and better energy resolution.

In addition to the signal and its fluctuations there is random thermal noise. The signal-to-noise ratio (SNR) is the amplitude ratio of the two. A discriminator level determines when a signal is considered to be present. If it is set high, there will be a low 'false positive' rate but also a decreased detection efficiency of real signals. If it is set low, real signals will be detected efficiently but there will be a significant rate due to noise. The SNR may be improved by increased amplification if the noise is not also amplified. Understanding the sources of noise is therefore important if signals are to be detected efficiently and unaffected by noise.

Acceptance and exposure

The largest factor in the efficient use of radiation is the acceptance of the detector array. Radiation that does not fall within the sensitive area of the array is lost. In some methods the radiation flux is confined to a narrow slice. In others it is isotropic. In the latter case an ideal detector array would envelope the experiment, subtending a solid angle of 4π steradians. Large reductions in the radiation dose experienced by patients are achieved by increasing the detector size and the solid angle to match the optics of the administered flux. Cheaper technology is allowing significant improvements in efficiency to be made. However, methods that rely on collimators and pinhole optics within the detector are by their nature inefficient.

As in solid angle, so also in time, the period of detection should be matched to the period of radiation exposure. This is a significant source of inefficiency in some current medical imaging procedures.

Precision and resolution in time and space

The error on the time (or position) of a recorded signal is limited by the signal pulse shape and its statistical fluctuations. The pulse shape in time is the impulse response of the detector and its electronics[2] folded with the distribution of electron arrival times to simulate the observed pulse shape. The actual value of the precision depends on the way in which the measured position is determined by the pulse. For example, it might be the peak, the mean or the leading edge of the pulse.

In practice the precision is not as important as the resolution given by the pulse shape itself. The resolution is the 'occupancy' of the signal. This determines how close a further signal can occur and still be resolved as being separate. Some detectors are disabled for a time following a signal.[3] In other detectors subsequent signals simply add on top of one another generating 'pile up'. A linear device would record a signal amplitude which was the sum of the unresolved signals. In most applications the resolution is the useful measure at the analogue level in both time and space. The element of a digital sampling of the analogue image is typically matched in size to the resolution. Detectors may record resolved data in one, two or three dimensions, and in time. A single digital element of a two dimensional picture is termed a pixel. A digital element in a three dimensional image is called a voxel.

6.1.3   Charged particle detectors

It matters in what form detectors provide image data and how they are stored. Photographic data are cheap, of high quality and convenient in a low technology environment. Electronic methods are often preferred because they can be linked straight to a computer. Circumstances and resources determine which detector is used.

Photographic emulsion

Photographic emulsion was the first radiation detector of charged particles. The photographic plates that Henri Becquerel left in a drawer recorded the radiation emitted by the specimens of uranium sulphate placed on top of them. A photographic emulsion contains fine grains of silver halide. These are metastable. The activation energy required to

[2] Corresponding to the signal generated by a single electron or photon at the detector input.

[3] This is known as the dead time. For example the Geiger–Müller tube has such a dead time of tens of milliseconds during which the tube is insensitive.

change the state of the silver can be provided by the energy deposited along the tracks of charged particles, or, just as effectively, by X-rays or light. The detector response, the blackening of the film following development, is a non-linear function of the energy deposition. The extent of this non-linearity is called the gamma of the film. It has rather weak response at low exposures but saturates at high exposures, similar to the dose curves for biological radiation response (see Figs 6.18 and 8.26).

Photographic film gives fine spatial resolution in two dimensions of the location of the radiation but gives no time information. The constituents of photographic emulsion, iodine, bromine and silver, have high atomic numbers and high photoelectric cross sections well into the X-ray energy region. Thus photographic emulsion is a reasonably efficient position-sensitive detector for photons as well as charged particles. It is used in this dual role as a robust cheap cordless radiation monitor which does not require calibrating. Traditionally film badges were carried by all radiation workers to record the time-integrated radiation dose to which they had been exposed. By attaching windows of different absorbers to the surface of the film badge, information about the nature and spectrum of the radiation dose could also be recorded. However, the non-linearity of response is a drawback and in many modern laboratories electronic dosimeters with better dynamic range are now used in preference.

Scintillation counters

Detectors based on recording scintillation light were used from the earliest days of nuclear research. The experiments on Rutherford scattering which detected the deflection of alpha particles by gold nuclei were based on observations of the flashes of scintillation light emitted when the alphas hit a screen coated with zinc sulphide. A more modern example is the phosphor coating on the inside face of a cathode-ray tube or conventional television display which emits light when hit by the electron beam.

The three characteristics that are important in the choice of scintillation material for a charged particle detector are the spectrum of the scintillation light, the photon flux and the scintillation decay time. The spectrum needs to be matched to the quantum efficiency of the detector to be used to detect it.[4] The photon flux when multiplied by the detector quantum efficiency of the photon detector gives the number of carriers that determines the overall efficiency. The characteristic decay time of the scintillator or phosphor determines the time resolution of the detector.

All materials scintillate to some extent but the effect of self-absorption is usually significant. In a dense material the requirements of transparency and emission at the same wavelength are incompatible unless the excited state decays to a state other than the ground state. The best scintillators are often doped materials where the emission is dominated by the contribution of a small concentration of another atom or molecule. Table 6.1 shows data on a few standard scintillator materi-

[4]Quantum efficiency is the probability of a signal for each incident photon. This is a function of the frequency.

Table 6.1 Properties of various scintillators. ε is the photon flux (relative to NaI) and τ is the decay time.

Name	Type	kg m^{-3}	ε	τ ns
Anthracene	Org. solid	1.25	0.43	30
Pilot B	Org. plastic	1.03	0.30	1.8
NE 213	Org. liquid	0.87	0.34	3.7
NaI(Tl)	Inorg. cryst.	3.67	1.00	230
CsF	Inorg. cryst.	4.1	0.05	5

als. The fluxes are quoted relative to sodium iodide (NaI) doped with thallium (Tl).

Because the scintillation emission is incoherent and isotropic it is difficult to extract fine spatial information from such detectors. A modern variant of the scintillation detector uses bundles of optical fibres formed from scintillator. However, there are two difficulties. Scintillation light is emitted isotropically and therefore the fraction of the flux trapped by the narrow angular acceptance of the fibre is small. Further, such fibres can only be of limited length because of the self-absorption problem. An advantage is that the signals may be matched into bundles of non-absorbing optical fibres which may be read out remotely with segmented photomultipliers or photodiodes.

Gas-filled detectors and the Geiger–Müller tube

The basic geometry of the classic gas-filled detector is shown in Fig. 6.5. A central fine anode in a cylinder of gas is enclosed by an outer cathode. It may be used in three different ways, depending on the gas with which it is filled, the diameter of the central anode and the applied voltage.

Its most robust form is the Geiger–Müller tube. In this case the anode is a stout wire of diameter $\sim 500\mu$m raised to several hundred volts. The cylinder is filled with a noble gas such as neon. There is no amplifier A. Charged particles passing through the gas create free electrons and positive ions which move towards the central anode and outer cathode respectively. The field rises as $1/r$ as the electrons converge on the anode. They therefore gain in energy as they move inwards. There are no soft vibrational or rotational modes for a neon atom and therefore the electron energy continues to rise until it is high enough to ionise the neon. Finally the electrons hit the anode with several electronvolts of energy. This process is sufficient to generate a number of ultraviolet photons at the anode. This gas is transparent in the soft UV and these photons generate fresh electrons by the photoelectric effect when they are absorbed by the surrounding cathode. For the UV discharge to propagate, the number of UV photons multiplied by their conversion efficiency to create a fresh electron simply has to exceed unity. The increased number of new electrons is attracted by the anode and a discharge is built up which spreads through the whole volume. The entire stored energy $\frac{1}{2}CV^2$ of the cylinder as a charged capacitor is dumped, and the signal saturates. It takes a significant time for the electrons and ions to clear and the external circuit to recharge the capacitor in readiness for another signal. While such a detector gives little quantitative information, the Geiger–Müller tube is widely used as a cheap and rugged radiation monitor in the field.

Gas-filled ionisation detector

The ionisation detector is a variant of the gas-filled counter, Fig. 6.5. It is a linear and quantitative device. The pure noble gas is replaced by a mixture containing a molecular gas with UV absorption length

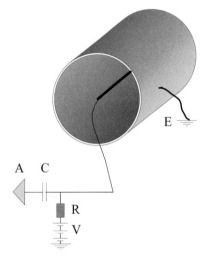

Fig. 6.5 A cylindrical gas counter with central anode at high voltage V and earthed outer cylinder E. The signal is readout by an amplifier A through a decoupling capacitor C.

which is short compared with the dimensions of the detector. Incident radiation ionises the gas as before releasing free electrons of charge Q. The electrons drift in the radial electric field as before, but they now lose energy in collisions with the molecular gas, keep cool and do not create further ionisation. The resulting current pulse in the external circuit is detected by an amplifier through a decoupling capacitor. This pulse is proportional to the rate at which the electrons move down through the potential difference V and begins as the electrons start to move. The total energy of the pulse $QV = \int IV\,dt$ can be related directly to the energy released in the gas QW/e where W is the value of the mean energy per ionised electron for the gas concerned.[5] Ionisation counters play a key role in measuring absolute radiotherapy doses. The important point is that the signal amplitude does not depend on an unknown detector gain requiring calibration. However, because the signals are not amplified they are not large. The integrated signal is only linearly dependent on the choice of voltage V. The anode diameter should not be too small and a value of 100–200 μm might be chosen.

The ionisation detector has no amplification. This is satisfactory when the signals are large and heavily integrated so that the bandwidth is small and thermal noise is reduced. For measuring absolute radiotherapy doses these conditions are satisfied. However, for small fast signals amplification is required.

Gas-filled proportional detectors

If the voltage V is increased substantially to 1000–2000 volts with the anode diameter small, 20–50 μm, amplification of the electrons by ionising collision occurs within a few diameters of the anode. Photons and metastable ions play no part as they are quickly quenched by the molecular gas. The amplification of the signal can be up to 10^6 and remain linear, which is why the detectors are named proportional counters.[6] The gas is typically 80–90% argon or xenon with an added polyatomic gas such as CO_2 or hydrocarbons.

Proportional counters can provide good spatial precision by deploying an array of separate anode wires with separation of a few millimetres within a chamber, as shown in Fig. 6.6. However, the energy resolution is limited by fluctuations of the small number of carriers, that is by the modest energy loss and the large value of W, the energy deposited per ionisation electron, which is about 20 eV.

Two reasons for the small signal from a gas-filled ionisation detector are the low density and the large value of W. Liquid ionisation counters can be used to increase the density by a factor of a thousand, but the purity required to achieve the necessary free electron lifetime in liquids is not practical outside a research laboratory.[7]

Proportional counters may also be used as photon detectors. Such photons need to enter the counter and be absorbed. Thus the photon energy range extends from the top of the absorption band of the chamber window to the top of the absorption band of the gas. This favours the

[5]This is not true if there is any electronegative component in the gas which can attach to free electrons and remove their contribution to the signal. Such negatively charged ions, like the positive ones, would contribute in the end to the signal. However, they arrive too late to be seen as part of the pulse because their mass and mobility in the gas are four orders of magnitude lower than the electrons. Oxygen is the main electronegative gas of concern.

[6]At large energies and gains the avalanche amplifier becomes non-linear simply because the magnitude of the charge in the avalanche itself becomes a significant fraction of the static charge on the wire which is driving the avalanche.

Fig. 6.6 A section through a multi-wire proportional chamber with thin aluminised Mylar windows acting as cathode and an array of fine anode wires, each providing gas amplification and a separate channel of electronic amplification (not shown).

[7]Liquid argon calorimeters have been used for many years but they require cryogenic as well as high purity conditions.

use of dense high Z gases such as xenon.

Solid state detectors

Detectors based on semiconductors provide improvements in both density and W. The density is increased by a factor of a thousand, and the mean energy per carrier which is closely related to the semiconductor bandgap is less by a factor of order 20, $W \sim 1$ eV. Relative to a gas this is an increase of 2×10^4 in the number of carriers per unit volume. On the other hand there is no amplification like the cascade around the wire. There is some benefit in using germanium rather than silicon with its smaller bandgap, although this is at the expense of running the detector at liquid nitrogen temperatures. Like the liquid argon calorimeter, the germanium counter has remained in the physics laboratory.

Detectors based on arrays of pixels in amorphous silicon are in widespread use not only to record images as such, but also for the more exacting task of monitoring absolute radiation doses for radiotherapy. Such arrays are now used to provide a combination of good spatial resolution with immediately available readout.

Other technologies exist with very much smaller values of W, and therefore better energy resolution. However, like the germanium counter, these need to be operated at a temperature T, such that $k_B T \ll W$. The superconducting tunnel junction technology (STJ) requires liquid helium temperatures but its resolution has already brought benefits in infrared astronomy.

6.1.4   Electromagnetic radiation detectors

The spectrum of electromagnetic radiation shown in Fig. 1.3 extends over 17 orders of magnitude. It is instructive to discuss how radiation may be detected over such a range. There is one consideration that divides the spectrum in two and causes a qualitative change in the way in which electromagnetic radiation is treated experimentally. In the hard part of the spectrum defined by

$$\hbar\omega \gg k_B T \tag{6.2}$$

thermal noise permits individual quanta to be detected. In the remaining soft part individual quanta cannot be resolved above background noise. At room temperature $k_B T$ is 0.025 eV, which is in the infrared. In research laboratories the boundary may be pushed substantially lower by the use of cryogenic temperatures. An overview of the basis of detection in each range is summarised in table 6.2.

Low energy photon detection

In the soft region the wavefunction of the photons behaves as a classical field because the number of photons is large. Signals may be filtered by selecting a very small range of direction (using an aerial or dish)

Table 6.2 An overview of electromagnetic wave detection in different energy ranges.

Region	Detector options
Radio	aerial/dish and resonant circuit
Infrared	calorimeter, room temperature and cooled (STJ)
Optical	photoelectric (PMT), photoexcitation (Ge, Si), photochemical
Ultraviolet	photoelectric (PMT, gas), photoexcitation (Si), photochemical
Soft X-ray	photoelectric (gas), photoexcitation (Si), photochemical
Hard X-ray	shower with generation of secondary optical photons
γ-ray	shower with generation of secondary optical photons

[8]The heat capacity of all crystal lattices falls rapidly at cryogenic temperatures. The benefit of cryogenic temperatures is both to reduce the thermal background and to increase the temperature excursion generated by the signal energy input to the calorimeter.

and a very small range of frequency (using a narrow band resonant circuit or other filters). The brightness of any source within that small acceptance is likely to exceed the thermal background, because the latter is reduced by both acceptance factors. If the bandwidth is large the signal may be detected as a temperature rise in a cryogenic calorimeter of low heat capacity.[8] If the bandwidth is small the signal may be detected coherently as a normal radio signal amplitude.

Photon detection in the optical energy range and above

In the hard X-ray and γ-ray range each photon has enough energy to generate a shower itself. In the optical to soft X-ray region individual photons can be detected but an external energy source is required for amplification.

Hard photons may deposit energy, and thereby be detected, by any of the interaction processes discussed in section 3.3. Photoelectric emission of an electron from an alkali metal surface has the lowest threshold and this is combined with external amplification in the photomultiplier tube (PMT). This is the classic workhorse of photon detection in or near the optical range. A schematic diagram of a PMT is shown in Fig. 6.7a. The primary photon enters the vacuum of a photomultiplier through a thin window (not shown) and strikes an alkali-coated electrode, emitting a photoelectron. This electron is accelerated onto a second electrode (or dynode as it is called) by a positive voltage, where on impact more electrons are ejected because of the increased kinetic energy. With several such stages a large amplification is built up. The dynodes are attached to a high voltage divider chain with high impedance elements R. Electrons are emitted in a cascade as they gain energy striking dynodes with more positive voltages and releasing extra electrons. The final cascade current generates a voltage pulse across the smaller input resistor r between the

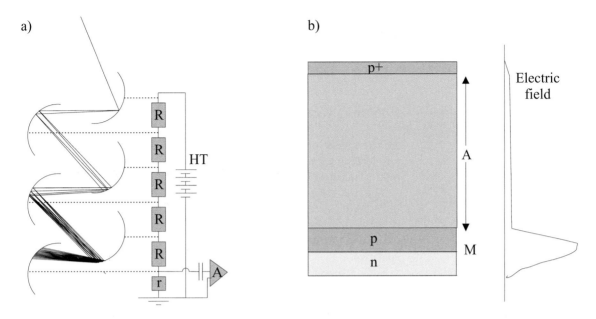

Fig. 6.7 a) A diagram of a vacuum photomultiplier, showing on the left the ladder of dynodes (electrodes) and the development of the cascade, and on the right the voltage divider chain and signal readout discussed in the text.
b) A schematic diagram of an avalanche photodiode. A photon enters from above through a thin metallised layer and is converted in the thick absorber layer A. A modest electric field transports carriers to the amplification region M where further carriers are excited in the high electric field.

last dynode and ground. This pulse is detected by a decoupled amplifier A. With modest capacitance C and very high values of R, the bandwidth of the divider chain, $\Delta\nu = 1/(RC)$, can be kept low. Consequently the noise associated with this chain is absent at high frequencies, and the signal can be looked at with very high bandwidth without loss of SNR. The photon energy threshold is low because the electron work function of an alkali metal is lower than for other materials. In the absence of signal the current is small, and a gain of order 10^6 is normal. The efficiency is high. The amplification is linear provided that the charge in the avalanche is small compared with the stored charge on the dynodes.

PMTs are now commercially available with arrays of anodes (8×8) each affording spatial resolution on a scale of 6×6 mm^2.

Available solid state photon detectors typically have p- and n-doped semiconductor layers separated by an intrinsic layer, a PIN device.[9] Photoexcitation of carriers in the thick intrinsic layer enables charge to flow. Devices are usually based on silicon and can be made in arrays with pixels on the scale of a millimetre or two. Layers are biassed by a few hundred volts. This is the avalanche photodiode (APD). Figure 6.7b is a generic diagram showing an absorber region and a multiplication region. Gains of 20–100 are achieved. Noise levels are a consideration for the APD, unlike the PMT. Sensitivity to wavelengths in the range 300 to 1700 nm depends on the choice of semiconductor.[10] Further materials such as CdTe are under development. Other developments include

[9] The terms *p-doped, n-doped* and *intrinsic* are explained briefly in the glossary and in texts on electronics.

[10] Si, Ge and InGaAs.

Table 6.3 Transparent electromagnetic shower calorimeter materials. X_0 is the radiation length in m. ε is the integrated light output relative to NaI (100%). The density ρ is in kg m^{-3}.

Material	X_0	$\varepsilon(\%)$	ρ
NaI	0.0259	100	3670
CsI	0.0185	45	4510
Bi_4GeO_4 (BGO)	0.0112	15–20	7130
$CdWO_4$	0.0106	35–40	7900
$PbWO_4$	0.0089	1	8280

[11]The number of radiation lengths required depends logarithmically on the photon energy.

the hybrid photodiode (HPD) which combines vacuum and solid state technology. Up-to-date information on these may be found on the Web.

Photon detection above pair production threshold

At the higher energies photon detectors work in two steps. In the first an electromagnetic shower is produced. In the second the scintillation light generated by this shower is detected by PMTs, APDs or HPDs.

The first stage calls for a dense high-Z transparent medium. By choosing a material with a sufficiently short radiation length the shower will be compact and may be contained within a small segment of a fine-grained calorimeter.[11] The segments may be isolated by reflective foil or coatings.

Some transparent materials with short radiation lengths that are available are given in table 6.3.

6.2 Analysis methods for elements and isotopes

Imaging and analysis of element concentration rely on the characteristic spectrum of fluorescent (or scintillation) X-rays. These spectra may be excited in several ways, by X-rays (XRF), electrons or protons (PIXE). Protons create less background than electrons. The scanning proton microprobe (SPM) may also be used to identify nuclei by mass using Rutherford back scattering (RBS). Unlike the relative concentration of elements, the relative concentration of different isotopes has a long memory and measurements are sensitive to age and changes in climate that remain preserved over long periods. Measurements of isotope concentration rely on differences in nuclear decay, differences in nuclear reaction cross sections, and differences in mass. Such mass differences may be measured by mass spectrometer. Specific examples discussed include ^{14}C, the ratio $^{40}Ar/^{40}K$, the ratio $^{230}Th/^{234}U$, and the oxygen isotopes of mass 17 and 18. The legacy of the natural nuclear fission of ^{238}U may also be used to date materials. Generally the radiation damage which is preserved in solids may be used either as a clock or as an irradiation record. Age determination requires that a clear time-zero point is defined. Systematic and statistical errors have to be controlled, and this determines which methods are reliable.

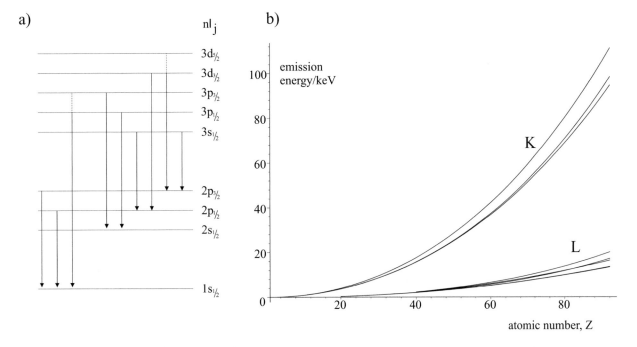

Fig. 6.8 a) The quantum numbers of the states involved in K and L shell emission. The notation s, p and d refers to $\ell = 0, 1$ and 2, respectively. To make the labels plain, the spacing of the M shell ($n=3$) states has been greatly exaggerated. b) The Z-dependence of K and L shell emission X-ray energies.

6.2.1   Element concentration analysis

Fluorescent X-ray method

The energy of the X-ray emission lines is directly related to the atomic number Z of the material, as determined by Mosley in 1913. The energy of an inner electron vacancy in shell n is given by

$$E_{nlj} = Z^2 \alpha \frac{1}{(n - \sigma_{nl})^2} + A \boldsymbol{l} \cdot \boldsymbol{s}. \qquad (6.3)$$

σ_{nl} is the correction for the shielding effect of other electrons on the energy of the nth shell with orbital angular momentum l, and A describes the spin–orbit splitting according to the value of $j = l \pm s$. These energies, E_{nlj}, are seen as the ionisation 'edges' in the absorption spectrum and, for all but the lightest elements, are independent of the chemical state which only affects outer electrons. The lines of the emission spectrum are *differences* of these energies, emitted as an electron drops down from a state to fill an inner shell vacancy.

The presence of individual elements may be imaged non-destructively by exciting such an emission spectrum in situ. This is a major advantage over chemical analysis. A plot of the energy of such X-rays associated with vacancies in the K shell ($n=1$) and L shell ($n=2$) is shown in Fig. 6.8b as a function of Z.

The combinations of lj angular momentum quantum numbers that

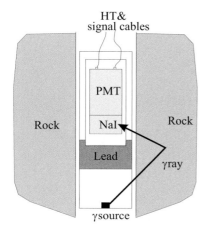

Fig. 6.9 A γ-ray fluorescent detector deployed to analyse rock in a bore hole.

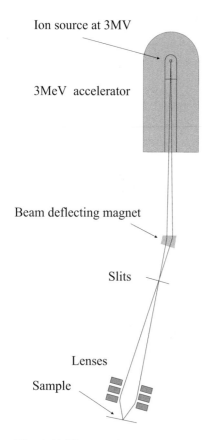

Fig. 6.10 The scanning proton microprobe.

arise for $n=1,2,3$ are shown in Fig. 6.8a. The resulting fine structure in the spectra gives the three close K lines and five close L lines, as shown in Fig. 6.8b. The dipole selection rules, $\Delta j = 0, \pm 1$ and $\Delta \ell = \pm 1$, restrict the transitions to those shown. Transitions from higher n states are allowed but the wavefunctions have smaller overlaps and so the matrix elements are small. Transitions from states with higher j appear stronger on account of the statistical weight $2j + 1$.

Analysis with XRF

For X-ray fluorescence a nuclear γ-source with energy above the K edge of the elements of interest is required. These γ produce K and L shell vacancies in the elements of the target material which then emit their characteristic spectra. The energy resolution of the X-ray detector is most important. In favourable cases element concentrations of a fraction of a part per million can be detected. The concentration value itself is determined by comparing with standards, with care taken over absorption and scattering. The overlap of the L lines of higher Z elements with K lines of lighter elements is simply handled by fitting a superposition of the emission spectra of elements checked with standards.

XRF is routinely used for the analysis of elements from sodium to uranium from the ppm up to the % level. Applications cover geology, forensics, planetary exploration, materials science, environmental sampling and industrial process monitoring. An example showing how the rock in a geological bore hole may be analysed as a function of depth is shown in Fig. 6.9. In a forensic context firearms may be identified and explosives traced. Other materials can be matched, for instance to those at a crime scene. No special sample preparation is required, the exposure can be carried out at ambient pressure and the analysis is immediate.

The element discrimination of XRF is outstanding but the spatial imaging is limited. Although X-rays cannot be focused they can be tightly collimated if the source is sufficiently bright. A synchrotron source (SR) can be used to provide such brightness. Spatial resolutions of order 1 mm square with high brightness and depth penetration can be achieved with SR. However, radioactive γ-ray sources are often cheaper and more convenient, as SR sources are still only available in specialised laboratories.

Scanning proton microprobe (SPM)

Alternatively, the characteristic emission X-rays may be excited with a bright spot from a finely focused proton beam. The beam can be focused down to a spot of 2–3 μm which can then be scanned across the sample. Electron beams are less suitable as they scatter significantly. They also generate bremsstrahlung which forms a background that swamps the fluorescent signals from all elements except those with the highest concentration. Figure 6.10 shows a diagram of the scanning proton microprobe. An electrostatic generator with terminal voltage of 3 MV produces a fine continuous beam which is focused by a set of quadrupole lenses onto the sample in vacuum. The proton-induced X-ray emission (PIXE) signal from the sample is detected as in XRF. As in XRF only a very small sample is required. The advantage of PIXE is the fine spatial resolution. However, only a thin surface layer can be analysed because the protons do not penetrate far, but if a surface layer analysis is required this is not a drawback. A more serious disadvantage is that the sample must be put under vacuum and the measurement must be done in a special laboratory.

Analysis by Rutherford back scattering (RBS)

The spectra of light elements do not extend to the X-ray region and therefore XRF and PIXE cannot be used. In these elements all electrons are outer electrons and are influenced to an extent by the chemical environment. These light atoms are most suited to analysis by Rutherford back scattering which was discussed in principle in chapter 3. An incident proton or light ion of energy E is observed scattered in the backward hemisphere with energy E'. The difference between E and E' is sensitive to the ratio M/μ, where M is the target nuclear mass and μ is the mass of the incident ion. By the application of simple non-relativistic kinematics it may be shown that at $180°$,

$$E' = E\left(\frac{M - \mu}{M + \mu}\right)^2, \tag{6.4}$$

see question 6.6.

With a suitable detector the SPM may be used for RBS measurements. By changing the gas in the ion source at the accelerator terminal, beams of helium, carbon or oxygen nuclei in place of protons can be made. This decreases the mass ratio M/μ and extends the use of the method to target nuclei of intermediate mass. Experimentally the detector for the back-scattered ion cannot be at $180°$. However, as Fig. 3.10b shows, the Rutherford angular distribution is relatively independent of angle for angles greater than $120°$, so that a detector anywhere in the backward hemisphere subtending a reasonable solid angle will serve.

The dE/dx energy loss of the incident ion, both as it enters the surface layers of the sample and as it leaves, plays a major role. Consider the three plots in Fig. 6.11 of the spectra from a sample composed of two nuclei, one light and one heavy. In a) the sample is very thin so that there is no effect of dE/dx. Scattering by the heavier nucleus causes little change in the incident energy E, while scattering by the light nucleus reduces the energy to E'. In b) the sample is thicker and scattering by the nuclei at the surface has the same result as in a). However, back-scattering from nuclei in progressively deeper layers shows extra energy loss due to dE/dx. Relative to those at the surface the ions scattered from the back of the slice have:

a)
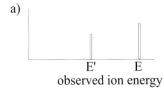
observed ion energy

b)
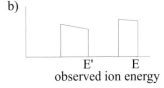
observed ion energy

c)
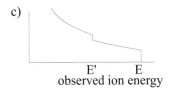
observed ion energy

Fig. 6.11 a) An ideal RBS spectrum from an infinitesimally thin target film composed of two nuclear species, a heavy nucleus giving a back-scattered energy E and a light nucleus giving energy E'. b) The same, but with a thicker target slice. c) The same, but with a sample slice so thick that some back-scattered ions stop.

◇ loss of ion energy penetrating to the back;

◇ smaller recoil loss but with larger probability when they get to the back, since the Rutherford cross section increases as the incident energy falls;

◇ extra loss of ion energy on the return to the surface, since dE/dx increases as $1/v^2$ as v the ion velocity falls.

These effects are responsible for the energy-dependent profiles in b). The contributions of different nuclei can ride up on one another. In c) the sample is so thick, that some ions stop before they get back to the surface.[12]

Because of the sharp depth dependence and fine focus, density distributions can be plotted in depth in thin layers with good resolution in all three dimensions.

6.2.2 Isotope concentration analysis

The previous section showed how the concentration of trace elements may be imaged or determined. The composition, the relative concentration of different elements, is a clear indicator of origin. Archaeological artefacts can often be linked to the actual mine or quarry of origin, sometimes very far away. However, the relative abundance of different elements is subject to change with every chemical or physical process. Elements and their compounds are separated from one another, not only by chemical change, but also by crystallisation and dissolution, by evaporation and condensation, by freezing, melting and boiling. Consequently the memory associated with the relative abundance of different elements is rather volatile.

The relative abundance of different isotopes of the same element has a much longer memory. These ratios can be measured precisely, and their variations in space and time form a record of valuable information. The relative isotopic abundance is only changed by nuclear reactions or by rather small mass-dependent effects that affect isotopes slightly differently. Nuclear decay reduces the concentration of an isotope, and irradiation and nuclear reactions may increase it. The isotope concentration is a faithful record of the combined action of these effects, and in this section we give a few examples of such records, how they may be determined and how they have been interpreted.

Separation of isotopes by mass dependence

Heavier isotopes have smaller random velocities than lighter ones in thermal equilibrium in gas or liquid phases, and lower vibrational frequencies in solids or in molecules. These differences give rise to slightly unequal chemical reaction rates, diffusion rates and concentration gradients in a gravitational field. The effects are largest in light elements where the fractional difference in mass between isotopes is largest.

The most pronounced differences between isotopes are their nuclear properties. Otherwise isotopes are analysed in the laboratory using con-

---

[12]In Fig. 6.11 we have neglected the effect of fluctuations in dE/dx, the straggling, which round the sharper features in the spectra.

ventional mass spectrometers, or special purpose ones for very small concentrations. Separation of isotopes in large quantities is based on gaseous diffusion, centrifuges or lasers that exploit the effect of mass dependence.[13]

[13]See section 6.5.2.

Distribution of stable isotopes

In the natural world stable isotopes may change their relative concentration as a result of phase changes. In the process of crystal growth, for example, the mass difference gives rise to a very slightly different rate of accretion. Biological activity can show a similar slight preference.

An important example is the concentration of oxygen isotopes in the oceans. Oxygen has three stable isotopes of mass 16, 17 and 18. The majority of atoms are ^{16}O. The concentration of ^{18}O is about 0.2%. A notable variation in this ratio is said to be due to the preferential freezing of ^{16}O, leaving a slightly higher concentration of ^{18}O in the liquid phase.[14]

[14]We note that this effect is reported to have the opposite sign to that which might have been expected.

During glacial epochs when a significant fraction of the water on Earth becomes locked up in ice sheets the concentration of ^{18}O in the ocean rises. Shell fish and new ice formed at this time then have a higher concentration as shown by analysis of marine deposits and ice core samples. The difference in concentration between glacial and interglacial epochs is about 2 parts per million.[15] The timing of these climatic changes has been compared with changes in the Earth's magnetic field, the various clocks associated with radioactive decay, pollen counts, and longterm changes in the Earth's solar orbit and the tilt of its axis. These all contribute data to the discussion of the climatic history of our planet.

[15]A change of 0.2 parts per million is said to be just measurable and equivalent to a change in world-wide mean sea level of 10m.

Neutron activation analysis

Isotope ratios may be determined by using their different nuclear properties. If a sample is placed in a nuclear reactor isotopes behave differently according to their neutron absorption cross sections. Typically one isotope may absorb a neutron with a large cross section in a reaction

$$A(N, Z) + n \rightarrow A(N + 1, Z) + \gamma, \tag{6.5}$$

while other isotopes do not. Such differences tend to be large, since neutron affinity is influenced by a preference for paired nucleons. If the sample is now removed from the reactor the product nucleus $A(N+1, Z)$ is typically unstable, with a characteristic and decay mode. Observation of the decay and its rate can be used to determine the concentration of that isotope. The measurement is most reliably calibrated by repeating the same procedure with a standard, a known quantity of the isotope in a similar physical form.

The scanning proton microprobe can also be used to distinguish isotopes by proton-induced reactions. There are many variations involving nuclear activation analysis which are described in some detail in the book by Alfassi.

Isotope detection by observation of decay

Isotope ratios may be determined by the detection of the radioactive decay of one particular isotope. For a sufficient counting rate this requires a certain minimum mass of the isotope that is proportional to the half-life. The errors of such measurements are usually dominated by the statistics of the observed decays. With the exception of ^{14}C most naturally occurring isotopes have far too long a half-life for this to be a realistic method.[16]

[16]See question 6.2.

Carbon has two stable isotopes of mass 12 and 13, and one of mass 14 with a mean life, $\tau = 8267 \pm 60$ years. The value of this is fortunate as it gives good overlap with typical timespans of archaeological interest. The presence of ^{14}C in the environment is entirely due to neutrons created in the upper atmosphere by cosmic rays. These react with nitrogen,

$$n + {}^{14}N \rightarrow p + {}^{14}C. \tag{6.6}$$

This ^{14}C then behaves almost indistinguishably from the other carbon isotopes. In particular it is present at a concentration, ρ_0, of 10^{-12} relative to ^{12}C in all living matter. As soon as matter dies and stops exchanging carbon with the environment, this concentration starts to fall due to the decay[17]

[17]The decay energy is only 160 keV. This is sufficiently small that, for it to be detected with reasonable efficiency, the carbon must be inside the detector. This may be done by filling a proportional counter with the carbon in the form of CO_2 or CH_4.

$$^{14}C \rightarrow {}^{14}N + e^- + \overline{\nu}_e. \tag{6.7}$$

At some time t later, the decay rate is measured, the current concentration ρ determined, and the value of t deduced from

$$\rho(t) = \rho_0 \exp(-t/\tau). \tag{6.8}$$

The method assumes that the production rate of ^{14}C has been constant in the past. This has been checked by measuring the ^{14}C concentration in tree rings whose sequence acts as a digital clock, phase locked to the seasonal year. The neutron flux is found to fluctuate slightly due to the solar cycle and to have been seriously affected by the era of nuclear weapon testing. Otherwise the production rate has been rather constant until recently.[18]

[18]The release of extra old carbon into the atmosphere by industrial activity, the source of the global warming problem, has tended to lower the ^{14}C concentration. The period of weapon testing has raised the production rate of ^{14}C in recent years, more than compensating for the release of old carbon. This makes dating easiest for modern materials. Indeed the successful identification of fake bottles of 'vintage' whisky or Armagnac has caught a few scoundrels unawares!

[19]Further details are described in the book by Aitken.

Carbon dating by accelerator mass spectrometry (AMS)

The large statistical error associated with the small number of ^{14}C nuclei that decay in the counting time T can be reduced by a factor of order $\sqrt{T/\tau}$. An accelerator mass spectrometer can detect all the ^{14}C atoms present, not just those that decay during the measurement. A by-product is that the ratio $^{13}C/^{12}C$ can also be measured. This is independent of date but forms a record of the fractionation effect, the extent to which the sample has been enriched or depleted of heavier isotopes by processes other than decay. The ^{14}C ratio may then be corrected for this.[19] Here we discuss the physics behind the accelerator mass spectrometer (AMS).

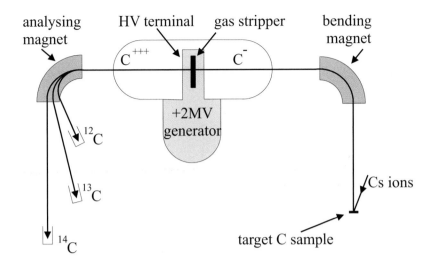

analysing magnet | HV terminal | gas stripper | bending magnet

C^{+++} C^{-}

+2MV generator

^{12}C

^{13}C

^{14}C

Cs ions

target C sample

Fig. 6.12 The layout of an accelerator specifically designed to measure ^{14}C.

The mass spectrometer shown in Fig. 6.12 has particular features designed to enhance its sensitivity to ^{14}C. Only a very small quantity of carbon is required, and the measurement for this is compared against several samples of known age on a carousel.[20]

A conventional mass spectrometer would suffer from backgrounds of ^{14}N and chemical fragments like $^{12}CH_2$ that have nearly the same mass as ^{14}C. This spectrometer reduces these backgrounds in two steps. Firstly it uses a negative ion C^{-} in the initial acceleration stage. N does not form a stable negative ion. At the positive terminal four electrons are stripped off (on average) and the C^{3+} is accelerated back down to ground potential.[21] The second stage of background reduction occurs because the chemical binding of a molecular ion such as CH_2^{-} could not survive such a stripping process intact. For each carbon isotope the relation between the kinetic energy given by the charge and potential, and the momentum as it passes through the magnets, depends on the mass. The momentum and the B-field determine the angle of the trajectory, so that the three isotopes follow separate trajectories and are counted in separate counters. Further, the energy deposited in the counter is well-defined for a carbon atom, but will be different for a background atom that reaches the ^{14}C counter spuriously.

These checks permit errors to be reduced such that samples as old as 50,000 years may be dated provided that they have not been contaminated. It is left to question 6.2 to calculate what this implies in terms of statistics and contamination tolerance.

Radiocarbon dating has become a mainstay of modern archaeology. This acceptance has been achieved by a particularly careful application of physics, including calibration techniques and consideration of statistical and systematic errors. Public relations have been well handled and the technique has added to the popular perception of science.

Amongst well-known results the date of the Turin Shroud was determined to be within AD1275–1290 at 68% confidence level, and within

[20]Such known reference samples are called standards.

[21]A stripper is a region of low density gas which knocks off electrons. Roughly speaking, all loosely bound electrons whose Fermi velocity is smaller than the ion velocity will be removed.

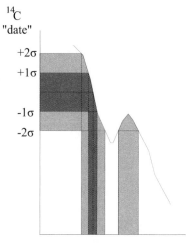

¹⁴C
"date"

+2σ
+1σ

-1σ
-2σ

historical date

Fig. 6.13 An illustration of how a normally distributed ^{14}C date is mapped by a calibration curve based on dendochronology into probability bands on an axis of historical date.

AD1260–1310 or AD1355–1385 at 95% confidence level. The two possible ranges at 95% confidence level arise from an inflection in the calibration curve that maps a single carbon measurement with Gaussian errors into a probability distribution in date with more than one peak. This is illustrated qualitatively in Fig. 6.13. Anyway these dates are consistent with an origin not too long before the Shroud was brought to the West by the Crusaders.

Other results of interest are frequently reported. In 1991 a perfectly preserved frozen body was discovered in the Alps, complete with clothes, tools, food and belongings. Accelerator mass spectrometry of ^{14}C was used to measure the age of this exciting find at 5300 years.

Dating by measurement of daughter concentration

There is another way in which the radioactive decay of an isotope may be used as a clock. The integrated number of decays that have occurred in time t may be inferred from a measurement of the ratio of decay product to parent element concentrations. If the decay rate is known, this gives an estimate of the period t over which the daughter nucleus has been accumulating.

There are two questions: 'What defines when the clock starts?', and 'Are there any circumstances in which the daughter nuclei have left the sample, thereby reducing the apparent time and slowing the clock?' The answers to these questions are related. The clock starts when the sample is formed with no daughter nuclei present. The clock is slowed if daughter nuclei diffuse away or are leached out of the rock sample. In some cases the daughter nuclei may themselves decay, but that is a known process whose effects can be corrected precisely, statistics permitting.

A good example is the decay of ^{40}K to ^{40}Ar for which the branching ratio is 10.5%. ^{40}K is widespread in nature and forms a fairly constant 0.0118% of natural potassium, with a lifetime of 1.2×10^9 years. This is sufficiently long that the rate of production of argon is effectively constant. The clock is reset whenever the argon is released. This occurs whenever rock is formed, whether deposited in water as a sediment or cooled from a molten lava flow. In theory, once the rock is deposited or cooled from the liquid state, the argon is trapped. In practice allowance has to be made for both a pre-existing concentration and losses of gas by diffusion, especially if the rock has been heated significantly since it was formed.[22]

An effective experimental procedure is the argon-argon technique. The rock sample is placed in a nuclear reactor to convert a fraction of the normal ^{39}K to ^{39}Ar. The sample is then heated and the ratio ^{40}Ar/^{39}Ar measured with time as the sample outgases. This method compensates for the speed of outgassing, gives a signal relative to the K content and indicates samples which have lost Ar from surface layers in the time since formation. Such samples give a 'young' reading as such surface layers outgas, but can give an asymptotically correct age as gas from deeper in the sample is released. Alternatively the trapped argon

[22]See the note in section 36 for further discussion of this point.

may be released from a single grain of a milligram or less with a laser. Errors vary with the K content. Age estimates with errors of $\pm 10^3$ years for rocks with high K content have been confirmed by radiocarbon dating for ages up to 3×10^4 years. The dates of early hominid remains have been determined at 1–2 million years with errors of 1–2% by this method.

Uranium series dating

Errors in dating based on parent and daughter isotopic concentrations arise when concentrations are affected either by other processes or when the initial concentration is uncertain. The absence of a significant initial argon concentration in rock is a good example where uncertainty is reduced small. Another is based on the low solubility of thorium salts compared with those of uranium.

As shown in table 6.6 the naturally occurring and long lived isotope ^{238}U gives rise on decay to ^{234}U with a half-life of 2.45×10^5 years. This in turn decays to ^{230}Th with a half-life of 7.7×10^4 years. It follows that, starting with rock deposited from solution at time $t = 0$, the concentration ratio of ^{230}Th/^{234}U rises from zero as the daughter concentration builds up. Thus the calcite of a stalagmite, for instance, has a ratio that is a direct measure of its age. The ratio eventually saturates after a number of lifetimes of ^{230}Th as dynamic equilibrium is established.

The experimental method involves the chemical separation of U and Th, followed by the identification of the two isotopes of interest by their characteristic α-decay energies.[23] Statistical errors may be reduced by the use of mass spectrometry of the isotopes instead, and the age range extended back to 5×10^5 years.

There are other methods too. The reader is referred to the book by Aitken.

[23]This is actually not too easy as the energies are close, 4.8 MeV and 4.7 MeV respectively.

6.2.3 Radiation damage analysis

In solids the record of past radiation may be recorded in the form of the frozen atomic, molecular and crystal-structure damage. Such a record is 'wiped' by processes such as heating, melting or recrystallisation. It may be recorded in the form of visible lattice damage, electronic excited states or paramagnetic electrons, with some overlap between these.

Fission track dating

The decay of the naturally occurring ^{238}U usually proceeds by α-decay to ^{234}Th. However, in a fraction 0.5×10^{-6} the decay process is spontaneous fission. The fission products are highly charged with a low velocity. Their dE/dx, or LET, is exceptionally high, and they stop in about 10 μm. The locally disrupted crystal structure may be dissolved preferentially by etching a polished face, making the track location visible under a microscope. Because the half-life is so long, a count of the density of these pits relative to the ^{238}U concentration is linearly dependent on the time since the crystal was last annealed. The count is normalised by observing the density of pits found on an identical sample which has been annealed and irradiated with neutrons in a reactor. This process reveals the location of induced fission by ^{238}U. Using the ratio of densities takes care of the parent uranium concentration and details of the counting procedure used.

Only certain minerals are suitable and the counting procedure is tedious. The technique is not practicable at densities below 100 cm^{-2}.

Dating by thermoluminescence

Thermoluminescence depends on the density of excited electrons frozen into a solid. These metastable excited electronic states are created by the general flux of radiation incident on a sample, which includes the background γ-rays from neighbouring rock and the external cosmic radiation, as well as local α- and β-activity. In a liquid or gas these do not persist, but in certain solids the relaxation time is very long, though a rapidly decreasing function of temperature. The record may be 'bleached' by light in surface layers as well as by heat. Below the surface the density of trapped excited electrons may accumulate. If such samples are exposed and heated, the light emitted by de-excitation is a measure, either of the age of the sample if it has been held in a constant radiation environment, or of the integrated radiation flux if the sample is of known age.

As the temperature of the sample is raised very slowly, light is emitted by progressively more deeply bound electrons. The second stage of the measurement requires the sample to be irradiated with a known dose, and then heated again slowly as in the original measurement. The ratio of thermoluminescent fluxes emitted indicates the ratio of the historical dose to the calibration dose. If the annual dose can be estimated then the age may be deduced. Spurious effects may be eliminated by observing the ratio as a function of the temperature. The details are described by Aitken.

The technique is particularly important for dating pottery and kilns. The $t = 0$ event is the date when these were last fired and cooled.[24] The dominant error tends to come from the uncertainty in the annual dose experienced by the sample.

These radiation-induced electronic states also involve unpaired electrons which give rise to paramagnetic properties. A measurement of the amplitude of the electron spin resonance (ESR) is another technique that may be used to indicate radiation exposure or date.[25]

[24]This firing and then cooling gives rise to a further effect that is also used in dating. The thermal cycling takes magnetic grains in the clay above and then below the Curie temperature, thereby freezing information on the Earth's B-field strength and direction at precisely the same time. This has been used to deduce information about dates and about the behaviour of the Earth's field. Data from lava flows as well as kilns and ceramics are discussed in the book by Aitken.

[25]ESR is sensitive to the integral of radiation exposure. So if the date is known, the radiation exposure can be deduced, as in Hiroshima studies and tests of food irradiation. Conversely if the rate of irradiation is known and steady, a date may be determined.

6.3   Radiation exposure of the population at large

Radioactivity may be measured as the rate at which nuclei decay. Radiation exposure is measured as the absorption of energy per unit mass. The response of living tissue to a radiation dose is essentially non-linear. Doses that are local and of short duration are more damaging than the same dose less densely distributed either in space or time. This may be simply understood in terms of healing processes. These are essential to the successful use of radiotherapy. The linear no-threshold model (LNT) naively assumes that a given radiation dose spread over a large population causes the same damage as when concentrated on a few individuals; this model is not sustainable, though frequently quoted. Such quotations have generated significant public misunderstanding. We give the radiation exposure of the UK population and discuss its various sources, namely radon gas, gammas from radioactive rocks, internal sources, secondary cosmic rays, medical doses and other human-made contributions.

6.3.1   Measurement of human radiation exposure

SI units

To discuss radiation in the environment we need first a basis of objective measurements. In SI there are units for three different quantities that may be measured in radiation health physics.

◇ The activity of a radioactive source is the number of disintegrations per second. One disintegration per second is 1 becquerel (Bq). This is not the number detected or counted, but refers to the whole source. A earlier unit, the curie, was equal to 3.7×10^{10} Bq.

◇ The unit of absorbed dose or deposited energy is the gray (Gy). A radiation dose of 1 Gy is equal to 1 joule per kg. A earlier unit, the rad, was equal to 0.01 Gy.

◇ The unit of equivalent dose for biological damage is the sievert (Sv).[26] For X-rays, γ-rays and electrons the equivalent dose in sieverts is defined to be equal to the absorbed dose in grays. For other types of radiation the equivalent dose in sieverts is greater than the dose in grays by a factor, w_R, the radiation weighting factor. Some values of this factor are given in table 6.4.

Biological equivalent dose

The operational definition of the sievert is that 1 Sv should represent the same hazard to tissue whatever the form of the ionising radiation. What determines the factors in table 6.4? First we shall justify, qualitatively at least, the use of these values. Later we shall come to see that actual biological damage has a more complex relation with absorbed dose which is not describable by a single factor.

[26] An older unit is the rem where 100 rem equals 1 Sv.

Table 6.4 Some values of the radiation weighting factor w_R for different kinds of radiation.

Radiation	Energy	w_R
X-rays and γ-rays	any	1
Electrons	any	1
Muons and muons	any	1
Protons	>2 MeV	5
Alpha/light ions	any	20
Fission fragments	any	20
Neutrons	<10 keV	5
Neutrons	10–100 keV	10
Neutrons	0.2–2 MeV	20
Neutrons	2–20 MeV	10
Neutrons	>20 MeV	5

The response of biological tissue to radiation damage is non-linear and the reason for this may be understood as follows. Compare the following situations in which two cells are exposed to a certain total absorbed radiation dose.

⋄ One cell receives all the energy and the other none, with the result that one cell dies and the other survives.

⋄ Both cells receive half the dose each, with the result that each has a fair chance of effecting a repair. As a result both have a better than even chance of survival.

With a non-linear response due to repair mechanisms and some biological redundancy, the question of the uniformity of a delivered dose is critical.

Now we examine the entries in the table. The energy loss associated with photons and $\beta = 1$ charged particles is thinly spread along the paths of the individual charges or quanta. This maximises the opportunity for biological repair.

However, a non-relativistic charge, whose rate of energy deposition along its path is enhanced by a factor $1/\beta^2$, delivers energy much more locally. Relative to irradiation by electrons, material irradiated with 1 Gy by such charges is traversed by a smaller number of more densely ionising tracks. Those cells that lie near a track get a large dose, from which recovery is unlikely, while those that are far removed are unaffected. Due to the non-linear effect of repair this radiation will do more damage per gray than electrons or γ-radiation. In brief the damage depends on the LET as well as the specific energy dose.[27]

Why do neutrons carry a heavy penalty per gray? A neutron, being a neutral particle, does not deposit energy at all until it has a collision. It does not collide with electrons. It can only undergo nuclear collisions. Products of such collisions typically consist of slowly moving nuclei or non-relativistic recoil protons. These have a high LET and a high w_R. Very low energy neutrons are moving too slowly to create recoils that ionise.[28] They have a low w_R. More energetic neutrons generate recoil protons of similar energy and therefore have similar w_R.

Timing of radiation dose

Life on Earth has developed under conditions in which it has been continuously exposed to ionising radiation from a number of different sources. The average exposure of the population in the UK is 2.7 mSv per year. There are significant fluctuations on this figure depending on location.[29] At the other end of the scale it has been found that a radiation dose to the whole body of the order of 5 Sv is fatal to life in 50% of cases.

Life has evolved and thrived in the gap between these two extremes. However, there is an inconsistency here, for the dimensions of the background rate and the fatal whole-body dose are not commensurable. We need a timescale. In other words, we have to answer the question 'Over what time should we integrate the radiation dose rate in order to compare it with the fatal dose?' Arguably, understanding the importance of

[27] This is particularly evident for multiple charged ions such as alpha particles or the highly ionised nuclear fragments emitted in the nuclear fission process. They get a very heavy rating in table 6.4 because they have a high LET.

[28] This will occur for neutrons with less than atomic electron velocity, that is with energy less than about 10 eV $\times m_N/m_e$. This is about 10 keV, in broad agreement with the table entries. Neutrons with less energy deposit it in the form of lattice vibrations or phonons which represent heat and cause no damage.

[29] However, there is rather good information that overall it has not changed much in the last 50,000 years. Looking back 10^9 years the radiation environment due to radioactivity in the Earth would have been more intense by a small factor. Since that time radiation levels have been declining slowly with a combination of the known exponential decay factors of the long-lived parent nuclei involved.

this question and the question of non-linearity discussed earlier are the two most important questions for radiation health physics.

The brief answer is that the timescale that is important is the cell replacement time or the repair time. This determines over what period the human body remembers radiation exposure. Over a longer time repair and replacement can make good the damage of irradiation. Over a shorter time damage is integrated. From a century of experience with radiotherapy and studies of cell recovery in vitro this time is known to be on the scale of days. It varies for different parts of the body and different organs, and also varies with age. To be very conservative we take this time to be a month. That is we assume that a radiation dose spread over a period of a month causes the same damage as if the dose is delivered in a single episode.[30]

[30] Clinical experience with radiotherapy treatment planning suggests that this may be an order of magnitude too long. See chapter 8.

Collective dose and the linear no-threshold (LNT) model

In assessing the radiation hazard to a whole population it is a natural temptation to multiply the radiation dose by the size of the population so exposed. This is what is known as the collective dose. Thus a million people receiving an additional dose of 1 mSv makes an extra collective dose of 1000 man Sv. The mistake is to suppose that the biological damage, for instance the number of additional cancers, is linearly proportional to this collective dose with no constant term. This supposition is called the linear no-threshold model (LNT). However, it is quite simple it is wrong. Biological systems do not respond linearly in this way.[31]

Laceration is another kind of biologically disruptive damage. If I cut my finger badly, in the absence of any coagulation, blood would be lost to the point that death would follow. This does not happen because the blood does coagulate and seal the wound, and in due course the loss is replaced. For this reason, although a major amputation may cause loss of life, a large number of people losing even a moderate amount of blood is not life-threatening. It makes no sense to represent the danger as proportional to the number of people multiplied by the loss of blood that each suffers. Otherwise every blood-donor clinic would be responsible for a large number of deaths every day.

The same argument applies to the effect of a moderate dose of ionising radiation on a large population. A whole body dose of 5 Sv delivered to each of 100 people can be expected to result in about 50 mortalities. Because of the non-linearity an extra dose of 0.5 mSv delivered to a million people results in no fatalities, although the collective dose is 500 man Sv in both cases. Exposure of an individual within a factor ten of a fatal level causes some permanent damage and results in scar tissue, in a radiation exposure just as it does in laceration. However, much larger margins of safety, the equivalent of avoiding pin pricks, make no sense in the real world. Life has evolved to recover completely from pin pricks and the effects of modest levels of radiation.

The effect of radiation doses on very large populations have been studied to check whether there is any additional response to radiation damage

[31] For any threat to which life has evolved a protection or repair mechanism the normal response is an S-shaped curve, as shown in Fig. 6.18. Such non-linearity provides complete, not proportional, protection provided that the repair mechanism is not overloaded, for example in space and time - hence, in the case of radiation, the dependence on LET and the exposure period. We should expect such non-linearity for any threat to which life is exposed in the natural environment and has therefore had the opportunity to evolve some protection mechanism. This includes ionising radiation.

Table 6.5 Some high statistics studies of the effects of radiation.

Location	Pop.	Ref.
Hiroshima and Nagasaki	120,000	p. 196
Urals accident	30,000	p. 191
Radon		
in China	80,640	§§
in Europe	various	p. 191
Chernobyl	various	p. 199

§§www.hps.org/publicinformation/ate/q1254.html

that is linear. Table 6.5 lists some of these. In a group of a million people there are about 2000 cancers that occur each year, predominantly from ill-advised life styles, but also naturally. Thus the statistical error on the number of cancer cases in a sample of this size is about 50. There are systematic effects as well. The panic, social and economic upset that followed Chernobyl caused fatalities through family breakdown, drink, smoking and despair. These were not consequences of the physical effects of radiation and contribute large systematic uncertainties. The various reports on Chernobyl, and on Hiroshima and Nagasaki, may be found on the NEA and RERF-LSS websites referenced at the end of chapter 1. A most readable account is given in the book by Henriksen and Maillie. The conclusion is that there is no evidence to support the LNT model. Use of the LNT model and collective dose is easy but dangerous. It has been the cause of the exaggeration of the threat posed by ionising radiation.

6.3.2   Sources of general radiation exposure

Data for the UK population

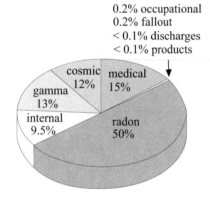

0.2% occupational
0.2% fallout
< 0.1% discharges
< 0.1% products

Fig. 6.14 The composition of the average radiation exposure of the UK population (2.7 mSv). Of this 84% is natural and 16% is artificial. [Data from HPA-RPD-001(2005).]

The average annual radiation dose to the UK population is 2.7 mSv. Fig. 6.14 shows the various sources of this radiation and how they make up this average. However, some sectors of the population receive more than others.

The greatest source of radiation is from natural radioactivity in the rocks of which the Earth is composed. Most materials, unless they have been very highly refined, also contain these radioactive nuclei in measurable quantities. All of the alpha radiation, most of the beta radiation and some of the gamma radiation is absorbed within the rock or other material and does not escape. The only products of these radioactive sources that contribute to the human dose are radon gas and gamma rays. Together these make up 63% of the average dose, but there are geographical locations where the exposure is five and more times greater than the average.

Background radioactivity is also present within our own bodies. In this case the dose includes the alpha radiation with its high weight (shown in table 6.4). The internal source of exposure represents some 9.5% of average dose. This is the only source for which little than can be done to reduce the level. Cosmic radiation contributes 12% of the average dose. This derives from the radiation that shines on the Earth from outer space. However, its effects are by no means uniform. The largest source of radiation over which we have substantial control is its use in medicine where it forms 15% of the average. In the following we study each of these sources and also those that contribute to the remaining 1% of the average.

Table 6.6 The two most important radioactive series with the half-lives of their components. In some cases there are alternative decay routes that have been omitted.

^{238}U			^{232}Th		
^{238}U	α	4.5×10^9 years	^{232}Th	α	14.1×10^9 years
^{234}Th	β	24.1 day	^{228}Ra	β	5.75 years
^{234}Pr	β	1.17 min	^{228}Ac	β	6.13 h
^{234}U	α	2.45×10^5 years	^{228}Th	α	1.91 years
^{230}Th	α	7.7×10^4 years	^{224}Ra	α	3.66 day
^{226}Ra	α	1.6×10^3 years	^{220}Rn	α	55.6 s
^{222}Rn	α	3.82 day	^{216}Po	α	0.15 s
^{218}Po	α	3.05 min	^{212}Pb	β	10.64 h
^{214}Pb	β	26.8 min	^{212}Bi	β	60.6 min
^{214}Bi	β	19.8 min	^{212}Po	α	0.3 μs
^{214}Po	α	164 μs	^{208}Pb		Metastable
^{210}Pb	β	22.3 years			
^{210}Bi	β	5.01 day			
^{210}Po	α	138.4 day			
^{206}Pb		Metastable			

Radon and gammas from rocks

In nature there are four distinct non-intersecting radioactive decay chains, each starting from a heavy nucleus of lifetime comparable to the age of the Universe, and ending with one that is metastable with a lifetime long compared with the age of the Universe.[32] At the heads of these four radioactive series or decay chains are the nuclei ^{232}Th, ^{238}U, ^{235}U and ^{237}Np with half-lives of 14.05, 4.47, 0.70 and 0.002×10^9 years respectively. The elements that we see today were synthesised about 6×10^9 years ago and the ^{237}Np series has already decayed to completion with ^{209}Bi. Little remains of the ^{235}U series which has largely decayed to ^{207}Pb.[33] However, the other two decay chains continue to contribute to natural radioactivity. Details are given in table 6.6.

The other contributor to primordial natural radioactivity is ^{40}K with a half-life of 1.25×10^9 years. This decays either to ^{40}Ca (89.5%) or to ^{40}Ar (10.5%). The energy level diagram Fig. 6.15 shows that there are four decay modes in all. All of these are very heavily suppressed indeed.[34] Potassium is ubiquitous, contributing 2.4% by weight of all elements, but only a fraction 1.18×10^{-4} is ^{40}K.

Uranium and thorium are not as common as ^{40}K, but they are present in most unrefined materials at a low level and are concentrated in particular rocks. These release radon and emit gamma radiation, radon because it is a gas and may be inhaled and gamma radiation because it

[32]All nuclei with A greater than about 100 are unstable but their α lifetimes may be very many times the age of the Universe, so that their decay is never observed.

[33]Although it contributes little to natural radioactivity, the remaining ^{235}U is crucial to the feasibility of nuclear power.

[34]The reason for this suppression is the subject of question 6.4.

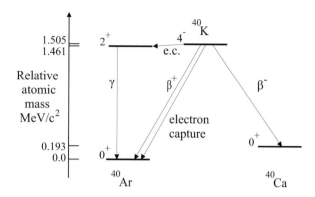

Fig. 6.15 The states with their energies involved in the decay of ^{40}K.

has a long range.

Radon is the heaviest of the noble gases. Two isotopes are produced in the radioactive decay chains, but the important one is ^{222}Rn with a 3.82 day half-life.[35] This is long enough for it to diffuse out of the rock where it is produced, at least from sites with depth less than 100μm.[36] Its geographical distribution is very uneven. More than two thirds of homes in England where the indoor radon radiation levels exceed 200 Bq m^{-3} are in Devon and Cornwall. On average this has the effect that the annual radiation dose for the population of Cornwall is 7.5 mSv rather than the average for the whole country of 2.7 mSv.

It is not difficult to minimise the exposure to radon. Because of its very high atomic weight of 222, it is some 15 times denser than air and accumulates on the floor.[37] As a result the concentration at head height is much less. The half-life of radon is only 3.82 days and so it tends to establish an equilibrium concentration on such a timescale. The radiation may be reduced by ventilating buildings, especially cellars, on a daily basis.

Internal sources

Potassium is present in the body as the main base ion of the intracellular fluid and therefore essential to the body's electrochemistry. The ^{40}K component is the main internal source of radiation, 70 Bq kg^{-1}. The corresponding radiation dose of 0.18 mSv per year is not subject to much variation. Other internal sources, predominantly ^{210}Pb and ^{14}C, add to make a total of 0.25 mSv per year.

Secondary cosmic rays

The Earth is illuminated by a wide variety of radiation from the Sun and elsewhere in space.

From the Sun we receive black body electromagnetic radiation which peaks in the optical region of the spectrum but extends down into the infrared and upwards into the ultraviolet. Both of these are largely absorbed in the atmosphere, although in the latter case this depends on the coverage of the ozone layer. Also emitted by the Sun are streams of

[35]The other one is sometimes called thoron but has a half-life of only 55.6s.

[36]These statements seem at variance with the observation that ^{40}Ar produced in the decay of ^{40}K does not diffuse out of the rock over periods of 10^6 years, as quoted in section 22. The ratio, a factor 10^8, is related to the recoil energy, see question 6.7.

[37]In the same way that nitrogen from a liquid nitrogen dewar falls to the floor, although that is only four times denser than air at ambient temperature.

charged particles. These are associated with sunspots, periodic regions of intense magnetic storms which appear sporadically peaking with an 11-year cycle. These charged particles are bent by the $\boldsymbol{v} \times \boldsymbol{B}$ force arising from the Earth's B-field. Those travelling most nearly parallel to $\boldsymbol{B}$ experience the least deflection and have the greatest chance of reaching down to the top of the atmosphere. This causes irradiation high in the atmosphere in the magnetic polar regions, particularly at the time of sunspot activity. The spectacular displays of the aurora borealis and aurora australis are the visible consequences. The increased ionisation of the upper atmosphere raises the plasma frequency and thereby has a major effect on the reflection and transmission of radio signals. The flux magnetically trapped in the Van Allen radiation belts is also increased, which is a hazard for Earth satellites and astronauts. However, none of this ionising radiation from the Sun reaches the ground.

Primary cosmic rays from outside the Solar System are composed mostly of protons with 13% alpha particles.[38] These have an energy spectrum falling as $E^{-2.5}$ extending up to the very highest energy. The flux is deflected by the Earth's B-field, depending on the energy and latitude. These cosmic rays interact in the atmosphere at a height of about 20 km, creating many particle species. The result is a 'shower' from a cascade of interactions. This shower peters out in the upper atmosphere. Its main products are more protons, charged π mesons and neutrons. The π mesons decay with a proper lifetime of 26 ns,

$$\pi^{\pm} \rightarrow \mu^{\pm} + \nu. \qquad (6.9)$$

The muons and the various unstable nuclei[39] produced are responsible for the environmental effects of cosmic radiation. Some of these muons, mean proper lifetime 2 μs, live long enough to reach the ground when relativistic time dilation is taken into account. These muons constitute the secondary cosmic rays detected at sea level. Their flux increases with magnetic latitude and also markedly with altitude. There is an increase of a factor of 2 in the cosmic radiation dose for each 2000 m increase in altitude. So at 12000 m at mid latitudes the dose is 0.004 mSv per hour or 35 mSv per year. Doses over the poles are correspondingly higher.[40]

Medical doses

There are three types of medical procedure involving ionising radiation.

⋄ External X-ray diagnostic examinations, both CT and conventional, and including dental.

⋄ Nuclear medicine (SPECT and PET), which involves the administration of a radioactive isotope to a patient, usually by injection (such diagnostic procedures are discussed in chapter 8).

⋄ Radiotherapy, although this contribution to the national radiation exposure may be seen in a different light and is not included in the dose figures discussed above (the exposure

---

[38] Necessarily there is also an equal flux of electrons to balance charge. In addition there is a small but interesting flux of high energy photons.

[39] For instance ^{14}C.

[40] A single flight from the USA to Europe yields the same extra dose as 10 dinners with meat containing ^{137}Cs at a level of 1000 Bq per kg. After Chernobyl the maximum level for food was set at 600 Bq per kg in most European countries. Such regulation is out of proportion.

a) Average X-ray dose per
   examination, mSv

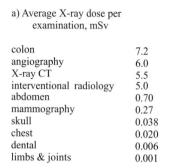

colon	7.2
angiography	6.0
X-ray CT	5.5
interventional radiology	5.0
abdomen	0.70
mammography	0.27
skull	0.038
chest	0.020
dental	0.006
limbs & joints	0.001

b) Proportion of total X-ray examinations

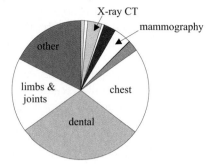

c) Proportion of dose, 0.33 mSv per person per year

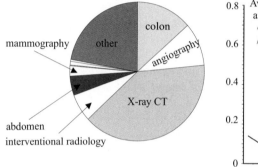

d) Average X-ray dose by age

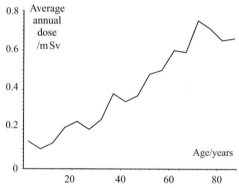

Fig. 6.16 The composition of the X-ray diagnostic radiology dose to the UK population. a) Dose per examination by procedure. b) Fraction of all examinations. c) Fraction of total dose (ordered and classified as in b)). d) Average dose by age (indicative). [Data from Hart and Wall, NRPB-W4 (2002).]

of staff during radiotherapy procedures is properly included within occupation-dependent doses).

The diagnostic X-ray dose accounts for an average of 0.33 mSv per head of the population, and Fig. 6.16 shows how this is made up. In addition there is a mean dose of 0.024 mSv from nuclear medicine. Three quarters of the latter dose involved the administration of ^{99}Tc. The number of such examinations was 360K in the UK in 1998 with an average dose of 3.3 mSv each.

The distribution of dose is far from uniform: 93% of the population receive a dose of less than 1 mSv from diagnostic radiology; 7% of the population receive 82% of the total collective dose. Dental examinations, chest and limb X-ray procedures involve very small doses. X-ray CT examinations, like the nuclear medicine procedures, carry a higher dose, close to the recommended annual limit, or the annual dose acquired by living in Cornwall. We shall consider the reasons for these higher doses in chapter 8. Many of those who receive higher doses are older and beyond child-bearing age. The age distribution, Fig. 6.16d, shows that those over 60 receive typically four times the dose of those under 20.

Radioactive consumer products and occupational doses

In the 1950s and 1960s every schoolboy wanted a watch with a large luminous dial. The radioactive constituent of the paint used on the dial and hands was a γ-emitter, such as radium. This gave an annual dose of about 1.3 mSv. The constituents now used are β-emitters whose radiation is contained. These are now marketed as 'beta-light' sources. Another consumer product in which radioactive material is used is the thoriated gas mantle. An efficient gas mantle should act as a black body radiator throughout the optical spectrum and have a high melting point. When heated to a high temperature it then emits a good white light. As a high Z material thorium has the required dense absorption spectrum. Thoriated gas mantles are still used, but gas lighting is very much the exception, and the dose is small. Smoke alarms are provided with a weak α-source (40 kBq ^{241}Am). However, no radiation can be detected outside the ionisation chamber. In summary, consumer products are no longer responsible for any significant dose.

The radiation exposure of certain occupations used to be significant. Now, carefully monitored and regulated on an individual basis, occupational doses are small compared with the variation of doses such as radon. Other contributions to the radiation exposure of the general population including the effect of atmospheric nuclear weapon testing and Chernobyl fallout are discussed later in this chapter.

6.4   Radiation damage to biological tissue

We discuss the range in scale of radiation damage to biological systems, and how such damage is related non-linearly to the radiation dose. Data on survival from exposure to high doses, the effects of low doses and the incidence of radon-induced lung cancer show that the health risks of low-level radiation have been significantly overstated. High doses are used to sterilise cell reproduction completely, and this has application to the sterilisation of hospital supplies and food.

6.4.1   Hierarchy of damage in space and time

On the left in Fig. 6.17 is shown a diagram of the structural hierarchy of healthy living tissue. On the right are shown the abnormal forms that can result from radiation exposure. This simplified picture is intended to show the relative scale at which different pathological conditions occur. Because there is a very large number of each lower scale structure in each higher scale form, there is significant redundancy in the system at each level. With this redundancy goes an ability to replace and repair when radiation-induced abnormalities occur.

Normal tissue   Radiation damaged tissue

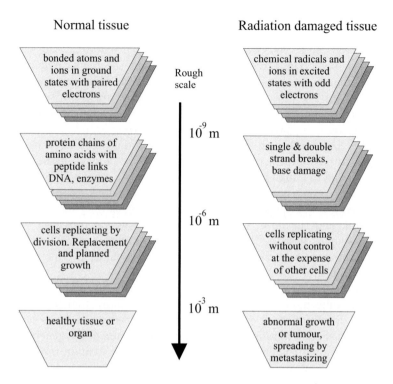

Fig. 6.17 The hierarchy of structures in living tissue with their errant radiation-induced forms to the right.

At each scale there is a corresponding time over which changes occur. The initial damage at the atomic level occurs at the time of exposure to the dose, or within nanoseconds of it. A dose of 1 mSv corresponds to the ionisation about 10^{15} electrons per kg. A large proportion of living tissue is composed of water and many of its mechanisms are water based, and so the effect of radiation on water is important. The primary products H_2O^*, H_2O^+ and e^- give rise to three reactive radicals: the hydroxyl radical OH, the hydrated electron e_{aq}^- and the neutral hydrogen atom H. The presence of any of these three may be detected by ESR. Within a few seconds such radicals react with other molecules in their immediate environment leaving breaks and changes in the structure of the large biomolecules. These are made of long chains of 20 different amino acid groups and peptide links. Within each cell there are the enzymes which carry out the work and the DNA which carries the blueprint or drawings for the cellular material and its operation.

The larger and more significant biomolecules, such as DNA, are repaired by enzymes. It has been shown that 90% of single breaks in a DNA chain are repaired within an hour of irradiation. Many damaged but less critical molecules simply become dysfunctional and are removed during the regular cycle of cell replacement by the garbage-collecting enzymes, a process called apoptosis. This takes place over a period of days or weeks. However, just occasionally, less simple errors induced in DNA may not be recognised as dysfunctional and may be replicated in the DNA copying process. These may then become the potential cause of longer term problems.

Factors affecting cell survival

Survival curves for cells show two marked effects that are important in radiotherapy.[41] The first is the oxygen effect. A higher oxygen concentration increases the mortality of cells by up to a factor 2. This is a problem for radiotherapy because a malignant tumour in an advance state is depleted of oxygen. However, with careful treatment planning, successive treatments can penetrate deeper into the tumour as normal metabolism recovers and the oxygen level rises to its normal level. The second effect relates to the timing of the cell reproduction cycle. A cell has to double its DNA in order to divide. The division stage is called mitosis. After mitosis there is the lowest redundancy in the DNA information in each cell and it is at this phase (1–2 hours in 24) that the cell is most radiation sensitive. *In vivo* cells are at different phases in their cycles. However, such questions may be studied with human cells grown *in vitro* in a tissue laboratory under controlled conditions.

[41] This is discussed at greater length in chapter 8.

6.4.2 Survival and recovery data

Damage at high doses

Very high whole-body doses cause catastrophic failure of biological mechanisms and early death with high probability and without significant delay. Unequivocal data on this are available, largely from animal studies but supported by available human data. The response is very non-linear. An example is shown in Fig. 6.18, in this case for rats. When humans or animals are irradiated the blood-forming organs in the bone marrow are immediately affected. After doses of 1 to 2 Gy the white and red blood cells decrease and the immune system deteriorates. If the dose is less than 4 to 5 Gy the bone marrow may recover after 3 to 4 weeks. The lethal dose LD_{50} is the whole body dose that induces 50% death within 30 days.[42] For rats LD_{50} is 7.5 Gy, for humans it is 3–5 Gy, with similar values for other higher mammals. Simpler organisms are more tolerant, for example for goldfish the figure is 23 Gy. Single cell organisms may survive much higher doses, such as 2000–3000 Gy. This is taken into account in the radiation treatment of food and the sterilisation of hospital supplies. Following exposure to a dose significantly above 5 Gy the reproduction of cells in humans shuts down. Systems with a high cell turnover fail first. Thus the gastrointestinal tract is an early casualty with symptoms of vomiting and diarrhoea.

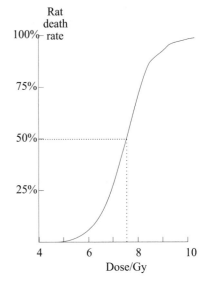

Fig. 6.18 The whole-body fatal dose curve for rats.

[42] This is called acute radiation death.

Effects of irradiation over long periods

The time lapse between the irradiation and the diagnosis of a cancer, called the latent period, may be many years. Our understanding of what happens during this time is still poor. There is no *in vivo* imaging with the required resolution and specificity, and the error rate itself is very small indeed. At a clinical level, the presence of cancer cells is often not diagnosed with certainty until a tumour many millimetres in size has developed. This is an unfortunately late stage at which to give therapy. The large number of malignant cells to be killed means that considerable damage is sustained by healthy tissue and there is also a significant chance that some malignant cells will survive and metastasize.[43]

[43]Metastasis is the stage at which malignant cells spreads around the body, typically through the bloodstream.

Damage due to low doses

[44]For example, below 100 mSv.

Low doses[44] do not cause cell death, at least not on a scale that has an impact on the survival of the organism as a whole. Low doses affect the cell or organism through errors in the information carried by the DNA that are mis-repaired or not repaired. Such errors or mutations also occur naturally and are a necessary source of diversity in a biologically competitive environment. Whether natural radiation is actually a beneficial and normal source of such mutations is unclear. The presence of mutations, whether natural or radiation induced, is not injurious.[45]

[45]Mis-repair also occurs in the case of modest levels of laceration. In that case we call it scar tissue, and it remains indefinitely as a source of imperfection.

Most mutations do not survive and are effectively sterile. However, the occasional mutation may be viable and also involve instructions to cells to divide and multiply without limit. This is a malignant cancer. It gives rise to tissue which drains resources and oxygen from the blood supply and grows without reference to the host body. The tumour created has an outer edge which grows anomalously fast and a centre which is starved of oxygen and nutrients. This centre becomes necrotic or dead. In later development the tumour metastasizes. The malignant cells break away to start secondary tumours elsewhere in the body carried by the bloodstream.

Exceptionally, relatively low doses may be concentrated by the body. The significant case is the absorption of ^{131}I by the thyroid. Following an accident this iodine may be ingested from the atmosphere, unless a simple treatment with regular iodine tablets is given. This was not dispensed at Chernobyl and there was a significant increase of thyroid cancer in consequence.[46]

[46]See section 6.5.3.

Particular attention has been paid to the effect of radon. As a noble gas it is airborne and absorbed by the body. It is an α-particle emitter

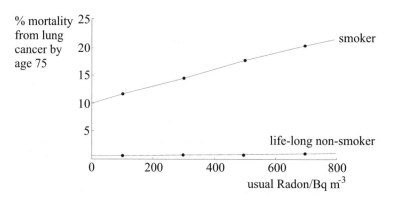

Fig. 6.19 The incidence of lung cancer for smokers and life-long non-smokers as a function of the radon count of their domestic environment. [Data and analysis from a compilation of 13 separate European case-controlled studies by Darby *et al.*]

with high LET and in some regions dominates the background radiation in homes over long periods. Large scale studies of the effect of radon in the lungs has been carried out in India, China and Europe. An increased incidence of lung cancer with radon exposure was found to be confined to heavy smokers. The European data on 13 studies in nine countries are described in the paper by Darby *et al* published in the British Medical Journal.[47] The results are summarised in Fig. 6.19. Their conclusion,

In the absence of other causes of death, the absolute risks of lung cancer by age 75 years at usual radon concentrations of 0, 100, and 400 Bq/m³ would be about 0.4%, 0.5%, and 0.7%, respectively, for lifelong non-smokers, and about 25 times greater (10%, 12%, and 16%) for cigarette smokers.

So smokers have a high chance of lung cancer and this may be doubled if they have been living in homes with a high radon count. However, given the sample size, 7148 cases of lung cancer with 14208 controls, the statistical errors are of order 1% and the figures for non-smokers are therefore consistent with zero.[48]

Other studies have been made of large populations contaminated by, and exposed to, unusual doses of β and γ radiation. In 1957 an explosion at a fuel plant in the southern Urals exposed 270,000 people in 23,000 km² to doses of ^{90}Sr. A recent IAEA study revealed no discernable changes in the subsequent incidence of cancer and other causes of death over more than three decades.[49]

Sterilisation of bacteria and viruses

The use of radiation to sterilise hospital supplies and surgical instruments is an unqualified benefit that has been in use since 1958. Bacteria and viruses are smaller biological structures that are more radiation resistant than higher biological forms. Doses of 20,000 to 40,000 Gy are used to ensure complete sterilisation. It is cheap and convenient to use a radioactive γ-source. It is essential that the irradiated material does not become radioactive as a result of the sterilisation process. Possible nuclear reactions of the form

$$\gamma + A \rightarrow n + (A - 1) \tag{6.10}$$

have a threshold associated with the binding energy of the last neutron. This lies between 5 and 9 MeV. An exception is the deuteron with a binding energy of 2.2 MeV although naturally occurring materials do not effectively contain any deuterium. Other nuclear reactions leading to radioactivity would require a charged particle to tunnel its way out through a Coulomb barrier and would be more unfavourable. Nuclear reactions of the form

$$\gamma + A \rightarrow \gamma' + A^* \tag{6.11}$$

may occur but these cannot make materials radioactive. Therefore if the γ energy is less than 2 MeV no radioactivity can be induced. The source that is usually used for sterilisation is ^{60}Co which emits γ-rays of energy 1.17 MeV and 1.3 MeV. This cannot induce radioactivity in any

[47] This study and others are to be found on the websites given at the end of the chapter.

[48] This author's observation.

[49] For details find the paper on the Web with the search keywords given at the end of the chapter.

material. With the advent of antibiotic-resistant organisms and evolving viruses this method of sterilisation has become especially important.

Fresh food can be sterilised in the same way. Typically a dose of 5,000 to 10,000 Gy is used. Both insect infestation and decay bacteria are sterilised. An advantage is that chemical preservatives are then not needed, and so the condition of the food is actually closer to its natural state. The chemical changes induced by irradiation are similar to those involved in the normal process of cooking although in most cases there is no change in taste. ESR can be used to measure whether food has been irradiated. Food irradiation is permitted by the World Health Organization and it is established that it is both beneficial and harmless. Regulations vary considerably throughout the EU and in some countries matters appear to have been fudged, whether out of ignorance or fear of public opinion. For example, it is not permitted in the UK except in special cases, such as for spices.

6.5   Nuclear energy and applications

The basic physics involved in nuclear fission and fusion is introduced. We then review applications from the harmful to the benign. We explain the principles and difficulties involved in the construction of nuclear weapons. Recent data on radiation exposure and its consequences are summarised, including the bombing of Hiroshima and Nagasaki, the atmospheric testing of nuclear weapons in the 1960s and 1970s, contamination incidents from the Soviet era, and nuclear power accidents such as Chernobyl. The principles of nuclear surveillance, methods of tackling nuclear contamination and the storage of waste are explained.

6.5.1   Fission and fusion

We start with an account of the nuclear physics involved. The binding energy of a nucleus of A nucleons with atomic number Z arises from a balance between the attractive nuclear potential and the repulsive Coulomb potential arising from the charge Ze. The attractive nuclear potential is spherical with a constant depth in energy and a volume proportional to the number of nucleons; thus the nuclear radius R varies as $A^{1/3}$. The nuclear binding energy is proportional to A less a correction term for the nuclear surface proportional to $A^{2/3}$. This is like a liquid drop where there is a correction to the latent heat due to the molecules at the surface, the surface tension energy. The dependence of the net energy per nucleon is illustrated in Fig. 6.20 as a function of A. It shows a broad minimum around $A = 56$ in the neighbourhood of iron between the less stable low A nuclei and the increasingly disruptive effect of the Coulomb term $Z(Z-1)e^2/(4\pi\varepsilon_0 R)$ at high A and Z. This is only a sketch because A and Z are separate variables. At a given A the lowest energy state occurs when $A \approx 2Z$. At least that would be the case were

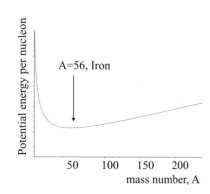

Fig. 6.20 A sketch of the potential energy per nucleon for a nucleus of mass number A as function of A.

it not for the effect of the Coulomb energy which favours neutron-rich proton-poor nuclei at high A. This excess number of neutrons has very significant consequences for nuclear fission.

Nuclei with A very much less than iron can release energy by increasing A by fusing. This is nuclear fusion. However, all nuclei have a positive charge and there is a high Coulomb barrier to be surmounted before the nuclei can get close enough to be affected by the extra attractive nuclear potential. For hydrogen the height of this barrier must be on the scale of

$$V = \frac{e^2}{4\pi\varepsilon_0 R}.$$ (6.12)

At a distance $R = 10^{-15}$ m, the nuclear radius, this is about 1 MeV. To penetrate such a barrier at a significant rate the hydrogen must be at a temperature much greater than 10^6 K with high density. Such conditions obtain in the interior of the Sun and this is its source of energy, known as thermonuclear fusion.

Nuclei with A much greater than iron can release energy by splitting. This is nuclear fission. The energy of nuclear fission comes from the reduction in electrostatic energy when a highly charged nucleus splits into two. This process absorbs nuclear energy by increasing the nuclear surface area, but the released electrostatic energy associated with the repulsion between the two nuclear fragments is much larger. In principle most heavy nuclei are unstable against fission. However, the increase in surface energy is large at small separations creating a potential barrier which inhibits the decay almost completely for all but the heaviest nuclei.

The spontaneous fission rate of any naturally occurring nucleus is very small. However, a thermal neutron reacting with an odd N nucleus such as

$$n + {}^{235}\text{U} \rightarrow {}^{236}\text{U}^*$$ (6.13)

creates a highly excited even N nucleus which prefers to decay by fission. The product nuclei cover a range in Z and are asymmetric in mass number A. The heavy fragment might be an isotope of atomic number in the range $Z = 48$ to 58, and the lighter fragment with corresponding mass between $Z = 34$ and 44. In each case the most stable isotopes have significantly fewer neutrons than the parent uranium. Some of the excess neutrons are emitted promptly at fission (about 2.5 on average), others are emitted after a short delay (a few per cent), while the remainder are responsible for the β-decay of the resulting neutron-rich isotopes.

Each of these processes has profound consequences. The initial flux of 2.5 neutrons drives the nuclear chain reaction, the delayed neutrons makes it possible to control a fission reactor, and the neutron-rich isotopes are responsible for the longer living radioactive waste products.[50]

[50]The following elements are plentiful among the heavy group, cerium, lanthanum, barium, caesium, xenon, iodine, tellurium. Among the light group are bromine, krypton, rubidium, strontium, zirconium, niobium, molybdenum, technetium, ruthenium.

6.5.2   Weapons and the environment

Nuclear weapons were developed in the World War II and its aftermath, the Cold War. Some devices are based on nuclear fission and others on thermonuclear fusion. There follows a brief sketch of the principles involved.

Nuclear fission bomb

A nuclear fission bomb relies on the rapid build-up of a neutron-induced chain reaction. Each nucleus of the fissile fuel emits 2 or 3 more neutrons when it fissions. These neutrons have a large cross section for inducing fission in further nuclei thereby releasing more neutrons and more energy. If the mass of fuel is small too many neutrons leave the surface for this build-up to diverge, a condition described as sub-critical. For an explosion to occur two or more subcritical masses have to be assembled into a critical mass for which the build-up does diverge. The technical problem is to assemble the extra mass sufficiently quickly. This may be achieved with a chemically driven implosion. A physical device comprises a number of concentric elements illustrated symbolically in Fig. 6.21 and listed below.

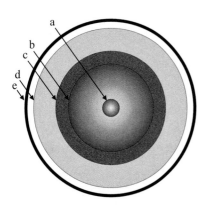

Fig. 6.21 A diagram illustrating the relation between the components of a fission-based nuclear weapon:
a, the neutron initiator;
b, the fissile fuel;
c, the tamper;
d, the chemical explosive;
e, the steel vessel.

◇ At the centre a neutron initiator with added tritium to trigger the rapid build-up of neutron flux just at the time of maximum compression of the fuel.

◇ A core of fissile material that must be assembled or concentrated very quickly into a critical mass. This critical mass for ^{235}U is about 20 kg and for ^{239}Pu about 6 kg.

◇ A 'tamper' material to reflect neutrons back into the core.

◇ A conventional chemical explosive to implode the fissile and tamper material with a physical shock wave.

◇ A steel vessel.

Timing is crucial. The critical mass must be fully assembled before the chain reaction is fully developed otherwise the assembly would be blown apart without developing its full power. The problem is to avoid the explosion starting prematurely, or 'fizzle' as it is known. The requirement is

$$\text{assembly time} < \text{chain reaction} < \text{mechanical dispersal}.$$

The chemical explosion is used to achieve assembly speeds in the range 1000–7000 m s^{-1}. Nuclear chain reactions that build up too quickly cannot be used for nuclear weapons. The start of the chain reaction build-up cannot be left to the rare spontaneous fission decay. The initiator is a neutron source made of a mixture of ^{9}Be and ^{241}Am which is an α-source. Neutrons are then made by the reaction

$$\alpha + {}^9\text{Be} \rightarrow \text{n} + {}^{12}\text{C}. \tag{6.14}$$

This neutron flux gives the chain reaction a 'push start' just at the right time as the assembly comes together.

Weapons-grade nuclear fission fuel

The main difficulty in making such a device, apart from the trigger mechanism just discussed, is obtaining a sufficient mass of fissile fuel. The only naturally occurring nucleus with a significant fission decay rate is ^{235}U. Its half-life is 7×10^8 years, so that only about 0.1% of that which existed when the Earth was formed survives today. Natural uranium is almost entirely composed of ^{238}U, with only 0.71% of ^{235}U. Any ^{238}U in a ^{235}U device would absorb neutrons at the expense of the build-up of the chain reaction. If ^{235}U is to be used, the ^{238}U must be removed. Separating these isotopes that differ by only 1% in mass is difficult. In principle it could be done in one of four ways:

◇ By mass spectrometry. This was attempted but proved unsuitable for the quantities required.

◇ By laser excitation. A tuned laser excites and ionises a chemical compound of one isotope but not the other. This is difficult even today but was not available in the 1940s.

◇ By centrifuge. Again this is a high technology solution which is difficult today but was not realistic in the 1940s.

◇ By diffusion. The compound uranium hexafluoride UF_6 is a gas. By building many stages of diffusion the nimbler molecule $^{235}UF_6$ can be separated from the slightly heavier and slower $^{238}UF_6$. This requires a very large plant and a large amount of energy.

In principle the centrifuge method is the same as the diffusion method but with the equivalent of a vastly enhanced gravitational acceleration. Such centrifuges require late twentieth century materials technology if they are to withstand the forces involved. In the 1940s such technology was not available and only the diffusion method was viable.

The other possible fissile material was ^{239}Pu. At the start of the Manhattan Project, as the bomb project was codenamed in the Second World War, this artificial isotope had only been produced as single atoms. To produce several kilograms a ^{235}U reactor had to be built as a source of neutrons to make ^{239}Pu by the reactions

$$\mathrm{n} + {}^{238}\mathrm{U} \rightarrow {}^{239}\mathrm{U} \rightarrow {}^{239}\mathrm{Np}\,(+\beta) \rightarrow {}^{239}\mathrm{Pu}\,(+\beta). \qquad (6.15)$$

The plutonium could then be extracted chemically. However, there was a further problem. The cross section for ^{239}Pu to capture a further neutron to make ^{240}Pu is high. Even a second neutron can be captured to make ^{241}Pu. ^{240}Pu fissions too readily and the presence of even a small concentration is sufficient to make a ^{239}Pu device fizzle. Consequently the ^{239}Pu has to be removed from the reactor rather frequently and before it has time to absorb further neutrons. This is a very inefficient use of the reactor fuel and some 10 tonnes of uranium must be processed to extract a critical mass of ^{239}Pu.

A plutonium test device called Trinity was detonated on 16 July 1945 in the New Mexico Desert. Of the two bombs dropped on Japan on

6 August and 9 August 1945, one used plutonium and the other uranium. As a result WWII ended on 15 August without the need for a land-based invasion of Japan.

Nuclear fusion device

There is a limit to the size (and power) of any nuclear fission weapon because of the difficulty of assembling a mass much larger than the critical mass within the time constraint. For a thermonuclear bomb there is no such limit. While the fission bombs dropped on Japan were equivalent to 15 and 22 kilotonnes of TNT, thermonuclear devices as large as 50 megatons were tested in the 1960s. Just as the fission device is 'lit' by a chemically driven compression, the fusion device must be lit by compression and heating from a fission explosion. To ignite, hydrogen must be held at high temperature and density for long enough for the nuclei to tunnel through the Coulomb barrier. Neutrons are needed to make helium and the fuel used is tritium (^{3}H) and deuterium (^{2}H) in forms such as lithium deuteride.

Legacy of Hiroshima and Nagasaki

There were about 429,000 people living in the cities when the bombs were dropped. Of these, 67,000 died on the first day and about 36,000 in the following four months. This number of fatalities exceeded the number in any subsequent nuclear incident in the next 60 years by more than three orders of magnitude. The number of survivors and the long subsequent timespan are unique. Studies based on the statistics of this data set are by far the best available and this will remain true.

Since 1950 the health of 283,000 people who survived has been followed. In particular the average dose of those exposed to radiation has been estimated at 160 mSv. However, there were considerable variations in dose which was a mixture of gammas and neutrons. The dose for individuals has been re-estimated several times: 86611 people received well established doses and another 25580 residents were outside the city and provide a control sample. Table 6.7 shows the most recent analysis of this cohort over the period from 1950 to 1990 in respect of death from leukaemia and from solid cancers.[51] There is no significant added risk of death from leukaemia for radiation doses less than 200 mSv. For solid cancers the background rate is higher and so the statistical errors are larger but there is no effect for doses less than 100 mSv; the increased risk is less than 1% for doses less than 200 mSv.[52] These limits are important. They show that the risks are too small to measure even with a sample of more than 100,000 people monitored over half a century following the largest-ever nuclear incident. These risks are utterly negligible compared with those arising from the use of fossil fuels.

There is data on the effect of the Nagasaki and Hiroshima bombs on children and pregnant mothers. Approximately 70,000 children have been studied where the parents received doses of about 350 mSv. Data on stillborn, child death, deformities, death before age 26, abnormal

[51]These are discussed in the book by Henriksen and Maillie. See also the RERF and LSS websites.

[52]To interpret these data we should look at them in terms of our previous knowledge (see chapter 5). This is that living tissue responds to radiation doses with an S-shaped dose–response curve. We know this in general from a) cases where the curve can be measured, b) understanding of response mechanisms, c) the observation that evolution has had the opportunity to find a defence mechanism. There is no reason to expect a linear response, and the data do not suggest it.

Table 6.7 An analysis of leukaemia and solid cancer deaths amongst the survivors of Hiroshima and Nagasaki between 1950 and 1990. The figures in brackets give the number predicted from the control sample. The excess gives the extra risk due to the radiation. This is shown per 10,000 people with estimated statistical errors. [Data from www.eh.doe.gov/radiation/workshop2005/presentations/neta.ppt]

Dose range (mSv)	Number	Leukaemia deaths	Extra risk per 10^4	Number	Cancer deaths	Extra risk per 10^4
0–5	35458	73(64)	3 ± 3	38507	4270(4268)	0 ± 20
5–100	32915	59(62)	-1 ± 3	29960	3387(3343)	15 ± 20
100–200	5613	11(11)	0 ± 10	5949	732(691)	70 ± 45
200–500	6342	27(12)	24 ± 10	6380	815(716)	155 ± 45
500–1000	3425	23(7)	46 ± 16	3426	378(262)	340 ± 60
1000–2000	1914	26(4)	120 ± 30	1764	326(213)	640 ± 100
>2000	905	30(2)	310 ± 60	625	114(58)	900 ± 170

number of chromosomes and sex ratio have been summarised up to 1991. These effects were sufficiently small that no genetic effects should be discernible for an accident on the scale of Chernobyl. It was also found that there was no cancer increase among the 2000 Japanese irradiated in the critical embryological stage of pregnancy.

Atmospheric nuclear tests

Testing of nuclear weapons in the atmosphere continued after 1945 with thermonuclear as well as fission devices and reached a peak in the mid 1960s. Following such a test some radioactive fallout reached the Earth's surface within the first month but much of the material thrown into the stratosphere was carried round the Earth and persisted for many years.

Tests low in the atmosphere drew surface material up into the fireball and generated the greatest radioactive pollution. Significant pollutants are given in table 6.8. The total release of ^{137}Cs and ^{90}Sr in testing was

Table 6.8 The major components of fallout from nuclear tests with their half-life and source.

^{95}Zr	64 day	Fission product
^{131}I	8 day	Fission product
^{137}Cs	30 years	Fission product
^{90}Sr	29 years	Fission product
^{14}C	5730 years	By n on ^{14}N

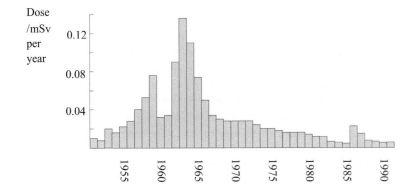

Fig. 6.22 The contribution of radioactive fallout to the average annual radiation dose of the population of the UK, 2.5m Sv, during the years of nuclear testing and the Chernobyl accident. [Data from NRPB report]

respectively 30 and 75 times the release in the Chernobyl accident. The average radiation dose in the UK due to fallout is shown in Fig. 6.22. The dose level started to drop when atmospheric testing was discontinued by international treaty. The small 'blip' in 1986 is the effect of Chernobyl. These annual doses are small. The maximum of 0.14 mSv is to be compared with an average total dose of 2.7 mSv.

In the immediate neighbourhood of a nuclear test much higher radiation exposures were recorded. For instance, in 1954 a low level thermonuclear test of 15 megatons exposed 23 Japanese fishermen who were 130 km away to very large doses in the range 2–6 Sv. They became nauseous and suffered β-induced skin burns. One died within 6 months but most were reported to be still alive 30 years later.

Nuclear surveillance

An important indicator of nuclear activity is the presence of the rare noble gas isotope ^{85}Kr in the atmosphere. This neutron rich isotope has a half-life of 10.8 years and its only source is nuclear fission. Its activity has increased from 1.0 to 1.4 Bq m^{-3} of air over the decade to 2004. Compared with the early 1950s this is an increase by a factor 10^6. Most of the atmospheric release occurs when nuclear fuel is reprocessed and its concentration is monitored by regulatory authorities.

Manufacturing nuclear weapons remains extremely difficult, compared with chemical or biological weapons. However, monitoring their production is relatively easy.

⋄ Plutonium production is not possible without operating a fission reactor with extremely frequent and energy-inefficient fuel changes. Except in the case of a heavy-water moderated reactor, in order to change fuel a fission reactor must be closed down. Such activity tends to be visible.

⋄ Heavy-water moderated reactors need a supply of heavy water. This may be obtained by the electrolytic enrichment of the small concentration of deuterium present in the hydrogen of natural water (0.015%). This requires large amounts of electrical power in quantities that may be seen by infrared satellite cameras.

⋄ Uranium enrichment by gaseous diffusion requires large industrial scale plant and high power input. These are generally visible. Modern methods using centrifuge or selective laser excitation require special materials and expertise which may be monitored. The enrichment required in the manufacture of fuel for power stations is quite modest and the plant required is easily distinguished from the scale of plant required to manufacture weapons grade uranium.

⋄ The materials and knowledge required for the neutron trigger may also be monitored. The security services noticed that Saddam Hussein was acquiring this technology prior to the First Iraq War.

These points have prevented the proliferation of weapons in the past in many cases at least. However, when weapons production is detected, political decisiveness is required to act on this information.

6.5.3 Nuclear power and accidents

Windscale and Three Mile Island accidents

There have been rather few major nuclear accidents. We discuss the three best known, Windscale (1957), Three Mile Island (1979) and Chernobyl (1986).

The Windscale case was caused by a build-up of unstable potential energy in the graphite moderator.[53] The result was a chemical fire in the reactor core. Radioactive debris was released into the local environment although the heat was not sufficiently intense to propel it into the upper atmosphere. The most serious immediate consequence was that 6×10^{14} Bq of ^{131}I contaminated the atmosphere, food supplies and local agriculture. This caused 260 cases of thyroid cancer, 13 of them fatal. Children were particularly at risk because they absorb more iodine than adults while they are growing. The highest thyroid dose was calculated to be 160 mGy. The full extent of the disaster was covered up until 1983. When revealed, the fact that this cover-up occurred seriously damaged the long-term credibility of official information about nuclear accidents.

The Three Mile Island accident was caused by human failure. The overheated and damaged reactor core did not release a significant quantity of radiation into the environment. However, there was a misunderstanding, a scare and a loss of public confidence in the nuclear industry to oversee the routine operation of nuclear power plants.

Chernobyl accident

The cause of the Chernobyl accident was a combination of human and management failure with a seriously inadequate reactor design.[54] An unregulated 'test' caused a loss of cooling water, then a fire of the graphite moderator. This was followed by a chemical explosion which blew the top off the reactor. The heat propelled the more volatile contents of the reactor high into the atmosphere. The accident occurred just as the Soviet Union was itself becoming unstable, and contributed to that disintegration as well as being partly caused by it.

Table 6.9 gives details of the activity released. There are the neutron-rich fission product nuclei and also the transuranic elements, neptunium plutonium and curium, formed by neutron absorption and β-decay from the uranium fuel. The volatile elements were all released into the environment, the xenon and most of the iodine, caesium and tellurium. Most of the heavy transuranic elements remained in the reactor. The hazard posed by each nuclide is determined by its lifetime as well as its initial activity. Chemical and biological affinity are also important. Thus xenon is not absorbed and decays in the atmosphere, but iodine is ingested and taken up by the thyroid gland.

[53] See section 3.2.4.

[54] There is an authoritative international OECD NEA report on the accident which may be found on their website; see the list at the end of chapter 1.

Table 6.9 Inventory of the core contents of the Chernobyl reactor on 26 April 1986 and the activity then released. [Source: table 1 *www.nea.fr/html/rp/reports/2003/nea3508-chernobyl.pdf.*]

Nuclide	Half-life	Activity Bq×10^{18}	Released %	Released activity
^{133}Xe	5.3 days	6500	100	6500
^{131}I	8.0 days	3200	50–60	~1760
^{134}Cs	2.0 years	180	20–40	~54
^{137}Cs	30 years	280	20–40	~85
^{132}Te	78 hours	2700	25–60	~1150
^{89}Sr	52 days	2300	4–6	~115
^{90}Sr	28 years	200	4–6	~10
^{140}Ba	13 days	4800	4–6	~240
^{95}Zr	65 days	5600	3.5	196
^{99}Mo	67 hours	4800	>3.5	>168
^{103}Ru	39.6 days	4800	>3.5	>168
^{106}Ru	1.0 years	2100	>3.5	>73
^{141}Ce	33 days	5600	3.5	196
^{144}Ce	285 days	3300	3.5	~116
^{239}Np	2.4 days	27000	3.5	~95
^{238}Pu	86 years	1	3.5	0.035
^{239}Pu	24400 years	0.85	3.5	0.03
^{240}Pu	6580 years	1.2	3.5	0.042
^{241}Pu	13.2 years	170	3.5	~6
^{242}Cm	163 days	26	3.5	~0.9

Table 6.10 The dose and fatality figures for those workers at Chernobyl who received an acute exposure. Of the total 238 a further 11 had died by 1998. [Source: table 12 *www.nea.fr/html/rp/reports/2003/nea3508-chernobyl.pdf.*]

Number of patients	Estimated dose (Sv)	Number of deaths
21	6–16	20
21	4–6	7
55	2–4	1
140	<2	0
237	(all)	28

Similarly caesium, being similar to potassium, is absorbed by the soil and enters the food chain. Strontium and barium are chemically like calcium and become absorbed in bone. Given their long lifetimes strontium and caesium are the most significant radioactive contaminants after iodine.

The medical casualties of the accident can be separated into acute cases, thyroid cancer cases and others. The fate of those who received an acute dose is given in table 6.10. Those who received the highest dose died, and there were 28 of these, in addition to 3 who were killed by the explosion. It is known that in the years up to 1998 a further 11 of the cases in the table died, many from unrelated causes.

There was also a large and significant increase in childhood thyroid cancers by a factor of 30, as shown in table 6.11. The latency is long and the numbers continued to increase up until 1993 but have since started to decline. While the dose fell rapidly with the 8-day half-life of ^{131}I, those who received that dose continue to be affected. The children concerned

Table 6.11 Data on cases of child thyroid cancers by year, then aged up to 15, showing the steep rise from 1986 as a result of the Chernobyl accident. The upper rows show the number of cases. The lower rows show the numbers normalised per 100,000 of the whole population. At the time the table was compiled 3 of these 1036 children had died. [Source: table 13 *www.nea.fr/html/rp/reports/2003/nea3508-chernobyl.pdf.*]

Year	86	87	88	89	90	91	92	93	94	95	96	97	98
Belarus	3	4	6	5	31	62	62	87	77	82	67	73	48
Ukraine	8	7	8	11	26	22	49	44	44	47	56	36	44
Russia	0	1	0	0	1	1	3	1	6	7	2	5	0
Total	11	12	14	16	58	85	114	132	127	136	125	114	92
Belarus	0.2	0.3	0.4	0.3	1.9	3.9	3.9	5.5	5.1	5.6	4.8	5.6	3.9
Ukraine	0.2	0.1	0.1	0.1	0.2	0.2	0.5	0.4	0.4	0.5	0.6	0.4	0.5
Russia		0.3			0.3	0.3	0.9	0.3	2.8	3.5	0.6	2.2	

are now adult. Up to 1995, 3 out of 1036 children with thyroid cancer had died. The incidence is high because the most elementary precautions were not taken by the population in the days immediately following the accident. There was also an increase of adult thyroid cancers.

Table 6.12 shows that the numbers of people receiving a dose of 100–200 mSv and more than 200 mSv are 1123 and 8124, respectively.[55] Using the Hiroshima and Nagasaki data (the fourth and last columns of table 6.7) we may estimate that an extra 40–120 of these may die from solid cancer caused by the radiation in the next 50 years; the number of extra deaths from leukaemia would be less than 5. In the Chernobyl data themselves there is no clear evidence for anomalous rates of leukaemia, genetic disorders or forms of cancer other than thyroid. Any such evidence would be buried by the statistical noise on the larger incidence of cancers that occur in any population. In addition there are systematic effects arising from the social dislocation and panic caused by the accident itself.

As a result of the accident land became contaminated by ^{90}Sr and ^{137}Cs. How can radioactive contamination over a large area be 'cleaned up'? It can be concentrated, stored and managed; it can be diluted and dispersed into the wider environment; or it can be left where it is. High levels of radioactivity may best be concentrated and managed, but over wide areas the only practical solution is dispersal, ensuring as much as possible that contamination does not enter the food chain. This means diluting the caesium with massive applications of potassium fertiliser to the land. This exploits the chemical similarity between caesium and potassium. Similarly the agricultural uptake of strontium and barium is minimised by a heavy application of calcium in the form of lime.

The same dilution technique should be used to minimise the incidence

Table 6.12 Distribution of estimated total effective doses received by the populations of contaminated areas (1986–1995) excluding dose to thyroid [Source: table 10 *www.nea.fr/html/rp/reports/2003/nea3508-chernobyl.pdf.*]

Dose mGy	Number Belarus	Russia	Ukraine
<1	133053	155301	
5	1163490	1253130	330900
20	439620	474176	807900
50	113789	82876	148700
100	25065	14580	7700
200	5105	2979	400
>200	790	333	

[55] These dose profiles are coarsely binned but there are large errors on the Hiroshima and Nagasaki data too. Whatever, the predicted fatalities are few compared with risks associated with climate change.

[56]It can be taken orally in the form of tablets.

of thyroid cancer. Following a radiation accident those exposed should take an excess of normal iodine for 2–3 weeks.[56] This excess reduces the uptake of radioactive ^{131}I which has a half-life of 8 days. After a few weeks the iodine has decayed and the thyroid risk is passed. Indeed the best advice in the event of a nuclear accident is 'don't go out in the rain', for rain concentrates the fallout, and 'take iodine tablets'. No such advice was given to the population affected by the Chernobyl incident.

Management of nuclear waste

[57]The expense of nuclear power depends in large measure on just how low such safe levels need to be. This is considered critically in chapter 10.

High level nuclear waste is managed and concentrated while low level waste is dispersed. The dispersal should be to safe levels,[57] so that there are no further cost, health or security implications. High level nuclear waste comes from weapon stockpiles, spent fuel elements, decommissioned reactors and used medical and industrial radioactive sources. Compared with high toxicity chemical waste the volumes of material are small. It is difficult but technically possible to 'burn up' nuclear waste. This is called transmutation and references to this may be found on the Web. In the absence of such a process high level nuclear waste must be stored securely and records maintained to avoid its inadvertent dispersal.

[58]This is about 1 GW-year. An equivalent coal power station would burn 2×10^6 tonnes of coal releasing 5.4×10^6 tonnes of CO_2 into the atmosphere (and 0.05×10^6 tonnes of SO_2, leaving 0.12×10^6 tonnes of ash). The mass of nuclear waste is smaller by a factor of 10,000. The environmental impact is small compared with the effect of the CO_2 on the climate.

The scale of the problem of waste from nuclear power stations may be judged from the following figures. In January 2005 there were reported to be some 440 operating commercial nuclear power stations world-wide with generating capacity of some 364 GW. A typical light-water-reactor power station generates 20 tonnes of spent fuel in the process of delivering 6.6×10^9 kWh of electricity.[58] This waste occupies 11 m^3. At full capacity each year the world's nuclear power stations would generate 8000 tonnes of spent fuel with volume about 4000 m^3. This is small.

The public perception is that the physical storage is a serious difficulty, and that no storage exists with an integrity on a sufficiently long timescale. That this is not correct can be illustrated by the Oklo reactor. In 1972 it was found that the ratio of ^{235}U to ^{238}U from this uranium mine in Gabon, Africa, was only 50% of the value elsewhere. The fission products found there have been used to show that, about 1.7 billion years ago when the isotope ratio was 3% and this rock was being formed, the ^{235}U went critical and induced fission occurred, moderated by water bound in the minerals. The energy generated was about 15 GW years, and over a considerable time a large fraction of the ^{235}U was burnt up, thus generating the anomalous uranium isotopic ratio found today. The important point for the waste storage question is that the tell-tale waste fission products have not moved significantly in 1.7 billion years. They remain quite secure. Indeed the Earth has always had a significant content of radioactivity, more or less widely dispersed. A few deep dry storage locations built by humans will be a tiny addition. In a few locations natural radioactivity is leached out of the Earth by groundwater. Far from being shunned by the public, fashionable spas[59] have been built at such places and have proved both popular and harmless.

[59]For instance, the Roman city of Bath, popular to this day.

In total, by the middle of the twenty-first century the accumulated high level nuclear waste requiring long term storage would be about 0.1×10^6 tonnes, or 5×10^4 m^3. This assumes nuclear power stations operating at 150 GW, with realistic levels of recycling.[60]

Many incidents have arisen from poor record keeping, especially in the less stable conditions that can occur in the Third World. In 1987 in Brazil a ^{137}Cs source of 50 TBq from an abandoned radiation medical therapy unit was broken open in a scrap-metal yard: 244 people were contaminated, 20 were taken seriously ill and 4 died.

In 1983 a radiotherapy device containing 15 TBq was taken from a warehouse where it had been stored. Two steel foundries in Mexico and one in the USA handled the scrap producing 4000 tonnes of radioactive steel that was used to make table legs and building reinforcements. Over 100 people received doses between 10–500 mSv, and some over 15 Sv over a period of several months. These instances show how nuclear material puts a heavy burden on society in the areas of record keeping, labelling, expertise, education, regulation and political continuity.

Only about 50% of the available fissile nuclear fuel is 'burnt' in a reactor before the fuel rods must be withdrawn, otherwise their chemical and structural integrity would become dangerous. Natural uranium is enriched somewhat (0.7% to 3.5%) for a light water reactor; 100 tonnes of such fuel will keep a 1 GW power reactor running for three years. When such a reactor is shut down, the primary chain reaction is halted and the fuel withdrawn. However, the daughter nuclear decay processes continue. After 10 days the fuel emits 60 MW, after 1 year 1 MW, after 100 years 40 kW. For this reason, for a number of years spent fuel must be concentrated, stored, managed and cooled. It is usually kept under water in large tanks. Since the mass is not that large this is quite practical. Currently 10% is reprocessed but this is only economic if the reprocessed fuel is needed. This may be carried out at Sellafield in the UK or at two plants in France. The advantages of keeping waste concentrated in one place are that its history and location are known. However, it remains a responsibility and a security risk. When it no longer needs active cooling it may be vitrified and set in concrete. Then it may be stored in dry caverns. The task of reactor de-commissioning will become somewhat easier with the advent of more advanced robotics. And significantly cheaper with more realistic safety assessments.

Modern nuclear power stations are much safer than the old Soviet design built at Chernobyl and 'burn up' a greater proportion of their fuel. In addition there is the reassuring expectation that power from controlled nuclear fusion will be available by the middle of the twenty-first century. Such power involves no chain reaction, no risk of explosion, and no creation of radioactive daughter products with long lifetimes. Making such power available to the Third World will be a challenge to future statesmen.

[60] In January 2006 there were 10 nuclear power stations under construction in OECD countries (Korea, Japan, Slovakia and Finland) with another 17 on order.

Read more in books

Detectors for Particle Radiation, K Kleinknecht, Cambridge (1986), a text on radiation detectors

Nuclear Physics, Principles and Applications, Manchester Physics Series, J Lilley, Wiley (2001), a textbook covering applications of nuclear physics

Chemical Analysis by Nuclear Methods, ed. ZB Alfassi, Wiley (1994), an edited compilation of articles on analysis techniques using nuclear methods

Tracing Noble Gas Radionuclides in the Environment, P Collon *et al*, *Annual Reviews of Nuclear and Particle Science* 54,39 (2004), a review exemplifying the information available from studies of isotope distributions

Medical Physics & Biomedical Engineering, BH Brown *et al*, IOP (1999), a standard text on medical physics

The Physics of Medical Imaging, ed. S Webb, IOP (1988), an edited compilation of useful articles, somewhat out of date but still frequently referenced.

Radiation & Health, T Henriksen and HD Maillie, Taylor & Francis (2003), a highly readable general text on radiation and health

Radiation and Radioactivity on Earth and Beyond, IG Dragonic, ZD Dragonic and J-P Adloff, CRC Press (1993), a wide-ranging descriptive book

Science-based Dating in Archaeology, MJ Aitken, Addison Wesley Longman (1990), a readable physics text on scientific dating in archaeology

Archaeology: Theories, Methods and Practice, C Renfrew and P Bahn, Thames & Hudson, 4th edn (2004), a textbook on scientific methods in archaeology

The Solid Earth, CMR Fowler, Cambridge (1990), a textbook on geology and geophysics

Look on the Web

Look at the book website at
   www.physics.ox.ac.uk/users/allison/booksite.htm

Find out how radiation is used to manipulate the behaviour of plastics, including shrink-wrap film, by searching on
   polymer memory electron beam

Keep abreast of developments in radiation detector technology
   PMT, APD, CCD, HPD, amorphous silicon, CdTe

Follow further developments in analytical and dating techniques
   XRF, SPM, PIXE, RBS, NAA, AMS

Check for nuclei with specific decays, lifetimes and other properties by consulting
   nuclear data BNL

Watch the growth in availability of synchrotron radiation sources (SR)
   synchrotron light

Study the known data on Hiroshima and Nagasaki, and on Chernobyl, by searching the websites of the authorities listed at the end of chapter 1

Take a look at papers on the incidence of radon-induced lung cancers and other studies
   bmj.com/cgi/content/full/330/7485/223
   www.hps.org/publicinformation/ate/q1254.html
   IAEA Urals soviet accident

Look up about the transmutation of nuclear waste
   transmutation nuclear waste

Read about sterilisation with radiation
   sterilisation radiation medical
   sterilisation radiation food

Questions

6.1 Calculate the number of ^{14}C atoms detected from a 'modern' 1 mg sample of carbon assuming 100% detection efficiency, (a) in an experiment counting the beta decays over a period of four weeks, and (b) in an accelerator mass spectrometer measurement.

6.2 Calculate the statistical error on the measured age of a 1 mg sample as determined by the methods of question 6.1 if the age is about 3000 years, and if it is about 40000 years. Calculate the systematic bias of the age determination caused by a contamination of 1% modern carbon.

6.3 The proportion of ^{14}C in living tissue is about 10^{-12} and its half-life is 5730 years. It decays to ^{14}N emitting an electron with maximum energy 156 keV. Calculate the associated internal radiation dose to the body in Sv making appropriate assumptions where necessary.

6.4 In the decay of ^{40}K, see Fig. 6.15, both possible daughter nuclei, ^{40}Ar and ^{40}Ca, have an even number of both neutrons and protons, and therefore $J = 0$. Consider the dominant decay mode

$$^{40}\text{K} \rightarrow\, ^{40}\text{Ca} + e^- + \nu.$$

The parent ^{40}K has angular momentum $J = 4$, so $\Delta J = 4$. Calculate the ratio of decay rates for this β-decay to that for the neutron (half-life 614 s, energy 0.78 MeV with $\Delta J = 1$),

$$n \rightarrow p + e^- + \nu.$$

Compare the value of this suppression factor with that for the EM decay of multipolarity $\Delta J = 4$ and wavelength λ of a nucleus of radius a, $(2\pi a/\lambda)^{2\Delta J - 2}$, discussed in section 8.3.1.
[The same angular momentum effect is responsible for the suppression, although in one case it is a weak nuclear process and in the other an electromagnetic one.]

6.5 A beam of 3 MeV alpha particles is incident on a flat surface. The energy spectrum of the back-scattered alpha particles consists of a continuous distribution up to 1.693 MeV, and a sharp peak at 2.766 MeV. What can be deduced about the structure and composition of the surface?

6.6 Derive the non-relativistic RBS relation for the mass ratio of target-to-beam

$$\frac{M}{\mu} = \frac{E + E' - 2\sqrt{EE'}\cos\theta}{E - E'}$$

in terms of the incident energy E, the scattered energy E' and the laboratory scattering angle θ.
Deduce the relation at 180° given by equation 6.4. For what range of θ is that an adequate approximation for the determination of M to within 10%?

6.7 a) The decay of ^{40}K to ^{40}Ar has a Q-value of 1.5 MeV. In most cases this energy is released as a γ-ray. Calculate the recoil kinetic energy of the residual ^{40}Ar nucleus.
b) In the α-decay of ^{226}Ra to ^{222}Rn the energy release is 4.87 MeV. Calculate the nuclear recoil energy of the residual ^{222}Rn nucleus.

Hence comment on the fact that the ^{40}Ar product may be trapped in rock for millions of years (allowing its formation to be dated) but that radon seeps out of rock on a timescale short compared with its half-life of 3.82 days.

Imaging with magnetic resonance

7.1 Magnetic resonance imaging   **207**

7.2 Functional magnetic resonance imaging   **221**

Read more in books   **230**

Look on the Web   **230**

Questions   **231**

in which the analysis, strengths and limitations of MRI methods in medicine are explored.

7.1 Magnetic resonance imaging

In MRI field gradients are used to encode spatial information into NMR. The fields of the gradient coils select a slice by excitation, and label a voxel within that slice by its phase and frequency. Because the signals are linear, a superposition from many voxels may be decoded by Fourier inversion. However, an image is never an exact copy of the related object. In MRI it may fall short in three ways: limited resolution, random noise and the presence of artefacts. As in other applications of Fourier optics, artefacts are generated as a result of incomplete sampling in k-space. MRI data acquisition, that is the sampling of k-space, can be made faster in three ways: the interleaving of data acquisition from different slices, the use of complete scans of k_x, k_y with a single excitation pulse, and the use of arrays of pickup coils. Each such method has its drawbacks. These are arguably the most important strategies, but there are others not discussed here.

7.1.1 Spatial encoding with gradients

Gradient coils

By applying different gradients to the B_0 field in the x-, y- and z- directions at different times, it is possible to determine a complete three dimensional map of the nuclear spin density in the region of interest. This region is called the field of view (FOV). Figure 7.1 shows how such a uniform gradient may be applied along the z-axis with a pair of coils. If the current in the two coils is in the opposite sense to one another, this generates a uniform gradient along z, the axis of symmetry. Departures from uniformity are minimised by adjusting the ratio of coil diameter to their separation, so that, on axis at least, departures are of order $(z/a)^4$ where a is the coil size. Uniformity is improved by making the coils large, and by arranging the geometry of further windings to ensure that higher polynomials contributing to the field variation are zero.

The coil geometry can conform easily to the bore of the NMR B_0 solenoid. A finite field gradient in z $(\partial B_z/\partial z)$ must still satisfy Gauss's

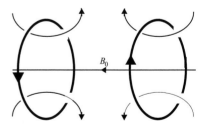

Fig. 7.1 Gradient coils giving a gradient of B_0 in the z direction, shown horizontal. The coils and current flow are shown with heavy lines. The B-field lines are shown lighter.

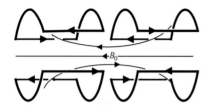

Fig. 7.2 Saddle-shaped gradient coils giving a gradient of B_0 in the transverse vertical direction. The four coils with their current flow are shown with heavy lines. The B-field lines are shown lighter. The geometry allows unrestricted access in the horizontal z-direction.

law,

$$\frac{\partial B_x}{\partial x} + \frac{\partial B_y}{\partial y} + \frac{\partial B_z}{\partial z} = 0, \tag{7.1}$$

and therefore unwanted radial B-field components must be present. However, although these will tip the *direction* of the field slightly as a function of z, they have only a second order effect on the *magnitude* of the field, and it is the latter to which the FID is sensitive. It is important to appreciate that the gradients are small. Typically the value of B_z is varied by up to 1 part in 10^4, while the chemical shifts in precession frequency correspond to about a part in 10^6. The x- and y-terms in equation 7.1 are therefore negligible.

In the x- and y-directions gradients of B_z can provided by sets of coils. However, the coil design has to be different because access to the magnet volume is required along the z-axis. So these coils have to conform to the cylindrical geometry of the main static B_0 solenoidal magnet. The required set of four saddle-shaped coils is shown in Fig. 7.2. The y-coils are similar to those for x, being related by a 90° rotation about z.

RF pulse shape and slice selection

Consider a problem in which there is just one voxel containing spins at position (x, y, z). How can we measure this position? The first step in position encoding is to select a thin slice in z by applying the initial RF pulse with a frequency ω in the presence of a uniform z-gradient. When this is done, only those nuclei in a narrow range of z will have the right Larmor frequency to be excited by the pulse, and therefore to contribute to the subsequent FID signal. Thus

$$\omega = \gamma \left(B_0 + \frac{\partial B_z}{\partial z} z \right). \tag{7.2}$$

By applying an RF pulse of a different frequency in the presence of a z-gradient field, different slices in z may be excited. If the observed amplitude of the FID is plotted out as a function of RF frequency, the z-value of the voxel occupied by the spins would be seen.

Before moving on to measurements in x and y we need to be a little bit more precise about this process for selecting spins in z. The thickness of the selected slice depends on the field gradient and the bandwidth of the RF pulse. In a uniform field gradient there is a one-to-one relation between z and frequency ω. So, to excite spins in a sharply defined slice in z, a pulse with a rectangular frequency distribution is required, as shown in Fig. 7.3b. The corresponding time modulation of the amplitude of the RF is shown in Fig. 7.3a. This is the Fourier transform of the rectangular pulse, and the same problem mathematically that occurs in Fraunhofer diffraction from a sharp slit. The functional form of the time modulation is

$$\frac{\sin(2\pi t/\Delta t)}{t}, \tag{7.3}$$

where the main peak is bounded by $\pm \Delta t/2$. The shorter the bunch modulation Δt, the larger the frequency slice $\Delta\omega$ and therefore the thicker

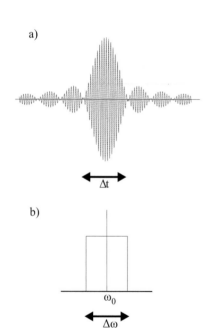

Fig. 7.3 a) The RF pulse in time showing the Larmor frequency modulated by a sinc function, as required to give b), the rectangular frequency distribution for sharp slice selection in z in the presence of a uniform B-field gradient in z.

the slice Δz. Quantitatively

$$\Delta t^{-1} = \frac{\Delta \omega}{4\pi} = \frac{1}{4\pi}\gamma\frac{\partial B_z}{\partial z}\Delta z. \qquad (7.4)$$

In this way, with a certain gradient the choice of central frequency and bandwidth determine the excited slice position and thickness respectively.

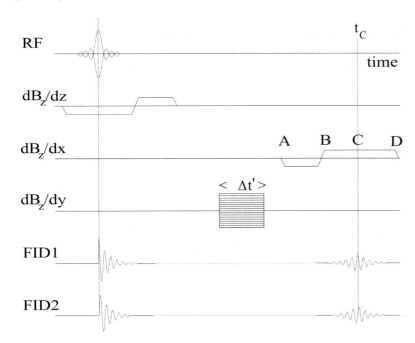

Fig. 7.4 A pulse sequence using all three gradient coils to encode the x, y, z spin density into the two orthogonal FID signals.

Gradient refocus in slice

To measure the position in x and y as well as z, coils providing independent gradients along all three axes are required. Figure 7.4 shows a simple pulse sequence as a function of time from left to right. On the first line is shown the applied RF waveform, on the next three lines the switched currents applied to the three gradient coils, and in the lowest two lines the detected FID signals picked up by coils in the xz- and yz-planes. Following the RF pulse, the spins within the slice will contribute towards a potential FID signal. However, the field varies within the slice because of the applied z-gradient. The spins in the different depths within the slice will get progressively dephased. To counteract this, not only is the z-gradient switched off, but it is reversed for a time, such that afterwards, all spins within the slice have seen the same time integral of B-field; that is, they will have precessed through the same angle, at least as far as the effect of the z-gradient is concerned. This is shown on the second trace of Fig. 7.4. This unwinding of gradient-related dephasing is called gradient refocusing. Unlike the spin echo discussed in chapter 2, this only compensates for the field variation coming from the applied gradients.

Spatial encoding by frequency measurement

The two FID signals shown in Fig. 7.4 are digitised and read out in the interval BD. During this readout time a steady gradient of B_z in the x-direction is applied. Therefore, if the single occupied voxel that we are considering lies in the excited slice, it will give a FID signal whose frequency indicates its x-position. The application of the reverse gradient during the period AB is such that there is a gradient echo at time t_c, in the middle of the readout interval.[1] Within this gate the FID traces generated by the two orthogonal magnetic fluxes due to the rotating spins in the particular voxel at x are

$$\cos\left[\gamma\left(B_0 + \frac{\partial B_z}{\partial x}x\right)(t - t_c)\right] \quad \text{and} \quad \sin\left[\gamma\left(B_0 + \frac{\partial B_z}{\partial x}x\right)(t - t_c)\right].$$
(7.5)

Linear superposition of signals

Now we come to a small but important step in the argument. Since the contribution of spins to the FID is linear, if we have a *distribution* of spins $f(x)$, the FIDs will be given by a simple linear superposition

$$F_1(t) = \int f(x)\cos\left[\gamma\left(B_0 + \frac{\partial B_z}{\partial x}x\right)(t - t_c)\right]\mathrm{d}x \qquad (7.6)$$

$$F_2(t) = \int f(x)\sin\left[\gamma\left(B_0 + \frac{\partial B_z}{\partial x}x\right)(t - t_c)\right]\mathrm{d}x.$$

We compare the equations (7.6) with the standard form for a Fourier transform, equation 5.3, in the form

$$g_1(k_x) = \frac{1}{\sqrt{2\pi}}\int f(x)\cos(k_x x)\,\mathrm{d}x \qquad (7.7)$$

$$g_2(k_x) = \frac{1}{\sqrt{2\pi}}\int f(x)\sin(k_x x)\,\mathrm{d}x.$$

We observe that with a suitable normalisation and definition of k_x these two functions (7.6) represent the real and imaginary parts of the complex Fourier transform of $f(x)$. The role of k_x in the trigonometric factors $\cos(k_x x)$ and $\sin(k_x x)$ is taken by $\gamma(\partial B_z/\partial x)(t - t_c)$. It is the spin density $f(x)$ that we want to determine. This is proportional to the inverse Fourier transform of the FID, $F_1 + \mathrm{i}F_2$.

Spatial encoding by phase

With the pulse sequence as discussed so far, the relative phase of all voxels at time t_c is zero. The remaining unmeasured coordinate y can be resolved by giving the signals from each voxel a phase shift between $-\pi$ and π that depends linearly on its y-position. We therefore apply a y-gradient $\partial B_z/\partial y$ for a fixed short length of time $\Delta t'$ before the readout gate as shown in Fig. 7.4. This results in an additional phase $\gamma y \Delta t'(\partial B_z/\partial y)$, linearly dependent on y. The multiple lines shown for the size of the y gradient in the figure are intended to indicate that FIDs are recorded for a whole *sequence* of different values of this gradient. The two orthogonal FID signals are related to the spin density $f(x, y)$ in the excited slice by[2]

$$F_1(t) = \int \int f(x, y) \cos\left[\gamma\left(B_0 + \frac{\partial B_z}{\partial x}x\right)(t - t_c) + \gamma\frac{\partial B_z}{\partial y}y\Delta t'\right] \mathrm{d}x\mathrm{d}y$$

$$F_2(t) = \int \int f(x, y) \sin\left[\gamma\left(B_0 + \frac{\partial B_z}{\partial x}x\right)(t - t_c) + \gamma\frac{\partial B_z}{\partial y}y\Delta t'\right] \mathrm{d}x\mathrm{d}y.$$

$$(7.8)$$

[2]The two independent FID signals could be read by two coils at right angles. In practice they are deduced from the magnitude and phase of the signals induced in the birdcage coil or other detector coils used.

Spatial reconstruction by Fourier transformation

These are the real and imaginary parts of the two dimensional complex Fourier transform of the spin density $f(x, y)$. To invert this transform and obtain $f(x, y)$ it must be measured everywhere in k-space. Then ideally

$$f(x, y) = \frac{1}{(2\pi)^2} \int \int (F_1 + \mathrm{i}F_2) \exp[-\mathrm{i}(k_x x + k_y y)] \, \mathrm{d}k_x \mathrm{d}k_y \qquad (7.9)$$

with $k_x = \gamma(\partial B_z/\partial x)(t - t_c)$ as before and $k_y = \gamma(\partial B_z/\partial y)\Delta t'$.

These two k-vector components differ only in that, for k_x, the gradient is fixed and the time t varies, while for k_y the gradient is varied in sequential scans and the time interval $\Delta t'$ is fixed. This distinction is peculiar to the particular choice of simplified pulse sequence we introduced in Fig. 7.4. For a general pulse sequence the (k_x, k_y)-vector is the (x, y)-derivative of the phase accumulated between the excitation of the transverse magnetisation and the read time,

$$k_x = \int_{\text{RF pulse time}}^{\text{read time}} \gamma\frac{\partial B_z}{\partial x}\mathrm{d}t \qquad (7.10)$$

$$k_y = \int_{\text{RF pulse time}}^{\text{read time}} \gamma\frac{\partial B_z}{\partial y}\mathrm{d}t.$$

Of course we cannot measure the transform at all values of (k_x, k_y). The best that we can do in practice is to sample the FID signal at a sequence of n_x discrete times in the interval BD and repeat the whole sequence for n_y discrete values of $\partial B_z/\partial y$. This restricted information generates the artefacts discussed in general in chapter 5.

7.1.2   Artefacts and imperfections in the image

Limited resolution

The reconstructed image may not show the fine detail of the object. This is related to the values of the Fourier transform at high k. If the signal is not recorded beyond $\pm k_{\mathrm{max}}$, detail on the scale of $1/k_{\mathrm{max}}$ will be absent from the image. In the simplified scan described in Fig. 7.4 better resolution in x would be achieved with greater values of k by increasing the length of the readout time $t - t_c$ (with the same gradients). However, that would increase the scan time, and also involve problems of signal loss by relaxation. In y more scans would be needed to sample at high k_y. In each case improved image resolution comes at the expense of longer scan time.

Random noise

The random thermal noise level of the first stage of electronic amplification is crucial, given the small signal amplitudes. This noise amplitude is proportional to the root of the bandwidth. Its effect may be reduced with multiple scans but, again, this comes at the expense of a lengthened scan time. Digital signal processing may be used to improve the SNR by general filtering as discussed in chapter 5, although this is only effective in so far as the characteristics of signal and noise are distinguished.

Wrapping in the image

Attempts to invert an incomplete Fourier spectrum result in artefacts in the image, as discussed in chapter 5. In particular, discrete sampling of the transform at points separated by Δk generates ambiguity. The reconstructed image is then normally treated as periodic, with a repeat length $2\pi/\Delta k$. To present a picture arbitrary decisions have to be taken during reconstruction, such as to present the endless periodic image, or to cut it somewhere. Prior knowledge may help, but the scan has no information. In the case of Fig. 1.2b there is prior knowledge that the end of the author's nose is connected to the rest of his nose and is not at the back of his head. So this picture was cut at slightly the wrong place. When pieces of the object space appear in the wrong place in the image they are said to be aliassed.

If there are measurements at n distinct values of k the spatial resolution in the image is limited to $1/n \times 2\pi/\Delta k$, where $n\Delta k$ in the denominator is the range k_{max}.

Overlaps in object space

In addition to ambiguities in the image, there are ambiguities in the object generated by the same mechanism. In the object space there is a field of view (FOV) which is the region of interest to be imaged. However, there are also the regions outside the FOV which can contribute to the signal and therefore to the image. Indeed there is no clear way to distinguish these. The choice that can be made is the granularity in k-space. This is normally chosen such that the width of the FOV, Δx, is $2\pi/\Delta k_x$ and the height of the FOV, Δy, is $2\pi/\Delta k_y$. This is sufficient to ensure a unique relationship between points in the FOV and in one period of the image. However, the contribution of each point in the FOV has the same phase factors $\exp(ik_x x)$ as a whole family of translated points $x + n\Delta x$ in the object space for all the discrete values of k_x. The same is true in the reconstruction. Just as the data reconstruct to a periodic image, signals from every translated copy of the FOV window are superimposed in the measured signal amplitude and therefore in the reconstructed image, as illustrated in Fig. 7.5. The situation is made worse by the fact that the fluctuations of the spins in the unwanted regions can contribute random noise to the image of the wanted FOV.

Note that here the problem is only in two dimensions because slices are reconstructed one at a time. Spins in object space outside the excited slice do not contribute.

If the mathematics of image reconstruction cannot help the ambiguity problem, can anything be done on the physics side to reduce this effect? Indeed it can. If the space around the FOV were empty of spins there would be no ambiguity. If a smaller pickup coil is used, magnetic flux detected from parts of the slice further away outside the FOV is reduced. Indeed a coil of a similar size to the FOV will have a good sensitivity to the FOV but a reduced sensitivity to objects in the regions outside the FOV and also to their contribution to noise. As discussed later, small pickup coils have additional advantages.

7.1.3 Pulse sequences

There are three simple properties that can be imaged, T_1, T_2 and the hydrogen spin density itself. In practice images are maps of spin density, more or less weighted by the local values of T_1 or T_2 (or T_2^*). The differences are determined by the pulse sequence. There are many options. The initial RF excitation may be a small angle rotation, a 90° rotation or a 180° rotation. Further choices involve the sequence of the gradient pulses, the timing and flip angle of subsequent RF pulses (including the use of spin echo) and the repeat time T_R. We cannot pursue here the details of all of the options. Instead we sketch the principal ideas involved, and show some examples.

T_1 and T_2 weighted images

If the initial pulse is a 180° inversion pulse, after a certain time the longitudinal magnetisation of a region with a particular value of T_1 passes through zero, as shown in Fig. 2.9b. Even a 90° pulse at this time would generate no FID signal from this region. Other voxels with different values of T_1 would generate a signal. This gives heightened contrast between regions with different values of T_1. Use of such a pulse sequence generates a 'T_1 weighted' image, such as shown in Fig. 7.6a.

To maximise the contrast of an image on the basis of T_2 the spin echo technique is used. Consider the effect of the choice of the echo time T_E. Figure 7.7 illustrates the FID amplitudes found in two regions with different values of T_2 but the same initial magnetisation density. It shows that there is no difference if T_E is small. Equally there is no

a)

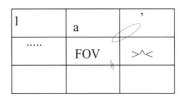

b)

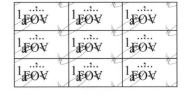

Fig. 7.5 An illustration of the effects of aliassing. a) A field of view (FOV) in object space surrounded by regions of no particular interest. b) The reconstructed image space in which the image is periodic and regions of the object space are superimposed on the FOV of interest.

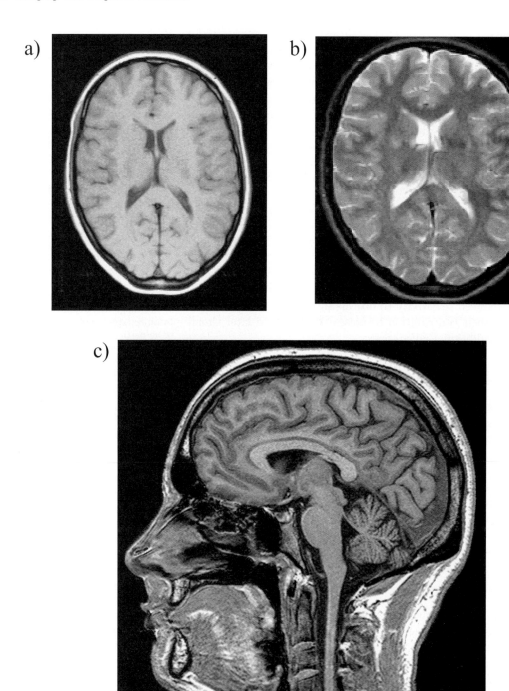

Fig. 7.6 a) Conventional T_1 weighted spin echo image, $T_E = 15$ ms, $T_R = 400$ ms, 1.5 tesla.
b) Conventional T_2 weighted spin echo image, $T_E = 80$ ms, $T_R = 2000$ ms, 1.5 tesla.
c) 3-D spoiled 'FLASH' (T_1 weighted) image, 625 μm in plane resolution, 1.5 mm slice thickness.
[Images kindly provided by the FMRIB Centre, University of Oxford.]

difference at large T_E because both signals have decayed away. If T_E is chosen between the two T_2-values there is a broad maximum in the contrast. This is illustrated as the curve of the difference between 50 and 150 ms in Fig. 7.7. An example of such an image is shown in Fig. 7.6b.

Figure 7.6c shows a high resolution image taken with a sequence called FLASH which uses a small flip angle.

Spatial resolution and timing

Given time, data more finely divided in k-space could be taken. If Δk is halved, the unique FOV is doubled in size, and the occurrence of ambiguities, ghosts or aliases correspondingly reduced. Alternatively the range k_{max} may be extended by taking more measurements deeper into k-space. Then the spatial resolution in the image is improved. But either of these options doubles the scan time required.

Scan time is valuable, and techniques that speed up scans are of the greatest practical importance. They permit increased throughput, enlarged FOV or improved resolution. The pulse sequence Fig. 7.4 can be shortened in a simple way. Because the behaviour is linear, the gradient pulses in the time between the end of the RF pulse and the start of digitisation at time B may be applied simultaneously rather than sequentially. But much more can be done to speed up scanning.

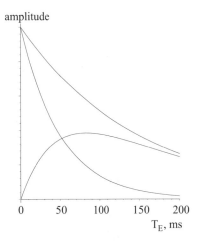

Fig. 7.7 The effect of varying T_E on the contrast between two regions with different T_2. The two decaying curves show the spin echo amplitude decaying with $T_2 = 50$ ms and 150 ms. The difference curve is shown assuming that the spin density in the two regions is the same.

Overlapping scans

The choice of T_R determines the longitudinal magnetisation at the time of the excitation pulse of the next scan. If T_R is long the spins will have completely recovered their thermal distribution. However, this may be a poor choice. For a start, waiting for thermal equilibrium to be reached takes precious time, and anyway the memory of the previous scan can be useful.

A different idea allows complete thermal recovery of the magnetisation of a slice without the penalty of lost time. As soon as the last remnants of the transverse magnetisation of one slice have decayed away, the excitation of a different slice can begin. This is in spite of the fact that the longitudinal magnetisation of the first slice is still recovering towards thermal equilibrium. With the B-field tuned to be on resonance at a different z, the spins in the first slice will not contribute to the FID, will still relax towards thermal equilibrium and will ignore off-resonance RF pulses. Later, after several further slices have been scanned, the pulse sequence can return to excite and scan the first slice again, by which time the longitudinal magnetisation will have reached thermal equilibrium. Meanwhile all the other slices, now off resonance, can be quietly recovering in their turn.

Echo planar imaging (EPI)

Every scanning strategy is a compromise between speed of acquisition and adequate signal-to-noise ratio. Having excited the magnetisation

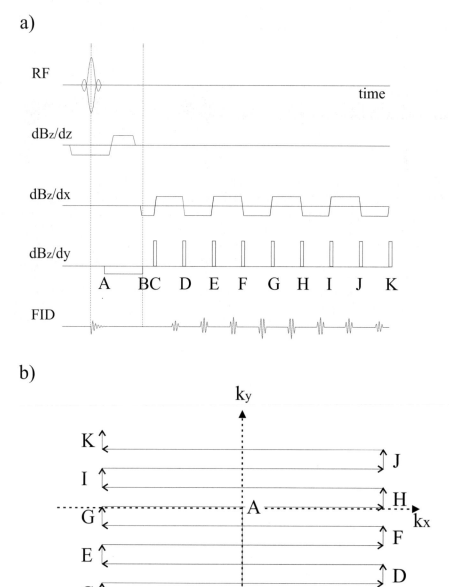

Fig. 7.8 a) An EPI pulse sequence and b) the related sweep in $k_x k_y$-space. The labels A, B,... indicate the k-values that correlate to specific times in the pulse sequence.

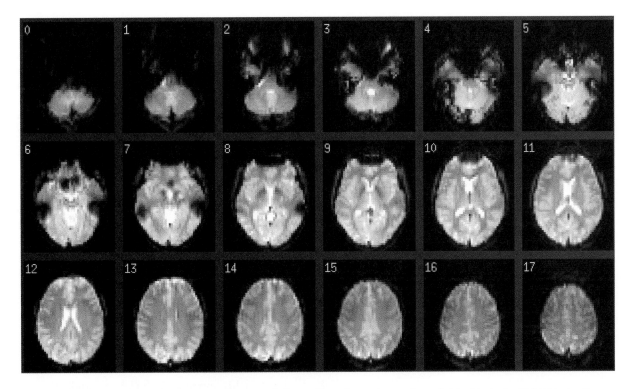

Fig. 7.9 Echo planar image of the brain. 18 horizontal slices collected over 2 seconds, each 64×64 pixels. Field 3 tesla. [Images kindly provided by the FMRIB Centre, University of Oxford]

from its value at thermal equilibrium with an initial RF pulse, a good pulse sequence will seek to scan as much of k-space as possible before signal amplitude is lost and the next scan is started. Echo planar imaging (EPI) is a sequence that scans the whole $k_x k_y$-space for each z-slice following a single RF excitation pulse. A typical EPI pulse sequence is shown in Fig. 7.8a and an example of a set of EPI images is shown in Fig. 7.9.

Figure 7.8b shows how this pulse sequence tracks through k-space with time during the long readout period. The sequence starts with a 90° RF pulse with slice selection by a z-gradient, followed by a reverse gradient, as in the previous discussion. At time A, immediately after the end of the RF pulse, a negative y-gradient is applied for a long enough time so that at B, k_y is large while k_x is still zero. Readout and digitisation of the FID signal begin at this point and continue without break for the rest of the pulse sequence. First a steady negative x-gradient is applied which sweeps the negative k_x-values at constant k_y. At time C a positive impulse is given to the y-gradient coils and the steady x-gradient is reversed. This sweeps the k-vector back towards positive k_x at constant rate. In fact, as the expression for each component of k shows, equation 7.11, the application of steady gradients gives a k-vector which sweeps across two dimensional k-space at constant velocity. As the Fig. 7.8 shows the regular sequence of small positive impulses on the y-gradient coils and the long alternating pulses on the x-coils sweep the k-vector back and forth across the whole k-region. In the middle of each k_x-sweep, it passes through zero. At this point all of the spins are in phase, except for the effect of y which is phase encoded. Such a point is described as a gradient echo or gradient refocus.

Many other rasters have been used for scanning k-space including a spiral sweep. However, the basic idea is the same and the benefits and disadvantages similar.

Advantages and drawbacks of EPI

The advantage of an EPI scan is speed, as shown by Fig. 7.9. For the scan to be fast, k must change rapidly. This means that Δk is large and the spatial resolution is limited.

Another disadvantage is signal loss. The loss of amplitude due to field inhomogeneities accumulates throughout the pulse sequence. These arise from three sources.

 ⋄ Intrinsic inhomogeneities of the static B_0 field. These can be mapped and vary slowly with distance. They are minimised by shimming and by working at the centre of symmetry of the magnet.

 ⋄ Dynamic effects due to eddy currents induced by gradient field changes. These can be reduced by shaping the pulse that drives changes to the gradient coil currents.

 ⋄ Static distortion of the field due to the magnetic susceptibility of the patient or sample. The longer range components of this effect may be reduced by adjusting the current in shimming coils. However, this is a time-consuming procedure and cannot compensate for local short range distortions.

The effect of a chemical shift is similar to a field distortion. Both produce a frequency change. Spins with a chemical shift will be displaced in the image depending on the phase shift that they accumulate during a long scan. A simple calculation of this effect is left to question 7.5.

7.1.4   Multiple detector coils

Encoding with gradients and multiple pickup coils

The birdcage coil which is designed to generate a uniform RF excitation pulse can also be used later in the sequence as the pickup coil. However, significant improvements in scanning speed can be achieved by the use of a dedicated array of small pickup coils instead. This involves an additional idea.

If the distance from a voxel to a coil is r and the coil is small, the contribution to the FID from the precessing magnetic moment in that voxel will fall off with the $1/r^3$ dependence of a dipole field together with

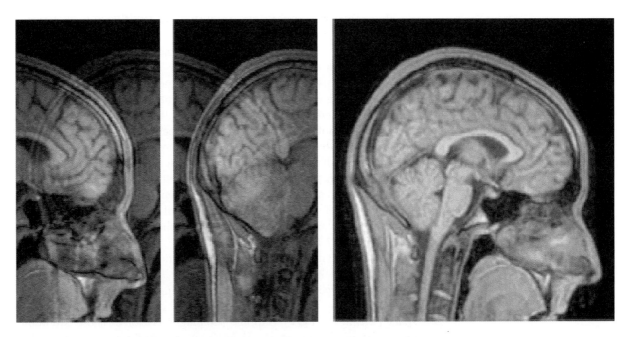

Fig. 7.10 Aliassed images (left and centre) from two local coils in an array are unfolded to one full image (right) without aliassing. [Images reproduced with kind permission of Jo Hajnal, *jo.hajnal@csc.mrc.ac.uk*.]

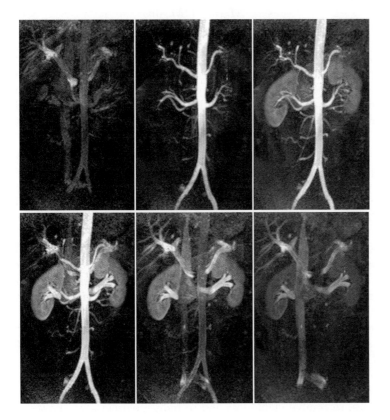

Fig. 7.11 Time-resolved images (4 s per 3-D frame with $R = 3$) of renal blood vessels taken following a bolus injection of a contrast agent. The resolution is $1.6 \times 2.1 \times 4.0$ mm^3. [Image provided by K. Pruessmann and M. Weiger, Institute for Biomedical Engineering, University and ETH Zurich, and reproduced by kind permission of John Wiley & Sons, Inc.]

simple angular factors.[3] So the amplitudes of the FID signals detected by an array of small coils depend on how close they are to the different voxels. In other words, there is useful information in the different signal amplitudes seen by the coils.

This may be used to solve aliassing ambiguities that arise when a large Δk is used. Figure 7.10 shows a simple example. An object is scanned and the FID detected by two small coils. If alternate phase encoding scans which would be required to reconstruct the whole FOV uniquely are omitted, the scan time is halved but the image reconstructed from data taken with each coil becomes ambiguous. The signals from the left and right halves of the object space will be superposed. In other words, the left and right halves of the image will each be the same superposition of the left and right halves of the object. However, the two coils see a *different* superposition, as shown in the left and centre images in Fig. 7.10.

If the sensitivity matrix of each coil to the pair of overlaid points in object space is known and is not singular, it may be inverted and a unique image reconstructed without aliassing. To be specific, consider one point in the image. This is the overlay of two points in the object space (within the FOV). Suppose the magnetisation at those two points are m_1 at r_1 and m_2 at r_2. If M is the sensitivity matrix, the coils a and b will detect signals

$$s_a = M_{a1}m_1 + M_{a2}m_2 \qquad (7.11)$$
$$s_b = M_{b1}m_1 + M_{b2}m_2.$$

Thus if M is known and can be inverted, the true m_1 and m_2 can be determined from the measured s_a and s_b. In principle this calculation has to be repeated for every pair of points in the FOV. This is an extensive computation but is not a limitation in principle. The resulting reconstructed image is shown on the right in Fig. 7.10 free of aliassing and still with the benefit of the reduced scan time.

The sensitivity matrix has to be measured. In practice the effective number of independent coils R is limited because the sensitivity matrix shows symptoms of singularity at R greater than 3 or 4. The matrix differs for different parts of the image, and in some places the sensitivity of the coils is the same, so that the method does not work there. In practice this means that the signal-to-noise ratio deteriorates. Thus, although in principle any number of coils may be used, the reduction in k-space scanning and time saving is limited to a factor 3 or 4. However, this saving can have a decisive effect on the images that can be taken when time is of the essence, for instance when a patient is required to hold his or her breath or when a bolus of contrast agent passes through blood vessels. An example of the latter, an angiogram, is shown in Fig. 7.11. These are complete high resolution images taken with $R = 3$ at 4 s intervals using a contrast agent. They give a graphical picture of blood flowing to and from the kidneys. Such parallel coil techniques can be used with EPI, MRS and other imaging strategies to image rapid changes, to combat signal loss and to reduce patient scan times.

7.2   Functional magnetic resonance imaging

Functional imaging methods based on paramagnetic contrast agents and the detection of blood oxygenation (BOLD) are described. It is explained how the flow and diffusion of proton spins may be measured. Such measurements relate to blood flow and the orientation and connectivity of nerve fibres. MRI may be extended to spectroscopy, such that hydrogen atoms in specific metabolites may be imaged, albeit with inferior signal-to-noise ratio and spatial resolution. The concentration of other species of magnetic nuclei may also be imaged, also with poor signal conditions. The risks, safety and installation configuration of a clinical MRI scanner are summarised.

7.2.1   Functional imaging

Paramagnetic contrast agent

Anatomical images do not answer all the questions. Clinicians need to know what the imaged tissue is doing. The simplest way to get data relevant to such questions is to inject the patient with a weak solution containing paramagnetic ions. The location of the ions is then observed through the large reduction in the relaxation time of all protons in the neighbourhood of any such ion. Such ions act as a contrast agent. Indeed we already saw an example of this in Fig. 7.11.

From the point of view of MRI there is a wide range of possible transition and rare earth ions. The usual choice is the gadolinium ion, Gd^{3+}, which has a very large magnetic moment, $7.98\mu_B$. The larger the magnetic moment chosen, the weaker is the solution with the required magnetic properties. The use of a weak solution minimises any toxic side effects. A concentration of order 10^{-4} mole Gd^{3+} per kg is usually sufficient.[4] The ion is bound in a chelate such as DTPA.[5] This chemical 'wrapper' for the ion is chosen to minimise toxic effects, to have the required selective uptake from the bloodstream in the different organs and tissue of interest, and to enable the effective fluctuation of local magnetic field direction experienced by proton spins in water molecules undergoing frequent interchange as part of the hydration state of the wrapped ion.

Images of tumours are often taken T_1-weighted because of the effect of paramagnetic ions on the local field direction. The abnormal vascularity characteristic of a tumour is seen as an enhanced accumulation of the agent relative to surrounding healthy tissue. In this way a tumour may show up with exceptional contrast, even when it is barely visible on a simple anatomical image.

To maximise the contrast an image is taken without the agent. The contrast agent is then injected into the bloodstream, and its distribution imaged as it is absorbed preferentially by pathological tissue. The agent will often leak out of the blood vessels and persist in the tissue. These areas are bright in a T_1-weighted image, as seen in Fig. 7.11.

[4] The SI unit mole in this instance is the amount of contrast agent that contains the same number of Gd atoms as there are in 12 g of ^{12}C. So 10^{-4} mole per kg means that the fractional mass of Gd ions in solution is $157/12 \times 10^{-7}$, where 157 is the atomic weight of Gd.

[5] Di-ethylene-triamine-penta-acetic acid.

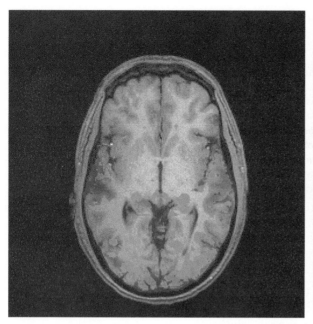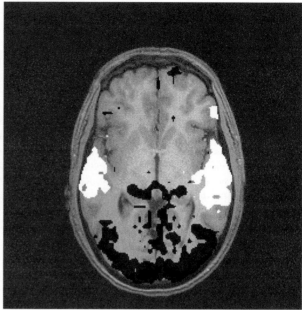

Fig. 7.12 The structural image on the left is also shown on the right with functional data overlaid. Black regions show brain areas responding to 40 s visual stimulation alternating with 40 s rest. White regions show brain areas responding to 30 s auditory stimulation alternating with 30 s rest. [Image kindly provided by the FMRIB Centre, University of Oxford.]

Blood oxygen level dependent (BOLD) scan

An ideal contrast agent is one that occurs naturally in the body, does not have to be injected, is not toxic and does not have to be excreted. Most fortunately such an agent exists. Red blood cells contain haemoglobin. As they carry out one of the functions of blood by transporting oxygen, the haemoglobin changes its form to de-oxyhaemoglobin at the point where it delivers the oxygen, indicating thereby the location of metabolic activity. Iron is a transition element and can change its valency. As a component of haemoglobin in blood, it can switch the diamagnetic oxygen-rich form (oxyhaemoglobin) to the paramagnetic oxygen-poor form (de-oxyhaemoglobin).

Muscular activity increases the return of venous de-oxyhaemoglobin to the heart. The vasculature of the muscle provides a reservoir of oxyhaemoglobin that is then replenished from the blood supply. However, in the brain this is not possible because there is inadequate muscle to provide storage. To protect the supply of metabolites and oxygen needed for brain function oxyhaemoglobin is therefore oversupplied by the blood flow even in the resting state. As a result the venous flow is only partially deoxygenated. Mental activity at a site in the brain requires metabolic activity there. Then, because of the poor storage, an increase in fresh blood flow is stimulated which actually overcompensates for the use of oxygen there. The result is to invert the expected effect. The return (venous) blood flow is actually even more oxygenated than in the resting state. This means that T_2 is increased at a site of mental activity, rather than decreased. Imaging mental activity with T_2 in this way is called the blood oxygen level dependent (BOLD) technique. An example of an image relating mental tasks to activity at particular points in the brain is shown in Fig. 7.12.

As this discussion suggests, the critical point is the change in the impedance of the microstructure (microvasculature) of the blood vessels in response to the activity. The rise time of this response at the onset of activity is 5–8 seconds. There is a similar fall time on cessation of activity. This response function can be determined and deconvolution used to achieve imaging with a response time of a few hundred milliseconds. The intrinsic PSF of the microvasculature response is 2–3 mm. Again some deconvolution is possible but is noise limited. There are also systematic effects coupled to the cardiac and respiratory cycles.

Faster signals of mental activity come from EEG and MEG but they have poor spatial resolution.

7.2.2 Flow and diffusion

Measurement of blood flow

Blood flow can be measured by exciting transverse magnetisation in one slice and then imaging a different slice, such that only those spins that passed from one slice to the other contribute to the detected FID. For example, if the first slice gets a 90° pulse and the second slice a 180° pulse at time $T_E/2$, only the spins that flow from one to the other in that time will give a spin echo FID signal after time T_E. In theory such measurements may be combined with BOLD measurements to extract the changes in blood flow and changes in oxygen usage. In practice blood passes through fine blood vessels with a spread of velocities, and these vessels change impedance in response to stimulation. The reader is referred to the book by Jezzard *et al.* for further information. Blood flow can also be imaged with ultrasound using the Doppler shift as discussed in chapter 9.

In a clinical context images of blood flow can show where infarcts[6] of blood vessels have occurred in stroke victims. These images can now be made available within hours of an actual attack.

Diffusion measurements

MRI can be used to image the diffusion of water molecules in tissue. This is of particular interest because it gives information on the direction of nerve fibres, and therefore on their general connectivity. Within a fibre molecules are free to diffuse axially. However, transverse diffusion is restricted by the fibre walls. The diffusion coefficient D is a symmetric second rank tensor[7] that relates linearly the components of the velocity

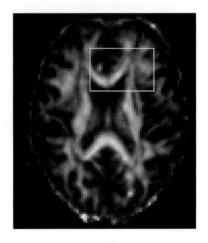

Fig. 7.13 Diffusion anisotropy map showing areas of the brain exhibiting high directional anisotropy in the diffusion coefficient of tissue water. The white box region is expanded in Fig. 7.14. [Image kindly provided by the FMRIB Centre, University of Oxford.]

[6] Obstructions.

[7] Tensors are discussed in chapter 4.

Fig. 7.14 An expansion of the region within the white box of Fig. 7.13 showing anisotropic diffusion. The direction of the principal eigenvector of the diffusion tensor is shown as a small line within each pixel. [Image kindly provided by the FMRIB Centre, University of Oxford.]

vector v to the components of the concentration gradient vector g. Thus

$$v_i = \sum_{k=1}^{3} D_{ik} g_k. \qquad (7.12)$$

The anisotropy of D is described by its irreducible traceless part with five independent components, like the stress or strain of a solid discussed in chapter 4. Diffusion measurements by MRI enable the eigenvalues and eigenvectors of the matrix describing D to be determined. For a cylindrical fibre this is axially symmetric with one large and two small eigenvalues. These represent data on the size and orientation of nerve-fibre bundles. This is a crude first step along the road to understanding how nerves are connected in the brain. A sample of data is shown in Fig. 7.13, with further detail in Fig. 7.14.

7.2.3 Spectroscopic imaging

Combining imaging with spectroscopic measurement

The crude distinction between fat, water, etc., based on measurements of T_1 and T_2 provides some contrast between different tissue types, as we have seen. However, it is a long way from a chemical analysis of what is occurring in the human body. What is really wanted is to image the density, not just of hydrogen atoms but of hydrogen atoms in specific molecules, thereby recording the presence and concentration of those particular molecules. This approach is called magnetic resonance spectroscopy (MRS). It comes with the prospect of rather large voxels or poor spatial information and is intrinsically slow and noisy. It covers not only hydrogen atoms at sites in specific classes of molecules, but also other NMR active nuclei with a workable NMR signal size and a specific chemical or biochemical story to tell, particularly phosphorus and fluorine.

Difficulties of spectroscopic imaging

MRS enables a start to be made on the non-invasive *in vivo* study of the chemistry of the functioning human body and brain. What is needed is a spectrum for each voxel. This is made difficult for a number of related reasons.

⋄ Even in theory, to measure the frequency of a 100 MHz wave with a resolution of 0.1 part per million (ppm) takes a time of order 100 ms. Thus each spectrum must be digitised for a long time.

⋄ A chemical shift of 0.1 ppm cannot be distinguished from a non-uniformity of 0.1 ppm in the applied B-field. Uniformity of the B-field is therefore crucial, and extensive use has to be made of shimming.

⋄ Since the frequency measured during readout is to be interpreted as the chemical shift, no gradient field is applied during this time. Both x and y are measured by phase encoding with an increased number of scans.[8] Figure 7.15 shows such a pulse sequence. So, for example, an MRS image with 100 pixels requires 100 scans, each with a different combination of values of k_x and k_y.

⋄ While the hydrogen atoms of interest may have a modest concentration, the spectrum is dominated by the very large peak due to water. Special techniques are employed to prevent this overwhelming the dynamic range of the electronics. Here is an example. With no gradient fields applied, a long RF excitation with small bandwidth matched to the water signal generates a 90° rotation. This is then dephased with large gradients, so that the magnetisation due to water is small. Now the pulse sequence of interest is applied and,

[8] In some sequences x or y may be selected by RF excitation with a gradient field, like the z slice selection.

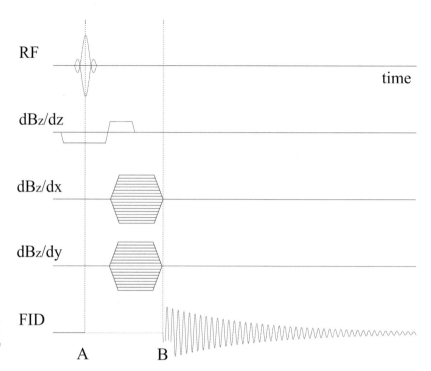

Fig. 7.15 An MRS pulse sequence. Both x and y are encoded by phase. The FID signal is digitised from time B onward, when all gradients are off.

although the longitudinal magnetisation of the water will recover, it will generate no FID provided that the last RF pulse is applied within a time short compared with its T_1.

The presence and concentrations of important metabolites in a sample can be measured *in vivo* by fitting its NMR spectrum as a linear combination of the spectra of known solutions of metabolites prepared *in vitro*. The spectra are even sensitive to the pH value, the acidity of the medium. An example is shown in Fig. 7.16 where peaks corresponding to some of the more important metabolites may be seen.

Few isotopes other than hydrogen generate a strong NMR signal. Even those with spin tend to have low concentrations in biological tissue, small isotopic ratios or small gyromagnetic ratios, as shown in table 2.2. Although there is no problem equivalent to the water peak, the magnetic moment at thermal energies is small on all counts. As a result data are usually taken for a few large voxels to get sufficient signal-to-noise ratio. Imaging of these is also referred to as MRS. The elements most frequently studied are ^{31}P and ^{19}F.

Phosphorus plays an important role in metabolism as various metabolites incorporating phosphate bonds are used by organs for short term energy storage. Fluorine, though not itself biologically interesting, has a large magnetic moment and is used as a tracer to study drug action. Among other isotopes, sodium is used in the academic study of cell action. Tracers labelled with the NMR-active isotopes ^{13}C, ^{14}N and ^{17}O are used in animal studies. Recent developments in hyperpolari-

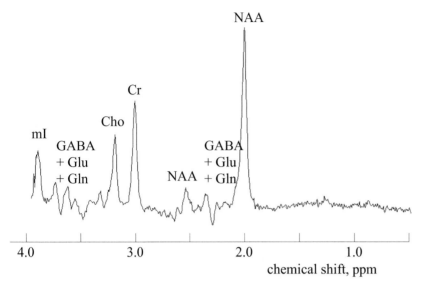

Fig. 7.16 Hydrogen metabolite spectrum collected using the PRESS sequence at 3 tesla. mI=myo-inositol, GABA=gamma butyric acid, Glu=glutamate, Gln=glutamine, Cho=choline, Cr=creatine, NAA=n acetyl aspartate. [Spectrum kindly provided by the FMRIB Centre, University of Oxford.]

sation have shown how signals from these isotopes may be significantly enhanced by dynamic nuclear polarisation (DNP).

In summary, the various techniques of MRI and MRS permit the non-invasive mapping of the flow of fuel, oxygen and services within the patient. This is the information that the clinician needs.

7.2.4 Risks and limitations

Benefits and resources

How effective is MRI as a clinical technique in terms of contrast, definition, speed, safety and cost? Contrast can be achieved by generating H-spin density images weighted by T_1 or T_2. This forms a rich and versatile basis for differentiation of tissue by anatomy or function. Contrast based on more specific chemical information (MRS) comes at the expense of a big increase in scan time and loss of spatial definition. Otherwise spatial definition is good to 1 mm and images may be obtained with a time resolution of a few seconds with some loss of definition in other respects. Although the cost of an MRI scanner is high, the availability of staff and know-how is more critical. Expensive facilities may lie fallow much of the time. Society is prepared to keep supermarkets open 24 hours a day, seven days a week. The need for scanners is more pressing. The availability of radiographers and clinicians would appear to be more critical than the availability of money to supply scanners, at least in the developed world. By the same argument, the widespread deployment of MRI in some parts of the Third World presents serious difficulties.

Safety

MRI appears superficially safe and there is no history of accidents, provided that standard safety precautions are followed. Nevertheless there are a significant number of safety concerns.

◇ The high static B-field itself is not responsible for any known well-defined physiological hazard. The current recommended limit is 4.0 T.[9] However, such a field is dangerous for anyone with a pacemaker or metal implants and MRI cannot be used at all for such patients.[10] It is interesting to note that in connection with overhead power lines and mobile phones there is public concern over microtesla fields!

◇ In the interest of speed the gradient fields would be switched as fast as possible. However, rapid switches of magnetic flux create large electric fields by Faraday's law. These in turn cause eddy currents which are felt as a tingling sensation in the skin and even give pain. This pain occurs at a lower threshold than any serious physiological effects. A maximum dB/dt of 20 T s^{-1} is permitted.

◇ The rapidly changing magnetic forces on the gradient field coils generate acoustic noise. They form a powerful loudspeaker system in a small volume. This noise and the claustrophobic sensation of being confined in a small tube for an extended period make an MRI scan a disturbing, even unpleasant, experience. Headphones with soothing music are not sufficient to overcome this. Significantly, conclusions on psychological research and clinical results are always qualified by the fact that, whatever else may have been experienced, the patient was being subjected to much noise in an oppressive environment. The noise level and spectrum depend on the pulse sequence and may be reduced by avoiding the resonances of the scanner as an acoustic cavity. However, much of the spectrum is around 1 kHz which is unfortunate for the human ear. Maximum permitted noise levels are specified.

◇ Fortunately most patients are not aware that they are lying enveloped by a liquid helium cryostat. One may wonder whether this would be allowed by the safety rules in a physics research laboratory!

◇ The maximum recommended level of RF heating in this frequency range for the general public is 0.08 W kg^{-1} to the whole body, averaged over any 6 minute period.[11] This is to ensure that no part of the body experiences a temperature rise exceeding 1°C. The maximum power that can be absorbed in MRI under saturation conditions is in fact less than 0.01 W kg^{-1}, safely below this level.[12]

[9] As discussed in chapter 1 and references there.

[10] The accidents that have occurred as a result of MRI have been in connection either with pacemakers or with the motion of ferromagnetic objects in the fringe field of the dipole magnet.

[11] See the references given at the end of chapter 1.

[12] See question 7.6.

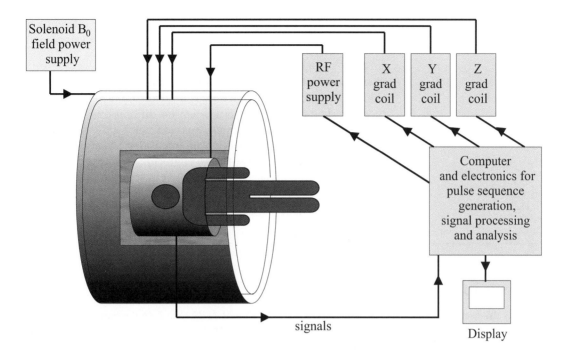

Fig. 7.17 A diagram of the components of a basic MRI scanner.

◇ The fringe B-field in the neighbourhood of such a magnet extends for many times the solenoid diameter and presents a serious hazard. Metal objects can be dragged by such field gradients with considerable force. In spite of the more sophisticated risks discussed above, the greatest danger from MRI is that of being impaled by a flying screwdriver or other loose metal object.[13]

[13]See question 7.7.

Summary of clinical equipment

In Fig. 7.17 we summarise schematically the various elements of equipment that make up an MRI clinical scanner.

◇ There is the large bore static superconducting solenoidal magnet with horizontal uniform B_0 field in the range 1.5 to 4.0 T.

◇ Inside which there is a set of x-, y- and z-gradient coils.

◇ Inside which there is an RF excitation birdcage coil giving 42.58 MHz T^{-1}.

◇ Inside which is the patient with an array of small detector coils to detect the FID (if the birdcage coil is not used for this purpose).

◇ The power supplies for the gradient coils and RF are controlled by the computer and electronics. The electronics also amplifies and mixes the FID signals with the reference frequency to give an 'audio' output, which is then sampled and digitised.

Not shown in the diagram are the B_0 shimming coils, the software and the team of people. The software is needed to control the pulse sequence, monitor the performance, carry out the calibrations, map the magnetic fields, invert the Fourier transforms, extract T_1, T_2, spin density, MRS spectra and many other tasks. The integrated team includes physicists, clinicians, hardware and software engineers.

Read more in books

The Physics of Medical Imaging, ed. S Webb, IOP (1988), an edited compilation of useful articles, somewhat out of date but still frequently referenced

Functional Brain Imaging, WW Orrison, JD Lewine, JA Sanders and MF Hartshorne, Mosby (1995), an edited volume by authors active in the field

Functional MRI, An Introduction to Methods, P Jezzard, PM Matthews and SM Smith, OUP (2001), an edited compilation of articles written by researchers in the field

Fundamentals of Medical Imaging, P Suetens, Cambridge (2002), a fully illustrated book with discussion of image processing

Look on the Web

Look at the book website at
  www.physics.ox.ac.uk/users/allison/booksite.htm

Read about a wider range of MRI pulse sequences than have been discussed here
  MRI pulse sequences

Follow the development of spatial encoding of MRI with multiple pickup coils
  MRI sense coil array

Find up-to-date examples of functional imaging using blood oxygenation
  BOLD imaging

Learn more about the use of MRI for diffusion, flow and perfusion measurements
  MRI diffusion and *MRI perfusion flow*

Look at examples of the clinical use of magnetic resonance spectroscopy
  proton MRS or *phosphorus MRS*

Consult the sites referenced at the end of chapter 1 on MRI safety matters, or search
  MRI safety

Find out about MRI magnet options
  MRI magnets

Questions

7.1 A voxel at position r has a magnetic moment m perpendicular to a static B-field about which it precesses with frequency ω. Show that the oscillating voltage induced in a small detector coil of area A at the origin has an amplitude

$$V = \mu_0 \omega A \cdot \nabla \left(\frac{m \cdot R}{4\pi R^3} \right).$$

7.2 Consider the signals induced by the magnetic moments of a pair of voxels at A and B in a pair of small detector coils at C and D where the four lie at the corners of a square ABCD. A uniform static B_0 field is perpendicular to the square. The normal of coil D points towards voxel A, and the normal of coil C points towards voxel B. Using the result of question 7.1 show that the ratio of signals detected in the two coils due to the precessing component of the magnetic moment of each voxel is $8\sqrt{2}$.

Hence discuss the spatial sensitivity matrix of coils C and D with respect to the positions A and B.

7.3 A water droplet is suspended in the field of view of an MRI scanner. The B_0 field at the origin of coordinates, the centre of symmetry of the MRI scanner, is 2.998015 T in the z-direction. An RF pulse of frequency 127.5987 MHz is applied at the same time as a z-gradient of B_0 is present. A FID signal is not detected subsequently unless this z-gradient field is 7.81×10^{-3} T m^{-1}. (After the RF pulse the z-gradient is reversed for a time such that all hydrogen atoms that are excited have a precession phase which is independent of z.) Two scans are made and two measurements of the FID signal are recorded, both in the presence of a y-gradient of the B_0 field of -12.2×10^{-3} T m^{-1}. In both cases the observed FID frequency is 127.5662 MHz. In the first scan no x-gradient field is applied; in the second an x-gradient field of -1.00×10^{-3} T m^{-1} is applied for a time of 100 μs before readout. In the second case the FID is measured to have an extra phase shift of 1.60 radians compared to the first case.

Draw the pulse-sequence diagram and explain how it works.

Determine (x, y, z), the position of the water droplet.

[The Larmor precession frequency for protons in water is 42.55747 MHz T^{-1}.]

7.4 A close sequence of water droplets as described in question 7.3 form a steady drip falling parallel to y from a single point. Suggest ideas for a pulse sequence to measure the mass of each drop, the spacing of drops and their vertical velocity with time.

7.5 If the hydrogen atoms of question 7.3 were subject to a chemical shift of 4 ppm, how far would they appear to be shifted in the image?

7.6 A paramagnetic salt is dissolved in water at 20°C. The concentration is such that in a magnetic induction of 3 T the proton spin relaxation time T_1 is 10 ms. Discuss the absorption of energy per kg of water in the steady state as a function of the average RF power level. Estimate the maximum power per unit mass that can be absorbed in this way. Comment on the relevance of this result to clinical safety considerations.

[A steady whole-body absorbed power level of 4 W kg^{-1} generates a temperature rise of less than 1°C in a healthy patient.]

7.7 A long thin screwdriver lies outside a magnet of central field B_0. It is drawn point first by the fringe field towards the magnet centre. Its density is ρ and its relative magnetic permeability is μ_r. Ignoring gravity show that when it reaches the centre of the magnet it has a velocity of about $B_0 \sqrt{\mu_r/\rho}$. Put in numbers to show that this might be on the scale of 1 m s^{-1} in a typical case.

7.8 A phantom is composed of two materials, A and B, with T_1 relaxation times 0.75 s and 1.3 s respectively. An NMR pulse sequence starts with a 180° RF pulse which is followed by a 90° pulse after time T'. Describe briefly the FID signals from A and B seen as a function of t. Find the ratio of the FID signal for B to the FID signal for A at time T' as a function of T'. For what choice of T' would the MRI contrast between regions of A and B be greatest?

Medical imaging and therapy with ionising radiation

8

8.1 Projected X-ray absorption images   233

8.2 Computed tomography with X-rays   241

8.3 Functional imaging with radioisotopes   246

8.4 Radiotherapy   256

Read more in books   264

Look on the Web   265

Questions   265

in which the uses of ionising radiation in medical imaging and radiotherapy are discussed and compared.

8.1 Projected X-ray absorption images

The source of X-rays generally used in imaging is bremsstrahlung emission by electron beams incident on high Z metal targets. Electrons are accelerated by a DC voltage or an RF linear accelerator. The efficiency, energy and angular distribution of X-ray production depend on the electron energy and the Z of the target. The X-ray energy spectrum is hardened by passage through a high Z absorber. The quality of the pinhole optics of the camera depends on beam source size, detector resolution and scattering. Image quality is affected by the statistical noise of the number of detected photons. The image contrast depends on the photon energy spectrum and the absorption edges of the regions of different Z within the object. For images of blood vessels and the digestive tract contrast is enhanced by administering iodine and barium as passive high-Z contrast agents.

8.1.1 X-ray sources and detectors

Conventional X-ray tube

From the time of their discovery X-rays have been used for imaging. The conventional X-ray examination, as used in medicine and dentistry, with a simple photographic plate is safe, cheap and widely available. In addition to the photographic plate as detector, a source of X-rays is needed. The initial electron beam can be focused to a small bright spot. But because it is not possible to reflect, refract or focus an X-ray beam, the straight line optics of the pinhole camera must be used to produce the image. Technically there are three possibilities. Intense highly collimated beams of X-rays are produced by electron synchrotrons and X-ray lasers. However, as yet these are too large and expensive to be available for clinical purposes. Secondly there is the production of X-rays using the more manageable linear electron accelerator with an RF power supply. These are used to produce X-rays for radiotherapy

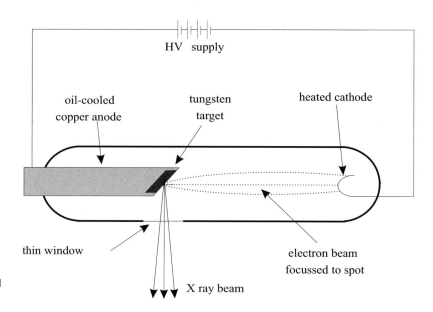

Fig. 8.1 The layout of a conventional X-ray tube.

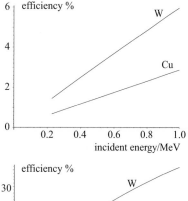

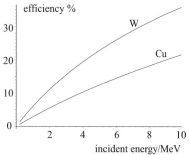

Fig. 8.2 The efficiency with which the energy of an electron beam generates bremsstrahlung radiation in a thick target as a function of the incident energy, calculated for tungsten (W) and copper (Cu). This simplified calculation tends to overestimate the efficiency, especially at low energy; see question 8.1.

where MeV X-ray energies are needed for their long range. For most imaging purposes, however, this increased energy is not helpful. For images using the contrast between the X-ray absorption of elements of moderate Z, a source based on a simple DC electron beam is sufficient. The power supply with terminal voltage up to 150 kV for such a source may use a diode array similar in principle to that in the Cockcroft–Walton accelerator with which they artificially split the nucleus in 1932.

A conventional low energy X-ray source of this kind is shown in Fig. 8.1. Electrons emitted by a heated cathode are accelerated in vacuum through a potential defined by the DC voltage supply (25–150 kV) and focused to a spot on the anode. The anode consists of a tungsten target (W, $Z=74$) set into an oil-cooled copper holder designed to conduct away the heat. The high Z maximises the production of bremsstrahlung relative to $\mathrm{d}E/\mathrm{d}x$ energy loss which is responsible for the heat. The process is very inefficient; the efficiency is calculated in question 8.1 using a simplified model and plotted in Fig. 8.2 for copper and tungsten. As well as having a high Z, tungsten has a very high melting point which increases the tolerance to the high thermal load. A rotating target can be used to spread the local heat load due to a tightly focused beam spot. The X-rays generated leave the vacuum through a thin window.

The raw spectrum emitted by such a target was discussed in chapter 3. It consists of a broad spectrum of bremsstrahlung with added emission lines characteristic of the target. The upper limit of the bremsstrahlung is given by the maximum incident electron energy. For a tungsten target the discrete lines are K_α, K_β, L_α and L_β at 59.3, 67.2, 8.39 and 9.67 keV respectively. For a lower energy source suitable for mammography a molybdenum target is used with a 30 kV terminal voltage. In either case the broad spectrum that emerges includes a flux of low energy photons which will be absorbed in any case. These increase the radiation dose

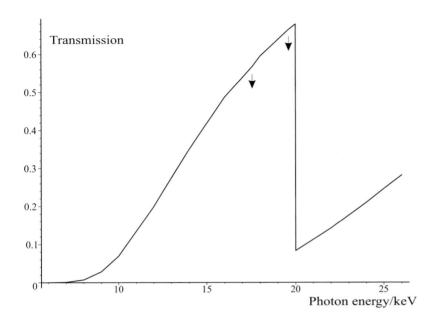

Fig. 8.3 The energy dependence of the X-ray transmission of a 30 μm thick filter of molybdenum. The arrows mark the position of the K emission lines of molybdenum.

without adding information to the image, and are therefore need to be absorbed with a filter. Such a filter has a pass band. Figure 8.3 shows the transmission for a 30 μm window of molybdenum which removes the soft photons and substantially reduces the hard ones using the molybdenum K edge. Used with a molybdenum target the K emission lines of Mo are transmitted. The X-ray spectrum is therefore determined by the incident electron energy, the Z of the target and the Z of the filter (which may be different).

RF linear accelerator and synchrotron sources

As shown in Fig. 8.2 and discussed in question 8.1, raising the electron energy increases both the radiation energy and the efficiency. However, high energy beams are less suitable for imaging because the contrast disappears. The photoelectric cross section falls to zero for all elements and attenuation becomes a function of electron density alone.

Higher energy beams are used for therapy and materials testing. Above a few hundred kilovolts the use of a DC accelerating voltage becomes less practicable. Such electrons are accelerated in bunches in a linear accelerator (linac) with RF power in a waveguide. At energies above about 500 keV the electrons become relativistic and the bremsstrahlung radiation that they create in the target becomes collimated in the forward direction[1] in a cone with angle of order $1/\gamma$ radians, where $\gamma = E/m_e c^2$. Therefore the target geometry shown in Fig. 8.1 is not suitable and the beam must be focused onto a thin transmission target which produces a bremsstrahlung beam in the forward direction as shown in Fig. 8.4. The spectrum is affected by the choice of target thickness as the electron beam loses energy. If the target is removed, the focused electron beam

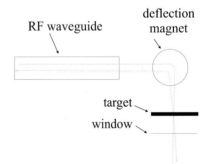

Fig. 8.4 An illustration of the components of an electron linac with steerable beam and an optional target producing a photon beam in the forward direction.

[1]This is because the radiation, which is nearly isotropic in the electron rest frame, is thrown forward by the Lorentz transformation to the laboratory frame.

itself exits though the thin window[2] and may be used for radiotherapy as an electron beam, as discussed in section 8.4.2.

In a synchrotron source the electrons radiate in the forward direction as a result of transverse acceleration perpendicular to an applied magnetic field. Because the radiation is produced in vacuum, there is no source of heating. The beam 'spot' size is finer and more intense than can be obtained in a bremsstrahlung beam. However, such sources are large, expensive and not yet generally available outside research laboratories.

Detectors for projected X-ray images

For many purposes the simple two dimensional photographic plate provides the resolution and ease of use that is required of a detector. Such data are not easily read electronically and have threshold and saturation effects that limit their dynamic range. Increasingly X-ray images are recorded electronically using CCD and other semiconductor technologies as these become cheaper and more widely available. Such images are more easily calibrated, stored and used quantitatively. This is important when relating images taken at different times or under different conditions.

8.1.2 Optimisation of images

Photon energy

[3]This is assumed normal to the xy-plane for simplicity.

The projected image in the xy-plane is the flux[3] $I(x, y)$ that penetrates the absorption density $\rho(x, y, z)$ due to the object and is successfully detected,

$$I(x, y) = \Phi(x, y) \exp\left(-\int \rho(x, y, z)\mathrm{d}z\right)\varepsilon,$$

where $\Phi(x, y)$ is the flux of the incident parallel beam along z and ε is the detection efficiency. However, this equation is too simple. The incident flux, the absorption and the efficiency all depend on the photon energy E in an essential way. At each energy the absorption $\rho(E, x, y, z)$ is the sum of the absorption due to each chemical element present, represented by the product of the cross section σ_i and number density N_i of each. Thus,

$$\rho(E, x, y, z) = \sum_{i,\text{elements}} \sigma_i(E) \times N_i(x, y, z). \tag{8.1}$$

A useful image gives information on the density N_i of elements of interest, albeit projected onto the xy-plane. For each element the dependence of the cross section $\sigma_i(E)$ on energy takes the form of absorption edges for different electron shells as discussed in section 6.2.1, and shown for a number of significant elements in Fig. 3.7. At an energy above the K edge the absorption cross section for each element falls progressively to zero, leaving only the Thomson scattering cross section[4] which is the same for all elements. Images taken at such energies are degraded

[4]The cross section *per atom* is Z times the Thomson (or Compton) cross section.

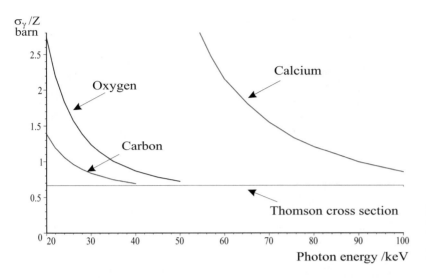

Fig. 8.5 The total photon cross section of carbon, oxygen and calcium, divided by Z, as a function of photon energy. All cross sections fall with energy to the Thomson value, 0.665 barn.

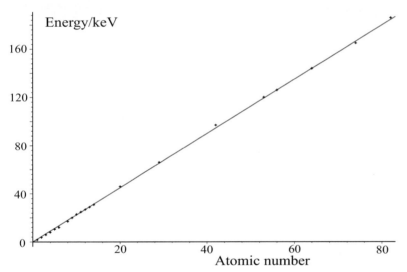

Fig. 8.6 The energy E_{1cm} plotted against Z, where E_{1cm} is defined in the text. Points are plotted for various elements and lie close to the plotted line, $E_{1cm} = 2.25 \times Z$ keV.

by scattering and are only sensitive to electron density, not elemental composition. The energy range of interest for X-ray medical imaging is shown in Fig. 8.5 on a linear scale.

Imaging with X-rays is therefore a matter of making measurements in an energy band in which the elements to be imaged have different photoelectric cross sections. Figure 8.5 shows that up to 70 keV there is a very large difference in the absorption of calcium and the normal elements of which tissue is composed. At energies up to 35 keV there is some discrimination between carbon and oxygen. We may ask at what energy each element becomes effectively transparent. In Fig. 8.6 is plotted the energy at which the absorption length of each element is 1 g cm^{-2} (in other words, 1 cm at the density of water) which fits well to $E_{1cm} = 2.25 \times Z$ keV. Assuming an energy dependence of the

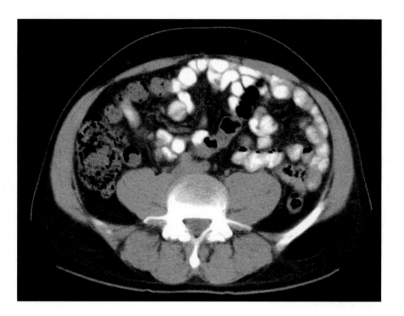

Fig. 8.7 An X-ray image of the intestine with contrast by orally administered barium. The image is of a transverse slice, reconstructed by tomography. The contorted shape of the gut can be picked out in the centre and upper part of the image as it crosses back and forth across the plane of the slice. The whole is contained within the ribcage which is seen in lower contrast due to the calcium. [Image reproduced by kind permission of Medical Physics and Clinical Engineering, Oxford Radcliffe Hospitals NHS Trust.]

cross section $E^{-2.5}$ in the region beyond the K edge, the energy at which the absorption length approximates the thickness of the scanned object may be estimated. Thus the need to penetrate thicker objects leads to a choice of higher energies. In this way the photon energies may be chosen in a range from 17 to 150 keV.

Noise

So much for the signal in an X-ray image, but what about the noise? In this case the dominant form of random noise is statistical. The number of radiation quanta n detected in a pixel of the image fluctuates with standard deviation $\sqrt{n}$. In a relatively coarse-grained two dimensional image the number of pixels is low and only a small radiation exposure is required to achieve an adequate image quality.

Sources of systematic imperfection of the image are the limited spatial resolution associated with the size of detector elements, the spot size of the X-ray source, and the background flux that comes from scattering of the X-rays. At higher X-ray energy the Thomson scattering cross section increases in size relative to the photoelectric cross section; thus scattering becomes more significant.

There is also an incoherent flux of scattered photons due to X-ray fluorescence and these photons are characteristic of the Z of the medium. However, in medical physics the fluorescence yields of low Z atoms are too small to be useful, but remain as a source of background.[5]

[5]Where higher Z atoms are imaged and higher radiation doses are acceptable these XRF signals are used, as discussed in chapter 6.

8.1.3   Use of passive contrast agents

Limitations of simple X-ray images

A simple X-ray image exhibits contrast in terms of the atomic number of the absorbing medium. This is very good for distinguishing tissue from higher Z metals such as iron, copper and silver. The ring on a finger, a coin swallowed by a child, a metal splint – all these show up very clearly. So too do bones through the calcium content. The higher absorption comes not from the linear dependence on Z of the electron density in higher Z atoms, as stated in some medical physics texts, but through the Z^2 dependence of the binding energy of the inner electrons. Lungs and air passages form images with good contrast because of the significant density difference. The presence of a growth in the throat and other airways is readily seen. Otherwise, foreign objects and bones apart, the composition of most tissue is a variable mix of carbon, hydrogen, oxygen with some nitrogen with rather uniform density. Absorption of X-rays by tissue is therefore rather uniform. Additional methods are needed to increase contrast, rather as the paramagnetic agents do for MRI.

Use of barium and iodine

The contrast agents most used with X-rays are iodine ($Z = 53$) and barium ($Z = 56$).[6] These are chosen for their high Z and low toxicity. Iodine is used as a contrast agent for blood vessels,[7] and barium for the digestive tract.

Barium is a divalent metal with a chemistry like calcium and strontium. The patient swallows a so-called barium meal, or, if administered through the rectum, receives a barium enema. The shape of features of the gut can be imaged as the barium gradually coats the stomach and the intestinal tract with its X-ray absorbing properties. Ulcers, growths and hernias can all be seen in high contrast. An example of such an image is shown in Fig. 8.7. The barium is inert and is excreted in the normal way.

Iodine is introduced to the bloodstream by injection and is removed by the action of the kidneys and passes out with the urine. It may be imaged as it spreads through blood vessels, but most examinations involve simply taking images before and after the administration of the iodine.

Digital subtraction techniques

To image the finest blood vessels special techniques are used. The images taken before the contrast agent is administered are compared with those taken afterwards, and these images are digitally subtracted to bring the blood vessels into maximum relief. Examples of the results are shown in Figs 8.8 and 8.9. However, if spatial resolution is not to be lost, the two images have to be registered with one another. This is difficult since some movement is unavoidable.

[6] These agents are the normal stable isotopes of iodine and barium, not the radioisotopes used in other applications considered later in this chapter.

[7] Images of blood vessels obtained in this way are called angiograms.

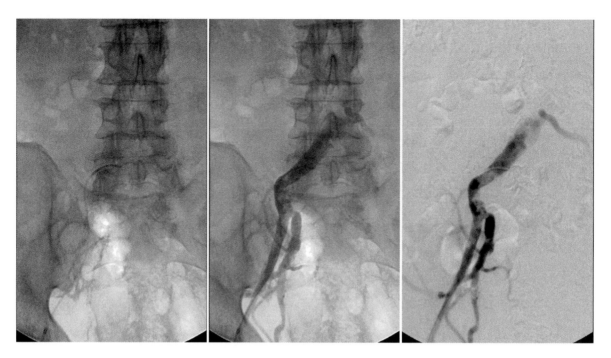

Fig. 8.8 Angiographic images of the pelvis and lower spine. On the left, an initial image. In the centre, the same after injection of iodine as a contrast agent. On the right, a digitally subtracted image formed from the other two and showing the right iliac artery (the main artery dividing from the aorta to supply the right leg). [Images reproduced by kind permission of Medical Physics and Clinical Engineering, Oxford Radcliffe Hospitals NHS Trust.]

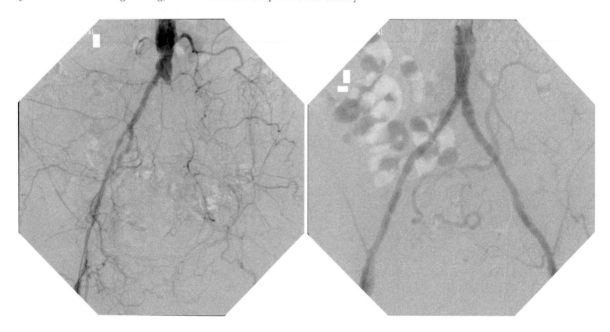

Fig. 8.9 Digitally subtracted iodine contrast images of the iliac arteries, on the left for an abnormal patient, and on the right for a normal patient. The abnormal patient has virtually no left iliac artery. [Images reproduced by kind permission of Medical Physics and Clinical Engineering, Oxford Radcliffe Hospitals NHS Trust.]

An ideal solution to this problem is to take two images in the presence of the contrast agent in rapid succession, one using X-rays just below the energy of the absorption edge of iodine, and the other just above the absorption edge. Ideal monochromatic X-ray beams are available from synchrotron radiation sources but useful results may also be achieved with filters. Similar studies have been carried out on the airways of animals using xenon as an inert contrast agent. The animals inhale the xenon and are illuminated by two monochromatic beams of SR, above and below the xenon K edge. As discussed above, such SR sources are not clinically available under normal circumstances.

The use of contrast agents is limited to the imaging of fluid passages. Further ideas are needed to improve the contrast of general images. The next step is to establish how three dimensional images may be made, free from the superposition of the contribution of those objects which lie above or below the plane of the object under study.

8.2   Computed tomography with X-rays

A three dimensional object distribution can be mapped as a series of two dimensional projections. With a sufficient number of projections this mapping process can be inverted and the three dimensional distribution reconstructed from the maps. The matrix to be inverted may be extremely large. If the projection axes lie in a plane, the reconstruction may be carried out one slice at a time. In this case the inversion is simpler and may be done using a Fourier method. Otherwise, to avoid the large matrix, a maximum likelihood method is used and the solution found by iteration.

The statistical noise in such images depends on the number of photons in each resolved image element. Consequently, for three dimensional image reconstruction the radiation exposure increases significantly with the image quality required, specifically with the fineness of the spatial resolution and with the statistical signal-to-noise ratio of each voxel.

8.2.1   Image reconstruction in space

Linear imaging by absorption or emission

When imaging by absorption the detected radiation flux depends exponentially on the density of absorber. So expressed in terms of the logarithm of the flux the behaviour of absorbers is additive.

In later sections we consider nuclear medicine, that is methods that involve imaging the emissions from administered radioisotopes, namely single photon emission computed tomography (SPECT) and positron emission tomography (PET). Then the measured flux from different regions itself is additive, and the intensity, not its logarithm, is the linear

[8] In nuclear medicine absorption of the emitted radiation within the object is a second order effect. This is because the photon energy is in the range 100 keV to 511 keV and the only process is Thomson scattering for which the correction required is small.

experimental measure.[8] With such a difference understood, the following discussion of 3-D linear imaging applies equally in principle to a spatial distribution of emitting sources as to a distribution of absorption.

Reconstruction in three dimensions by linear inversion

Consider an object described by the 3-D density $\rho(x, y, z)$, defined relative to axes (x, y, z). Data are taken relative to different sets of axes (s, v, t) at N different rotations labelled i relative to (x, y, z), such that, for each i, a 2-D projective density $\zeta_i(s, v)$ is measured integrated over t. We treat ρ as defined for a discrete set of voxels ρ_{xyz} and each ζ_i for a discrete set of pixels $\zeta_{i\,sv}$. The two are linearly related thus[9]

[9] In an oversimplified visualisation an element of matrix $C_{isv,xyz}$ is equal to 1 when the voxel at xyz projects onto the pixel sv at the angle labelled by i, and 0 otherwise. Actual values include partial overlaps.

$$\zeta_{i\,sv} = \sum_{\text{all } xyz} C_{i\,sv,xyz}\rho_{xyz}. \qquad (8.2)$$

The matrix C is very large indeed, but its elements are essentially geometric and known, and most of them are zero. In principle this linear relation can be inverted provided that the matrix C is not singular,

$$\rho_{xyz} = \sum_{\text{all } isv} [C_{i\,sv,xyz}]^{-1}\zeta_{i\,sv}. \qquad (8.3)$$

This condition will not be satisfied if there are too few different projections i or the granularity of the voxels in (x, y, z) is too fine relative to the pixels in (s, v).

This is a general statement of principle. In practice the problem cannot be solved so simply. A relatively coarse image with projective data 100×100, each recorded for 100 different projection angles, would require the inversion of a matrix with $10^6 \times 10^6$ elements to determine the 3-D picture. For data with finer resolution the numbers are correspondingly larger. There are two solutions to this difficulty.

[10] This use of Fourier transforms in image reconstruction is distinct from its use in MRI.

A simple solution is to take data projections in one plane only and then reconstruct slice by slice. Fourier analysis can be used to perform the inversion[10] as described in the paragraphs below.

For some imaging methods such a solution is not possible because the option to take data by slice is unavailable. Then an iterative method must be used to work towards a best-fit solution by the maximum likelihood method, described in outline in chapter 5.

Fourier inversion for slice images

Consider the family of projections $\zeta_i(s, v)$, restricted such that the v-axis is the same for all and is parallel to z, for instance. The planes of projection are illustrated in Fig. 8.10. Then the matrix C is partitioned into submatrices for which the values v and z are equal. At each value of z we have a 2-D transformation between (s, t) and (x, y) for each projection i at angle θ_i.

Figure 8.11 shows such a slice at constant z (and v) of an object density $\mu_z(x, y)$. The data $\lambda_i(s)$ are given as a series of projections onto

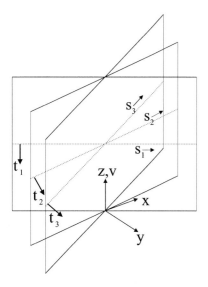

Fig. 8.10 An illustration of a family of projection planes with normals $(t_1, t_2, t_3, \ldots)$, all perpendicular to z.

axes $s(\theta_i)$ integrated in the orthogonal direction $t(\theta_i)$ at different angles θ_i relative to x. Thus

$$\lambda_i(s) = \int_{\text{constant } s} \mu_z(x, y) \, \mathrm{d}t. \tag{8.4}$$

To find $\mu_z(x, y)$ we form the 1-D Fourier transform of λ_i with respect to s,

$$\begin{aligned} P_i(k) &= \frac{1}{\sqrt{2\pi}} \int \lambda_i(s) \exp(-iks) \, \mathrm{d}s \\ &= \frac{1}{\sqrt{2\pi}} \int \int \mu_z(x, y) \exp(-iks) \, \mathrm{d}t \, \mathrm{d}s. \end{aligned} \tag{8.5}$$

But $s = x \cos \theta_i + y \sin \theta_i$ and $\mathrm{d}s \, \mathrm{d}t = \mathrm{d}x \, \mathrm{d}y$, so that

$$P_i(k) = \frac{1}{\sqrt{2\pi}} \int \int \mu_z(x, y) \exp[-ik(x \cos \theta_i + y \sin \theta_i)] \, \mathrm{d}x \, \mathrm{d}y. \tag{8.6}$$

We recognise this as $\sqrt{2\pi}$ times the 2-D Fourier transform of μ_z,

$$P(\boldsymbol{k}) = \frac{1}{\sqrt{2\pi}} \int \int \mu_z(\boldsymbol{r}) \exp(-i\boldsymbol{k} \cdot \boldsymbol{r}) \, \mathrm{d}^2\boldsymbol{r}, \tag{8.7}$$

where $\boldsymbol{k} = (k \cos \theta_i, k \sin \theta_i)$. If data are taken at enough angles to span the $\boldsymbol{k}$-space, $P(\boldsymbol{k})$ is determined.

Then the transformation can be inverted to find the source density in the slice $\mu_z(\boldsymbol{r})$, which is what we are trying to find,

$$\mu_z(\boldsymbol{r}) = \left(\frac{1}{2\pi}\right)^{3/2} \int \int P(\boldsymbol{k}) \exp(i\boldsymbol{k} \cdot \boldsymbol{r}) \, \mathrm{d}^2\boldsymbol{k}. \tag{8.8}$$

To reconstruct the full density in three dimensions this procedure must be repeated for each z-slice in turn. In practice each integral becomes a summation because the pixels and voxels are effectively discrete.

If the data projection axes do not lie in a plane this 2-D Fourier slice-by-slice method cannot be used. Then a maximum likelihood method must be used with a model using the principles described in chapter 5. This takes more computation but is more flexible in the use of scanning hardware and can incorporate the secondary effects of absorption.[11] As computational power becomes evermore easily available, such general methods become more attractive.

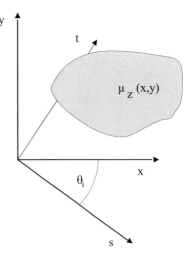

Fig. 8.11 A 2-D diagram of axes and projection directions for the slice $\mu_z(x, y)$ at constant z. The data form a series of projections onto the s-axis defined for each θ_i.

[11] In the Fourier method these have to be treated as a correction.

Reconstructed X-ray CT images

Examples of such reconstructed images free from the effect of overlying and underlying layers are shown in Fig. 8.12. Of course the images themselves are in three dimensions, and what is shown are just slices of the whole. Once the image of the whole has been constructed, whether from a sequence of slices or not, sections or slices may be constructed and viewed in any orientation. In Fig. 8.12 the slices are seen framed in the software display environment with its options to scan through a

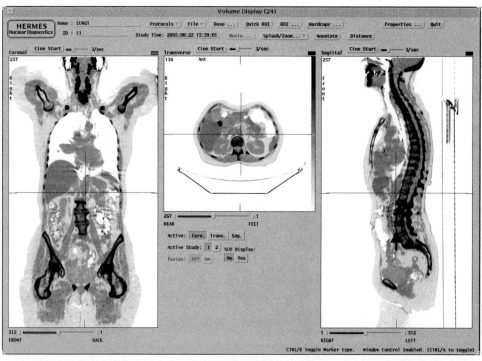

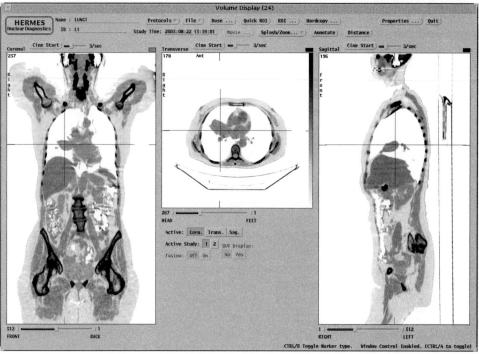

Fig. 8.12 Two sets of orthogonal slices (coronal, transverse and sagittal) from a complete 3-D tomographic X-ray whole-body image. In each case the superimposed crossed hair-lines indicate the position of the section shown in the other two slices of the set. [Images reproduced by kind permission of Medical Physics and Clinical Engineering, Oxford Radcliffe Hospitals NHS Trust.]

progression of slices as a cine at a selected speed. The provision of such a facility is a reminder of the extensive and time-consuming task involved in examining 3-D images and finding features that may be important.

The second row of slices shown in Fig. 8.12 are centred on a growth in the lungs of a patient which later images in this chapter will suggest is tumour. This growth is only visible in X-rays because it is seen in contrast to the air in the lungs. Tumours embedded in healthy tissue are not so simply seen with X-rays. This requires the functional imaging ability of nuclear medicine.[12]

[12]See section 8.3.

8.2.2   Patient exposure and image quality

Noise, spatial resolution and radiation exposure

The statistical signal-to-noise ratio in a data element is $1/\sqrt{n}$ where n is the number of counts in that data element. This noise is random and uncorrelated in different data elements. If an image is formed by taking linear combinations of such elements, as in a tomographic reconstruction, the noise is not increased but does become correlated between different voxels.[13] The SNR is still of the order of $1/\sqrt{n}$ where n is the number of counts detected in the voxel. Thus a picture with SNR=10 and 100×100 voxels in each of 20 slices requires 100 counts in each of 200,000 voxels.

[13]Such a linear transformation conserves information and therefore does not increase noise.

The radiation dose is related to the number of quanta absorbed by the patient rather than the number seen by the detector. However, as is generally true, the information per photon is maximal when, averaged over the exposure, the chance of absorption is 50%.[14] Therefore the radiation detected and the radiation absorbed by the patient should be of the same order.

[14]The proof of this statement is the subject of question 5.1.

The next question is 'How does the radiation dose scale with the spatial resolution Δx, when the same SNR is maintained?' If the number of voxels is increased, the radiation dose is increased in proportion. Thus if the resolution is improved by a factor two in all three dimensions, the radiation dose increases by a factor 8 to maintain the same statistics per voxel.

Consider the step from a single projected 2-D image to a 3-D image. The latter is composed of 20–100 slices or 2-D images at different z. So a 3-D image involves a dose increased by a factor 20–100 compared with a 2-D one.

It is not surprising then that recent improvements in image quality have tended to be accompanied by increases in the general medical radiation dose. However, this has been balanced by improved radiation detection efficiency the effect of which has tended to reduce the doses given.

8.3   Functional imaging with radioisotopes

By injecting a patient with a radioisotope, regions of high metabolic activity may be imaged by the anomalous concentration of the isotope there. For this to be effective most of the radiation from the decay should escape the body without attenuation or scatter, and the half-life of the decay should be matched to the duration of the procedure. Such a long decay time is only possible either with a highly suppressed electromagnetic decay or with a prompt electromagnetic process following a weak interaction decay. This may be achieved with activity giving one or two high energy photons. The procedure with single photon emission (SPECT) involves rather high patient dose and poor spatial resolution, although the contrast is exceptional. With two photon emission (PET) the shortcomings of SPECT are solved but the procedure is more expensive in resources and expertise.

8.3.1   Single photon emission computed tomography

Contrast in radiation images

X-ray absorption can only generate contrast in terms of density and atomic number. Bones, teeth and most foreign objects such as coins, needles or artificial joints show up clearly against tissue. The contrast between air and tissue is also adequate. However, since tissue, healthy or otherwise, has a fairly uniform density and atomic composition, intertissue contrast is poor except in the few cases where simple passive contrast agents can be used. As with MRI, the need is to find techniques that can generate contrast by activity. This allows functional images to be made, as opposed to purely anatomical ones. The simple solution is to administer a radioactive isotope carried by a suitable chemical which is absorbed differentially according to the local metabolic rate. In the region of a tumour where exceptional growth is occurring, the accumulation rate will be high. The image is formed by recording the distribution of the decay locations of the radioactive isotope. This technique gives images of exceptional contrast although the spatial precision is poor compared with CT or MRI.

Choice of radioisotope

There are severe constraints on the choice of a suitable radioactive isotope. The first requirement is that the radiation from the decay should escape from the patient without absorption. Also, in order to make a good image the radiation should suffer minimal scattering. Finally the energy deposition in the patient should be minimised. These eliminate radiation from isotopes that involve any decay by alpha or beta emission. Alpha radiation would be absorbed in a few millimetres. Beta radiation may penetrate a few centimetres but is subject to severe scattering. The same argument eliminates gamma emission with energy less than 70 keV which would be subject to significant absorption by calcium. A secondary requirement is that the radioisotope in the chemical form in which it is administered should not be toxic. This can be quite restrictive since most elements in the periodic table are toxic to a greater or lesser extent.

However, the tightest constraint on the choice comes from a consideration of the acceptable range of half-life for the isotope. The half-life must be long enough for the isotope itself to be prepared, the required chemical solution of it to be made and administered to the patient, the isotope to circulate in the body and be absorbed by function and the imaging data to be collected. The half-life must be short enough that the patient does not suffer the radiation dose of being radioactive long after the scan is over. The effect of a long half-life may be mitigated by excretion or exhalation from the body. A radioisotope with a half-life in the range between 10 minutes and a small number of hours is needed.

Generally lifetimes for gamma emission are much shorter than this. The rates of electromagnetic transitions depend on the energy of the transition and the multipolarity of the decay. Other factors appearing in the matrix element for the rate of the decay process given by equation 3.6 seldom affect the rate by more than one or two orders of magnitude at most.

The multipolarity of the decay categorises the decay rate according to the angular momentum and parity change of the nucleus, and the suppression factor that this generates if angular momentum and parity are to be conserved. The change of angular momentum of the nucleus, ΔJ, is the vector difference between the angular momenta J_i and J_f of the initial and final nuclei. Thus

$$|J_i - J_f| \leq \Delta J \leq J_i + J_f. \tag{8.9}$$

Without orbital angular momentum the photon can only carry away $\Delta J = 1$ unit of angular momentum as the photon has spin 1. The rest of the angular momentum change must be carried by the orbital motion of the recoil in the centre of mass. The field of the photon vanishes near the origin with an extra factor $1/r$ for every increase in this multipolarity, ΔJ.[15] Because of the reduced amplitude of the wave near the nucleus there is a suppression factor in the emission rate of order $(2\pi a/\lambda)^{2\Delta J-2}$ where a is the radius of the nucleus emitting the radiation and λ is the wavelength. Although such suppressed transitions occur, they are termed forbidden, doubly forbidden, and so on.

We are looking for a radioisotope with a γ-decay half-life of order 10^3 s and an energy of about 10^5 eV. A decay with $\Delta J = 1$ and a change in parity is termed an E1 or allowed transition and at this energy (100 keV) would have a half-life of about 10^{-13} s.[16] So we want suppression by 16 orders of magnitude relative to an allowed E1 transition. This suggests a triply forbidden transition ($\Delta J = 4$).

Surprisingly there are transitions that satisfy these requirements.[17] The best match is the excited state of an isotope of technetium, ^{99}Tc, a daughter nucleus produced by the β-decay of molybdenum. The parent ^{99}Mo is produced in the neutron-induced fission of ^{235}U or by neutron

[15]This is a general property of any wavefunction with angular momentum about the origin, and is familiar for the electron wavefunction with orbital angular momentum L in the hydrogen atom.

[16]Since most of the story may be understood in terms of the need for a large ΔJ, we do not dwell here on the effect of a parity change.

[17]There are no such transitions in the electronic structure of atoms because states of high J have high energy, and long-lifetime states are de-excited by collision anyway. (The small spin–orbit coupling effect raises the energy of high J states.) But in nuclei the spin–orbit interaction is large and negative giving some high J states at low energies. Thus there are transitions with ΔJ as high as 5. For example an excited state of ^{192}Ir* has $J = 11$ with a half-life of 240 years for decay to another excited state ($J = 6$) with an energy of 155 keV.

[18]Such long-lived γ-emitting nuclear states are called isomers. Sometimes they are distinguished from the ground state with a superscript ^{m}. We use a superscript *.

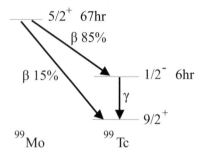

Fig. 8.13 A simplified diagram of the energy levels of ^{99}Mo and ^{99}Tc.

[19]This means that images depend on the approximation of point sources and are defined by straight line trajectories as they pass through narrow collimators.

absorption by ^{98}Mo. Anyway it is readily made in a fission reactor. The nuclear decay scheme is shown in Fig. 8.13. The $\Delta J = 2$ β-decay to the excited state ^{99}Tc* predominates over the decay direct to the ground state.[18] The $\Delta J = 4$ γ-decay to the ground state of ^{99}Tc with energy of 140 keV and a 6 hour half-life satisfies the requirements for the transition that we are looking for. Note that there are no accompanying emissions that would increase the dose to the patient.

The parent ^{99}Mo with its 67 hour half-life is manufactured at a reactor site on a substrate of aluminium oxide and shipped to the hospital by normal transport. The parent radioisotope and its substrate, both of which are insoluble, are immersed in a solution within lead shielding and held in the hospital radiation store. As it is formed, the daughter nucleus ^{99}Tc* dissolves in the solution. After 6 hours or so a concentration of technetium in dynamic equilibrium with the source is available and can be drawn off in solution as needed.

There are other sources of delayed gamma emission that rely on a nucleus that decays by electron capture (EC) with the required half-life leaving a daughter nucleus that decays promptly by emission of the γ of interest without suppression. The electron capture releases a neutrino but deposits no energy except for the kinetic energy of the nuclear recoil; ^{123}I is an example. However, ^{99}Tc* is the radioisotope used in the majority of clinical applications.

Camera and detector

The photons emitted from the patient have to be imaged in order to determine where the active agent has been absorbed. Just as with conventional X-rays, the only available optical imaging technique is the pinhole camera.[19] At an energy of 140 keV most materials are transparent and the highest Z materials must be used for collimators. Figure 8.14a shows that the photon interaction length in lead is about 0.4 mm so that a thickness of a few millimetres is needed in the construction of an effective collimator. The way in which this may be used to construct a camera, known as the Anger camera, is shown in Fig. 8.14b. At a given position of the camera only the decay γ-rays that pass through the collimator give a signal in the detector.

The detector itself consists of a large high-Z crystal to convert each γ into one or more electrons. The detection mechanism involves the creation of a large number of scintillation photons in a transparent dense crystal following photoelectric absorption, possibly with Compton scattering. The traditional crystal used is NaI, which gives a large secondary photon flux. By using a single large crystal the probability of containing all the energy of a photon passing the collimator is maximised. The scintillation light is detected by a number of PMTs around the crystal. The effect of scattered background signals may be reduced by recording the pulseheight of each photon and discarding those inconsistent with 140 keV.

The collimator and detector assembly as a whole is moved and rotated,

a)

b)

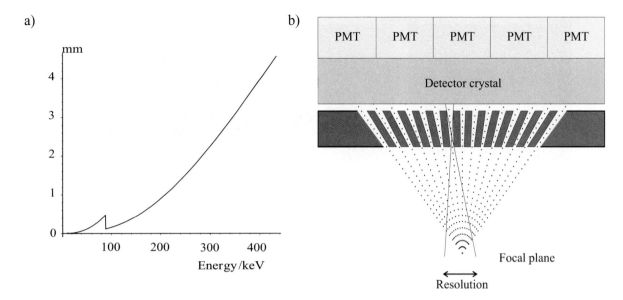

Fig. 8.14 a) The interaction length of photons in lead in mm as a function of photon energy in keV. b) The principle of a SPECT camera.

so as to scan the region of acceptance along and also around the patient. Referring to Fig. 8.14b, the patient lies in the focal plane perpendicular to the plane of the diagram. Each longitudinal slice is scanned by rotating the camera in the plane of the diagram. Other geometries are also used. According to the geometry of the projections there are changes to the reconstruction strategy as discussed in section 8.2. In more modern scanners the single large NaI crystal is replaced by an array of BGO crystals each with their own slit geometry and PMT. This enables more channels of data to be taken simultaneously.

The apparatus is necessarily heavy and has to be moved with precision which requires a very substantial gantry. This creates a claustrophobic environment for the patient, similar to other scan modalities with the exception of ultrasound.

Among other difficulties there is the effect of scattering. At this energy the scattering cross section of γ-rays is not small compared to photoabsorption. As a result typically 30% of the detected photons have scattered, and this represents a background flux in images.

The quality of SPECT data may be seen in Fig. 8.15. The strength and message of the functional signal given by SPECT is immediately evident. The MRI anatomical images shown in Fig. 8.16 show no significant signal in spite of their high spatial resolution. However the two modalities may be combined to give a fused image, if adequately registered, as shown in the bottom row of images in Fig. 8.15.

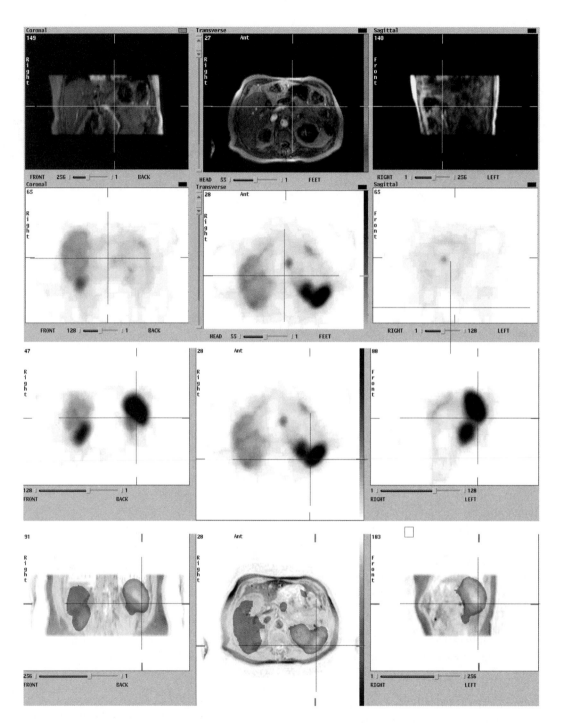

Fig. 8.15 Sets of coronal, transverse and sagittal image slices of a patient's liver. The first row is MRI; the second and third rows are SPECT; the bottom row is a fused MRI/SPECT composite image. The cross-hairs and slices in the lowest two rows intersect at the location of the strong SPECT signal indicating exceptional metabolic activity. This feature is not clear at the same location in the MRI image shown in Fig. 8.16). [Images reproduced by kind permission of Medical Physics and Clinical Engineering, Oxford Radcliffe Hospitals NHS Trust.]

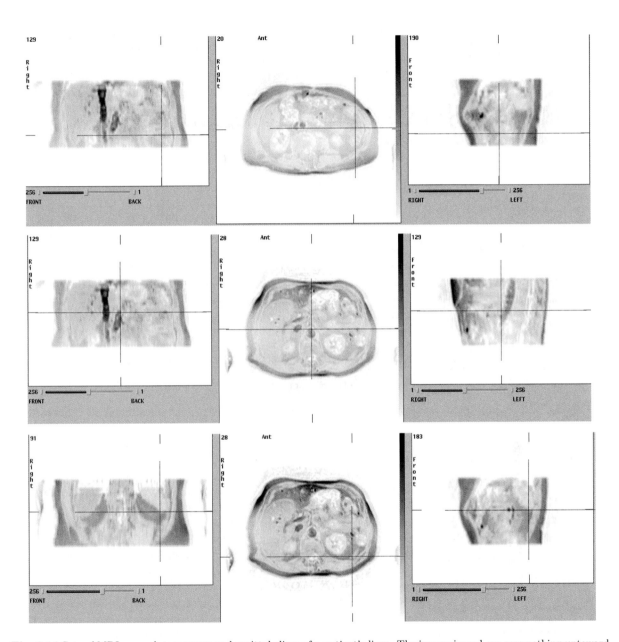

Fig. 8.16 Sets of MRI coronal, transverse and sagittal slices of a patient's liver. The inexperienced eye sees nothing untoward. The cross-hairs and slices in the lowest row are at the location of the high metabolic activity evident in the SPECT image, Fig. 8.15. [Images reproduced by kind permission of Medical Physics and Clinical Engineering, Oxford Radcliffe Hospitals NHS Trust.]

8.3.2 Resolution and radiation exposure limitations

A consideration of the geometry of the collimator slits shows the dilemma for SPECT. The radiation dose to the patient is determined by all the radiation, whether it can pass through the slits or not. The size of the slit determines the spatial resolution as illustrated in Fig. 8.14b. A fine spatial resolution requires fine slits with a small acceptance solid angle $d\Omega$. At any time the ratio of radiation incident on the patient compared with that used in the image is of order $1/d\Omega$. This is unfavourable. To keep the radiation dose to the patient low, wide slits and a coarse image are accepted.

In summary, SPECT is quite practicable and has very high functional contrast. But it has poor spatial resolution with acceptable doses. Since the scan is completed in a fraction of the 6 hour half-life of ^{99}Tc, the patient remains a source of radiation for some hours. An X-ray CT scan also uses collimators to define a small solid angle beam. However, in that case the beam is collimated between the source and the patient. The radiation to which the patient is exposed is already collimated. In SPECT, on the other hand, this is not possible and the collimators are placed between the patient and the detectors. Then the patient is exposed to the full uncollimated radiation, only a fraction of which is used.

8.3.3 Positron emission tomography

Self-collimating principle

The source of the difficulty with SPECT is the collimators. These necessarily discard most of the potential signal in order to achieve reasonable spatial resolution. Is there a way to determine the line of flight of the γ-ray with higher acceptance? The answer is yes, and the method is to use the radiation from positron annihilation as shown in Fig. 8.17.

A positron travelling through matter loses energy like an electron.[20] The in-flight probability of annihilation with an electron is small. Only very close to the end of its path where its velocity has fallen to $\sim \alpha c$ does it pick up an atomic electron and form a bound state of e^+e^- known as positronium. This often self-annihilates[21] with the emission of two γ-rays back-to-back each carrying half the available rest-mass energy of $2m_e c^2$. If both of these γ-rays are detected in coincidence we have the basis on which a line through the annihilation point can be constructed. The time coincidence and the energy constraint are used to reject random coincidences, scatters and 3-photon decay events.

Production of suitable radioisotopes

Positrons (β^+) are emitted by proton-rich radioisotopes that decay by weak interaction,

$$_{Z}^{A}\text{N} \longrightarrow {}_{Z-1}^{A}\text{N} + e^+ \nu. \tag{8.10}$$

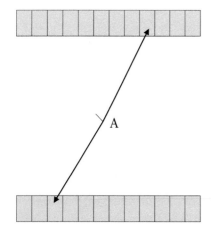

Fig. 8.17 In PET the detection of both back-to-back γ-rays coming from positron annihilation defines the line of flight without the use of collimators.

[20] We have seen that energy loss and scattering are independent of the sign of the charge in first order perturbation theory.

[21] Annihilation occurs in the $L = 0$ state in which orbital angular momentum does not prevent the electron and positron being at the same point. There are two such states that differ in their spin wavefunction, $^1S_0$ and $^3S_1$. The first decays to two 511 keV photons in 125 ps. The second decays to three photons in 142 ns, long enough for it to be perturbed by its environment. This effect is used in condensed matter studies but not in medical physics.

Table 8.1 The principal positron-emitting nuclei used in PET.

Nucleus	Production	Positron		Half-life
		Max. energy	Max. range	
^{11}C	$p+^{14}N\longrightarrow^{11}C+\alpha$	1.0 MeV	2 mm	20 min
^{13}N	$p+^{13}C\longrightarrow^{13}N+n$	1.2 MeV	3 mm	10 min
^{15}O	$p+^{15}N\longrightarrow^{15}O+n$	1.7 MeV	4 mm	2 min
^{18}F	$p+^{18}O\longrightarrow^{18}F+n$	0.7 MeV	1.4 mm	110 min

Such nuclei are quite plentiful and are relatively easily made. Some useful examples are given in table 8.1. This list shows immediately that such isotopes can be substituted for the stable nuclei of C, N, O, etc. in any pharmaceutical agent of choice and the problem of toxicity does not arise. As the nuclei are of low Z and the Coulomb barrier which forms an energy threshold for their production is low, protons from a cyclotron of about 15 MeV have sufficient energy to drive the production reactions given in the table. Because the β^+-decay is a weak process, the half-lives are normally quite long. We no longer need to find a factor 10^{16}. The positron is emitted with a range of energies up to a maximum which determines how far it can go before it stops and annihilates.

Spatial resolution of PET

Figure 8.18 shows a little more detail. If the isotope decays at D, the positron will have a range to its annihilation point A. The distance DA represents an error in the image. As the table shows, the smaller the positron energy, the shorter this range and the longer the half-life. For once the choices go the right way! We seek both a longer half-life and a shorter range.[22] An additional position error comes from the fact that the two γ-rays are not quite back-to-back. At annihilation the positronium is likely to have a velocity of the same magnitude as a typical atomic electron orbital velocity which is of order αc. So an angle $\Delta\theta \sim 1/137$ is to be expected. These two contributions to the spatial resolution are of order 2–3 mm.

The largest contribution to the spatial resolution of PET comes from the size of the detector elements. Detectors are composed of large arrays of small crystals, about 4×4 mm^2, of BGO or calcium tungstate chosen for their high Z and short radiation length. The segmented detector needs to record the position and contain the γ-ray energy, now 511 keV. The position of the photon may best be reconstructed as the centroid of the shower in the detector. Some of the practical problems may be judged by noting that the photon mean free path in lead at 511 keV is about 5 mm, greater by a factor 10–12 than at 140 keV.[23]

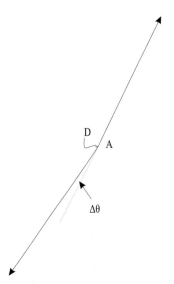

Fig. 8.18 The annihilation point (A) is a distance from the radioisotope decay point (D) depending on the positron range. The line of flight of the γ-rays is not quite back-to-back, angle $\Delta\theta$.

[22] Many PET images are taken with fluorine. Fluorine is not normally present in the body, but it can be incorporated into glucose as fluoro-deoxy-glucose (FDG) without toxic side effects. Since glucose has a high take-up at sites of metabolic activity, this is most fortunate for clinical oncological use.

[23] See Fig. 8.14a.

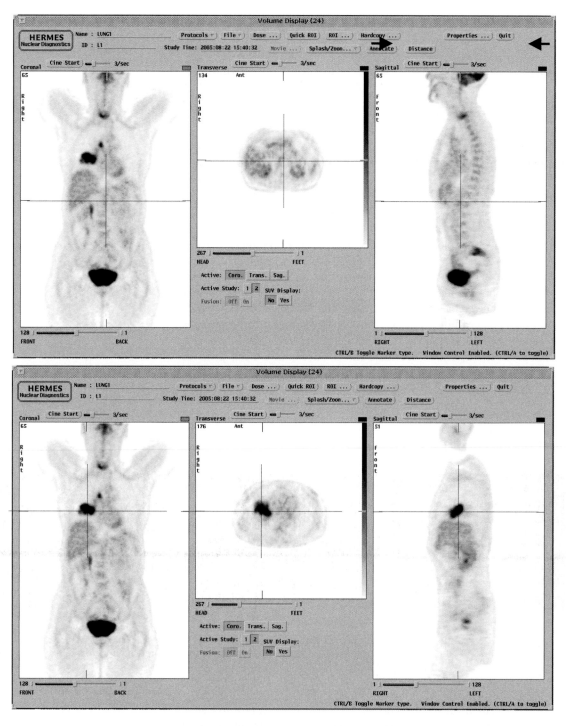

Fig. 8.19 Two sets of slices from a complete 3-D tomographic PET whole-body image. In each case the superimposed crossed hair-lines indicate the position of the section shown in the other two slices of the set. The PET signal fits the position of the anomalous growth in the lung evident in the CT scan of the same patient shown in Fig. 8.12. [Images reproduced by kind permission of Medical Physics and Clinical Engineering, Oxford Radcliffe Hospitals NHS Trust.]

Figure 8.19 shows an example of PET imaging. The effectiveness of the functional imaging is impressive. The spatial information is better than SPECT although worse than MRI or CT.

The major problem with PET is that the half-lives of the radioisotopes involved are only a few minutes, as shown in table 8.1. The most useful isotope ^{18}F has a mean life of 110 min. This makes it possible to transport an activated pharmaceutical from a hospital which has a cyclotron facility to a neighbouring one that only has a PET scanner. The shorter-lived isotopes have to be made on site at an accelerator adjacent to a highly automated hot chemistry laboratory. The reagents made in this laboratory are delivered to the patient by injection, circulated in the bloodstream and absorbed in the region of interest. The whole procedure must be completed within a few half-lives of the radioisotope chosen. Such facilities are available at a small but growing number of hospitals.[24]

[24]In 2003 there were 160 such sites in USA and 120 in Europe. Notably there were only 5 in the UK.

The principal benefit of PET with FDG is its sensitivity at flagging anomalous metabolic activity. The information that this brings to the task of monitoring the progress of cancer is reported to outweigh its high cost. A need for more interdisciplinary skilled staff to run such facilities persists.

The strong signals from a SPECT or PET scan may be fused with MRI or CT scans with their precise anatomical detail. Provided that the registration of the two is sufficiently accurate the fused image may then be used as a basis for therapy. In Fig. 8.15 fusion of MRI and SPECT images was shown for the liver. In Fig. 8.20 an instance of fusion of CT and PET images is shown for a case of prostate cancer with evidence for metastasis.

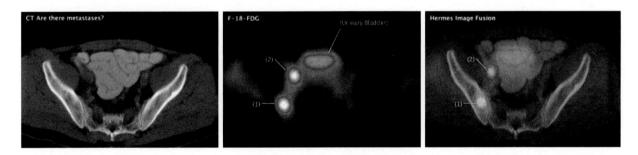

Fig. 8.20 PET–CT image fusion for a case of prostate cancer. Left, anatomical CT image. Centre, PET image showing anomalous take-up of FDG. Right, fused image showing precise locations of tumours. Note the increased glucose utilisation near the right pelvic wall (1). Hermes Image Fusion reveals that this hot spot is in the right ilium directly lateral of the sacroiliac joint. Lymph node metastasis right iliac region (2), with increased FDG-uptake. [Image reproduced by kind permission of Dr Richard P Daum, *info@rpbaum.de* and *www.zentralklinik-bad-berka.de*.]

8.4 Radiotherapy

Ideally the task of therapy is to deposit energy uniformly within the volume of a tumour and not to deposit energy outside in healthy tissue. The precise shape and location of the tumour are established using one or more of the imaging modalities. In radiotherapy the energy source may be a high energy collimated γ-ray or electron beam, an ion or proton beam, or a radioactive source. The radioactive source may be external or internal. Treatment by an implanted internal source is known as brachytherapy. In general it is not possible to match the radiation dose to the tumour volume, but the best compromise maximises the uniform irradiation within the tumour volume and minimises the dose to the surrounding tissue. The dependence of cell survival on dose is non-linear in both space and time. This is exploited to improve the effectiveness of treatment. Oxygenation and hydration also affect this dependence. Precise delivery of therapy relative to a diagnostic image depends on registration, choice of energy and scattering. The use of ion beams has advantages over γ-ray or electron beams in respect of focusing, scattering and the depth–dose profile.

8.4.1 Irradiation of the tumour volume

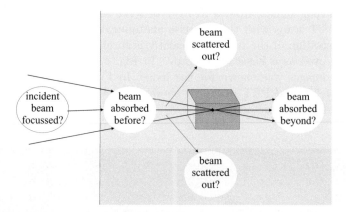

Fig. 8.21 A sketch of the problem of the radiotherapy of a deep-seated tumour. The volume to be irradiated uniformly is shown as the heavily shaded region within tissue which should be spared, shown lightly shaded.

The objective of radiotherapy (RT) is to deliver a fatal dose of many Gy to the cells in the tumour while minimizing the damage to healthy peripheral tissue. The problem is illustrated in Fig. 8.21. Energy may be absorbed by healthy tissue for any of the following reasons:

⋄ the radiation is not aimed at the target region, because of inadequate focusing;

⋄ the radiation is not aimed at the target region, because of alignment errors in registration;

⋄ the radiation is scattered out of the beam;

⋄ the depth distribution is such that radiation is absorbed in overlying tissue before it reaches the target;

⋄ the depth distribution is such that radiation passes through the target region and is absorbed in healthy tissue beyond.

The importance of these varies according to the therapy radiation source. For example, the depth distributions sketched in Fig. 8.22 are quite different for a charged ion beam and a γ-ray beam. Whatever method is used the predicted deposited energy distribution is calculated in 3-D; this process is called treatment planning.

8.4.2 Sources of radiotherapy

Brachytherapy

A solution adopted in the early days and still used in certain cases today is to implant a radioactive source at the site of the tumour. Low energy γ-sources, such as ^{192}Ir, ^{125}I (35 keV) or ^{103}Pd (21 keV), are used. The strong attenuation and the simple geometric fall-off of the flux with distance from the implant delivers a highly localised dose. Sources may be surgically implanted in accordance with the treatment planning procedure to give a well-defined dose distribution. The relative locations of the sources and tumour are imaged with ultrasound or MRI. The method is only in common use today in situations where the tumour is particularly close to a radiation-sensitive region, such as in the treatment of the prostate, which is very close to the rectum. Doses up to 100 Gy are given.

External radioactive sources

Beams of γ-rays from an external radioactive source may be used for RT. Usually ^{60}Co is used. This emits γ-rays with energy 1.2 and 1.3 MeV, and β-rays with 0.3 MeV. Such beams have drawbacks when compared with the high energy X-ray and γ-ray beams from accelerators discussed below. The radiation source cannot be turned off when not required. It has limited brightness as a point source providing irradiation in a narrow cone of interest. But it also has a high isotropic flux. This generates a high scattered background and requires extra shielding to protect staff. Such sources are used extensively for non-medical applications, but therapy with such sources is largely confined to Third World facilities.

Electron and photon beams

Most RT today is provided by photon or electron beams. The machines to provide such beams were discussed in section 8.1.1. The energy is chosen to provide the penetration depth required, and this depends on whether the tumour is on the surface or is deep seated. For surface tumours RT with an electron beam may be used.[25]

For deep-seated tumours RT is usually delivered by a beam of γ-rays of 1–20 MeV from an electron linear accelerator. Any low energy components of the spectrum would be absorbed rapidly in the surface layers and contribute no dose to the target volume. This effect is reduced by hardening the beam with high Z filters. In this range of photon

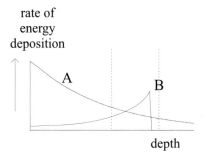

Fig. 8.22 Sketch profiles of the depth distribution from two sources of radiotherapy: A, a γ-ray beam, and B, a charged ion beam.

[25] As shown in table 3.1, in low Z materials the energy loss of electrons is predominantly non-radiative up to 70 MeV. Therefore, for electrons of less than a few MeV, the contribution of bremsstrahlung is small and they have a range of roughly 5 mm per MeV at the density of water.

energy, Compton scattering is the dominant form of energy deposition. The recoil Compton electrons deposit their kinetic energy close to the line of flight of the beam, but the scattered photons travel away at an angle of order $1/\gamma$ radians where $\gamma = E/mc^2$. Thus the effect of beam scatter reduces with energy. This angle is $\sim 1°$ at 20 MeV.

The transverse size of the irradiated volume depends on the beam size. The beam cannot be focused and its size is determined geometrically by the electron spot size on the target anode and by the sets of collimators which are shaped to match the tumour. The energy is deposited over a wider area than that defined by the collimators because of the finite spot size, the multiple scattering of the secondary electrons, and the effect of the scattered secondary photons. The longitudinal size of the irradiated volume in a given orientation is poorly defined by the range of the radiation. In the simple terms of Fig. 3.1, this is dominated by the exponentially distributed point at which each photon interacts. It results in a rather featureless longitudinal distribution of energy deposition. As a result substantial energy is deposited both in front of, and beyond, the target volume. By rotating and translating the beam relative to the patient, the dose to healthy tissue may be spread out. Because of the non-linear relation between dose and cell damage, this spread in the dose distribution results in an important reduction in damage to healthy tissue. But for this effect, photon beams would be quite unsuitable for deep therapy.

Ion beams

Each of the shortcomings of photon RT may be addressed by the use of ion beams. Such a beam might be a proton beam of variable energy up to about 200 MeV, which would have a range in water of about 20 cm.[26]

As shown in chapter 3 the rate of energy loss, dE/dx or LET, of an ion of charge Ze and velocity βc varies as Z^2/β^2. So for most of its trajectory such an ion loses energy relatively slowly, but the energy loss rises rapidly as its velocity falls at the end of its range. The increased dose, delivered to a small volume where the protons stop, is known as the Bragg peak (see Fig. 3.12).[27] By choosing the initial energy carefully to get the right range, most of the dose may be deposited at the required depth. No dose at all is deposited beyond the target region, and only a small fraction in the overlying tissue.

The use of a charged ion beam brings the following improvements to the transverse profile of the deposited dose.

- ◇ Unlike a γ-ray beam, a charged ion beam may be focused into the treatment volume very precisely using magnetic lenses. Spot sizes on the scale of 1 mm can be achieved.

- ◇ An ion beam passing through matter creates negligible secondary photons.

- ◇ An ion beam does not itself scatter significantly and remains tightly collimated compared with an electron or photon beam. In question 8.4 using equation 3.52 we calculate

[26]The density of tissue and water are almost the same. Since the range–energy relation for dE/dx is determined by density, we may treat water and tissue as equivalent.

[27]This is broadened somewhat by straggling.

that the RMS scattering angle for a proton beam passing through 10 cm of tissue is about 1°. Its large mass ensures that each proton continues on an approximate straight path even to the point where it stops.

Such therapy requires a substantial accelerator and there are already several such clinical facilities with more under construction. If light ions such as carbon are used the LET may be increased further. The energy distribution of the Bragg curve is then amplified further by the non-linearity. Such facilities are under development.

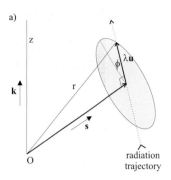

8.4.3 Treatment planning and delivery of RT

Inversion problem

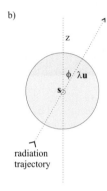

The radiation dose delivered in three dimensional space is the superposition of the radiation absorbed from a number of pencil rays. In a simplified picture we consider each such ray depositing energy with an absorption coefficient μ per unit density ρ. The geometry of a ray is shown in Fig. 8.23. Relative to the origin at O and the z-axis, each such ray may be described by a vector $\boldsymbol{s}$, the point of closest approach to the origin, and ϕ, the azimuthal angle of the direction of the trajectory in the plane perpendicular to that vector relative to the z-axis.

We write $\boldsymbol{k}$ as a unit vector along the z-axis, and the vector $\boldsymbol{s}$ in terms of its length s and unit vector $\hat{\boldsymbol{s}}$. Then a unit vector along the ray is

$$\boldsymbol{u} = \frac{(\hat{\boldsymbol{s}} \times \boldsymbol{k})\sin\phi + \hat{\boldsymbol{s}} \times (\hat{\boldsymbol{s}} \times \boldsymbol{k})\cos\phi}{|\hat{\boldsymbol{s}} \times \boldsymbol{k}|}. \tag{8.11}$$

The equation of the line of the ray is then

$$\boldsymbol{r} = \boldsymbol{s} + \lambda\boldsymbol{u}, \tag{8.12}$$

where λ is the distance along the ray from the point of closest approach to the origin. An element of flux may be written $\Phi(\boldsymbol{s}, \phi)\, \mathrm{d}^3\boldsymbol{s}\, \mathrm{d}\phi$. Then the energy deposition per unit volume of $\boldsymbol{r}$-space is found by summing over all elements of flux and values of λ to give a dose

$$D(\boldsymbol{r}) = \mu\rho(\boldsymbol{r}) \int \int \left[\int\int\int \Phi(\boldsymbol{s}, \phi)\, \delta^3(\boldsymbol{r} - \boldsymbol{s} - \lambda\boldsymbol{u})\, \mathrm{d}^3\boldsymbol{s} \right] \mathrm{d}\lambda\, \mathrm{d}\phi. \tag{8.13}$$

Looking at this relation we consider the task. The required dose distribution $D(\boldsymbol{r})$ is known. It is uniform over the volume of the tumour and zero outside.[28] We want to find the distribution of flux $\Phi(\boldsymbol{s}, \phi)$ that should be applied to achieve it. We have the relation 8.13 between the two.[29] In fact it is rather more realistic to consider small discrete elements of volume and turn these integrals into summations. Then the linear relation between the elements of flux Φ and the dose to voxels D can be expressed in matrices, as was the case for the general CT problem for imaging, equation 8.2. Then, as in equation 8.3, the matrix has only

Fig. 8.23 The geometry of a trajectory defined by the vector of its closest approach $\boldsymbol{s}$ to the origin at O and azimuthal angle ϕ in the plane normal to $\boldsymbol{s}$. The vector $\boldsymbol{r}$ is the position of an element of tissue on the trajectory. This geometry is seen a) as viewed in a general direction, and b) as viewed in the plane normal to $\boldsymbol{s}$ showing the azimuthal angle ϕ.

[28]In fact the irradiated volume is chosen to be slightly larger than the tumour itself to allow for imperfections of alignment or registration between the treatment and imaging coordinate systems.

[29]The 3-D δ-function in equation 8.13 reduces this 5-D integral down to a 2-D one. These may be thought of as the two angles, polar and azimuth for example, of the radiation that deposits energy in the voxel at $\boldsymbol{r}$.

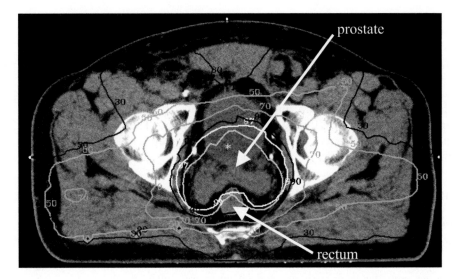

Fig. 8.24 A 2-D section of the dose contours (shown as a percentage of the peak value) calculated as part of the treatment plan for RT of the prostate. The contours are shown superimposed on the anatomical section. Note the re-entrant contours around the rectum. Such doses are delivered in a number of fractions spread over several months, as discussed in the text. [Image reproduced by kind permission of Medical Physics and Clinical Engineering, Oxford Radcliffe Hospitals NHS Trust.]

to be inverted to derive the elements of flux Φ required to deliver the elements of dose D. Mathematically this is atraightforward; the inversion is not singular.

But this attractive conclusion is flawed. Common sense tells us that we cannot irradiate a volume deep inside the body with such a superposition of rays without also irradiating the overlying tissue to some extent. What is wrong with our calculation? The answer is that such a computation describing a perfect treatment would require the use of negative as well as positive fluxes! With positive flux elements alone the distribution of delivered dose includes irradiation of tissue elements that we do not wish to irradiate.[30]

[30]These ideas may be extended by considering μ to be a function of λ to describe the range distribution in brachytherapy or the Bragg curve in ion beam therapy.

Intensity-modulated radiotherapy (IMRT)

We can still define the treatment planning task with the help of equation 8.13. We need to determine the flux elements $\Phi(s, \phi)$ that minimise the irradiation of the rest of the tissue while maintaining the required dose $D(r)$ inside the tumour. With the constraint that flux elements be positive this is no longer a linear problem and must be solved iteratively.

As available computational power has increased, the completeness and precision of planning calculations have improved. The optimisation of the treatment plan in terms of the model is driven by the delineation of the region to be irradiated and the sensitive regions to be spared. Numerical optimisation procedures can be set to work to find by iteration the set of fluxes that come as close as possible to the desired dose distribution.

Delivery of such elements of flux requires a beam that can be rotated, translated and modulated in intensity. In practice this involves irradiation at a number of angles with a parallel beam which can be modulated in shape by a 2-D collimator composed of dynamically controlled 'fingers'. The optimisation is then made in terms of the actual degrees of freedom of the particular therapy accelerator, its collimators and geometry. If there are restrictions on the beam angles available because of the position of the patient, these can be taken into account in the treatment planning. The irradiated volume may follow complex 3-D contours with re-entrant regions that reduce exposure to sensitive organs. This procedure is called intensity modulated radiotherapy (IMRT).

Because the plan is a 3-D problem it is not easy to visualise from 2-D representations. An example of a set of 2-D dose contours is shown superimposed on an anatomical image in Fig. 8.24. This treatment plan provided for the uniform irradiation of the prostate, minimising the irradiation of the rectum which is particularly sensitive and close by. This is achieved by five beam settings each with particular collimator configurations. A 2-D sketch, Fig. 8.25, indicates the general directions of the flux but it does not describe the 3-D nature of the problem and its solution, and does not show transverse shaping of the beam by the collimator fingers.

Figure 8.24 shows the difficulty rather clearly. A dose of many Gy is required to treat the tumour volume but the peripheral healthy tissue would receive as much as 50% of that over a wide area. The actual value of the dose is determined by the tolerance of the peripheral healthy tissue. Thus an improved plan would permit a higher peak dose to be applied to the tumour with a higher probability of successful treatment. The treatment for the plan shown in Fig. 8.24 called for a total peak irradiation of 60 Gy. To permit the peripheral healthy tissue a chance to recover this is spread over 20 fractions, as discussed later. This means that the healthy tissue is subject to 1.5Gy for each fraction, a significant dose which will leave scar tissue.

The details of treatment planning differ depending on the method and energy of irradiation. With γ-ray RT, the beam is attenuated slowly and a reduced flux reaches deeper layers. Compton scattering and electron–positron pair production must be included in the calculation of the dose distribution map. For brachytherapy and ion beam therapy the effects are different but a precise treatment plan is needed nonetheless. In every case allowance has to be made for systematic errors, particularly in alignment.

Fig. 8.25 A rough 2-D sketch of the directions of radiation flux required in the IMRT plan, Fig. 8.24.

Dose monitoring

During the delivery of RT the irradiated region should be precisely aligned with the tumour concerned. This alignment depends on the mapping of imaging information into the coordinate system of the planning and therapy delivery system. Depending on the location of the tumour this may require

◇ an RT delivery coordinate system defined with a set of low power laser beams relative to which the patient may be aligned;

◇ a set of visual marks on the patient's skin or implanted metal targets that can be related directly to the coordinate system of the scanned images and related treatment plan;

◇ a number of tightly defined mechanical restraints to which the patient is attached;

◇ a portal imaging system in which the transmitted RT flux is used to provide an image relating the treatment volume and to the physiology of the patient in real time.

The high photon energies used for RT show little contrast for any physiology properties apart from (electron) density.[31] Consequently such portal images have poor contrast and are not ideal for aligning the patient's physiology with the treatment plan.

[31] As discussed in chapter 3 such photons are far removed from any photoelectric absorption edge showing discrimination between elements.

The good spatial definition of ion beam therapy puts extra demands on registration which must be matched to it. Real-time imaging during therapy becomes increasingly important. When the spatial resolution of the beam is finer than the volume to be treated, the dose distribution may be delivered by scanning the beam on a raster. With such resolution the dose to the target volume can be increased, the dose to peripheral tissue reduced, and the success rate of the therapy increased.

The other monitoring function concerns the measurement of the delivered flux. In the first instance this comes from the settings of the X-ray source and the exposure time. The portal imaging device monitors this dose. Traditionally this was done with photographic plates and ionisation chambers, the former for the position information and the latter for the absolute calibration of dose. These are now replaced by amorphous silicon pixel arrays that provide fast readout of the portal image with good linearity.

8.4.4   Exploitation of non-linear effects

Cell repair

As discussed so far in this section it has been implicitly assumed that the effect of irradiation on the cells of a tumour is linear. However, we have seen already in chapter 6 that this is not the case. This is because cells can repair themselves if the DNA does not suffer multiple damage.

As a result, cell death is not a linear function of dose. The dependence of cell mortality on dose is shown qualitatively as curve A in Fig. 8.26. Such behaviour follows if cells suffering single hits can survive by repair. If in a given treatment peripheral healthy tissue gets half the dose, the cell mortality would be described by curve B in the figure. For a treatment corresponding to the vertical line, the healthy tissue would then suffer 10% cell mortality while the tumour cells would suffer more than 80%. In this illustration the non-linearity brings the benefit that the tumour cells suffer eight times the mortality although the dose is

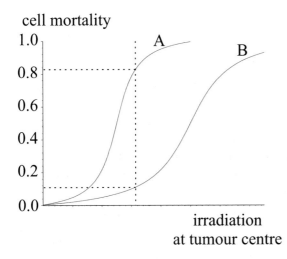

Fig. 8.26 The cell death in a tumour as a function of RT dose might be given by a curve A. If the peripheral tissue around the tumour receives half the dose, then its curve for cell death would be displaced to the right by a factor 2 (curve B), assuming that it is otherwise similar to the tumour. So a treatment described by the vertical dotted line would give a cell mortality for the tumour that was over 80% and a cell mortality in the peripheral tissue that was only 10%.

only doubled. Of course the dependence varies but the benefit of the non-linear effect of cell repair is general.

In choosing the level of irradiation the dose is increased as much as possible consistent with the survival of otherwise healthy tissue.

Fractionation

The repair mechanism takes time, and therefore it matters how the dose is delivered as a function of time. To maximise the non-linearity just discussed, the repair mechanism must be given time to work. Clinically the dose is described as being divided into a number of 'fractions'. These treatments are separated by recovery periods of duration chosen to match the cell reproduction time. This may be a day or two, depending on the tissue or organ concerned and the age of the patient. This protracted procedure is called fractionation. The need for this iterative treatment increases costs significantly and contributes to a miserable patient experience.

If the energy dose were better targeted there would be less need to appeal to the repair-related non-linearity to improve the relative effect on target and peripheral tissue. Thus while γ-ray treatment calls for 20 and more fractions, ion beam treatments need 2–4. The dreadful patient experience of RT is directly related to the need to recover so many times over a prolonged course of treatment. These improvements are the prospects held out by ion beam therapy.[32]

[32]Focused ultrasound therapy described in chapter 9 holds out similar prospects.

Oxygen dependence

Cell death at a given level of radiation dose is increased by a factor of 2 or more by the presence of oxygen. Thus oxygen levels in normal tissue are responsible for enhanced cell damage when compared with cells in tumours which are normally low in oxygen (hypoxic). The consequences of this unfortunate effect are reduced by fractionation. During the recovery periods the oxygen levels in the tumour start to return to normal as the runaway metabolic activity slows. As a result, later fractions benefit from oxygen levels in the target tumour that were absent when the initial fractions were given.

Further sources of non-linearity

Any source of non-linearity can improve the separation of the survival of peripheral tissue from the mortality of treated cells. The steeper the curves sketched in Fig. 8.26, the better will be the result.

In chapter 6 the non-linear response of tissue to highly charged slow ions was discussed. The biological damage caused with such ions is greater than with γ-rays, joule for joule, by an order of magnitude and more. This effect is crudely described by the radiation weighting factors w_R in table 6.4. This effect depends on the LET. Consequently as an ion slows and stops in tissue, not only does the deposited energy rise as $1/\beta^2$ described by the Bragg peak but the damage rises even faster due to w_R. This extra non-linearity is such that the number of fractions may be reduced and the dose to the tumour raised. The effect is expected to be even more pronounced with multiply charged ions such as carbon.

Read more in books

A Century of X-rays and Radioactivity in Medicine, RF Mould, IOP (1993), a book of early pictures and interesting authoritative history

The Physics of Medical Imaging, ed. S Webb, IOP (1988), an edited compilation of useful articles, somewhat out of date but still frequently referenced

Contemporary IMRT: Developing Physics and Clinical Implementation, S Webb, IOP (2004), a recent text on radiotherapy, see also earlier books on radiotherapy by the same author

Look on the Web

Look at the book website at
www.physics.ox.ac.uk/users/allison/booksite.htm

Search for further examples of images including the use of contrast agents with combinations of
CT, angiography, SPECT, PET

Look at manufacturers' websites and the images that may be downloaded, for example
GEhealthcare or *Hermesmedical*

Look on the Brookhaven database for nuclei that might be suitable for nuclear medicine
nuclear data BNL

For radiotherapy searching is a bit harder. Many websites are concerned simply to reassure patients and are short on factual information. Try some of
brachytherapy, radiotherapy treatment planning, IMRT

For prospects in radiotherapy
proton therapy, BASROC or *carbon therapy*

Questions

8.1 Consider the following simple calculation of the efficiency of bremsstrahlung production by electrons. If it is assumed that electrons have $\beta = 1$, the rate of energy loss with distance is $-E/X_0 - b$, where b is the rate of non-radiative energy loss. Show that the solution to this equation is

$$E(x) = (E_1 + bX_0) \exp(-x/X_0) - bX_0$$

for an electron whose energy is E_1 at $x = 0$.
Hence show that its range is $X_0 \ln(1 + E_1/(bX_0))$, and that the efficiency, the fraction of energy emitted as bremsstrahlung, is

$$(E_1 - Rb)/E_1 = 1 - X_0 b \ln(1 + E_1/(bX_0))/E_1.$$

[This simple model neglects K shell emission, angular distributions, scattering and self-absorption. By assuming that $\beta = 1$ it over-estimates the bremsstrahlung efficiency, especially at low energy.]

8.2 When $^{99}\text{Tc}^*$ decays to ^{99}Tc emitting a 140 keV γ-ray, the nucleus recoils. Calculate this recoil energy.

8.3 A patient is irradiated with X-rays in a CT scan. The 3-D spatial resolution is 1 mm and the statistical noise is 1%. Make a reasoned estimate of the minimum dose that the patient receives.

8.4 A 175 MeV proton beam passes through 10 cm of tissue. Using data in table 3.1 show that the RMS scattering angle of the beam is just over $1°$.
[Treat the tissue as equivalent to water and neglect the effect of energy loss by the proton.]

8.5 A radiotherapy beam of carbon ions with charge 6+ is to penetrate a depth of 20 cm in tissue and stop. Find the required incident beam energy.

Ultrasound for imaging and therapy

9

in which the use of ultrasound for medical imaging and therapy is explored, and the important roles of scattering and non-linearity are emphasised.

9.1 Imaging with ultrasound

The direction of objects that emit sound may be determined by passive imaging. In active imaging the observer emits a sound wave that it is reflected by the object. The detected signal is the convolution of the emitted wave train and the reflection coefficient as a function of radial distance. The resolution of the spatial imaging information depends on the wavelength used. In the non-destructive testing of materials both longitudinal and shear waves are used. Medical imaging uses longitudinal waves for which tissue materials differ little in mean properties (with the exception of bone, tooth and air). Contrast in images is made possible by differences in scattering from inhomogeneities, contrast agents, Doppler shifts by moving fluids, and harmonic generation.

9.1.1 Methods of imaging

Passive directional imaging

Sound is used for navigation and communication by humans and other creatures over distances large compared with a wavelength. The frequency range of 100–20,000 Hz employed by humans corresponds to a wavelength range in air of 3.3 m down to 20 mm for a sound velocity in air of 330 m s^{-1}. In water the velocity is five times higher and the wavelength five time larger at the same frequency. Sounds in this range are easily generated by bio-mechanical or electro-mechanical transducers.

In passive imaging the direction of sound emitted by an object is sensed by using several detector elements. For example, the use of two ears is sufficient for animals to detect the direction of sound sources in the horizontal plane.

If the wavelength is small compared with the size of the head, the head casts a sound 'shadow' so that the ear furthest from the source hears a weaker signal. Thus for humans in air at frequencies above 1500 Hz the relative sound intensity is used to determine direction. If the wavelength is large compared with the width of the head, the waves diffract around

9.1 Imaging with ultrasound 267

9.2 Generation of ultrasound beams 272

9.3 Scattering in inhomogeneous materials 280

9.4 Non-linear behaviour 290

Read more in books 305

Look on the Web 305

Questions 306

the head and there is no shadowing. However, the direction may then be detected by the difference in phase of the sound waves heard by the two ears. For the human head this phase difference gives directional information in air at 500 Hz and below. At frequencies around 1000 Hz there is little directional information. In water these frequencies would be five times higher.

Directional information can be improved dramatically using parabolic mirrors to focus the sound.[1] In espionage technology these principles are extended with large diaphragms and sensitive microphones. Similar passive detection methods are used in the sea to detect ships, submarines and creatures such as whales and porpoises. However, the passive use of sound gives little indication of range.

Active imaging and range by timing or frequency shift

In the active use of sound for imaging, a sound pulse is emitted and the echo of the reflected signal is detected. The timing of the echo gives the range of an object. The directional information is added either by emitting the pulse in a specific direction or by sensing the direction of the echo. In practice both may be used. Such active imaging is used less by humans than by other creatures, most famously bats who employ it to sense a complete 3-D image of their surroundings.

The traditional method is to emit a short pulse and to determine the range distribution of a number of objects by measuring the time-lapse spectrum of the reflected signals.[2] An alternative is to emit a continuous beam with a frequency that changes linearly with time. The difference in frequency between the transmitted beam and reflected signal at any time is proportional to the delay since the received beam was emitted. This frequency shift therefore encodes the radial distance to the reflecting surface. The phase as well as the amplitude of the reflected signal is measured by interference with a reference beam. This approach is called frequency scanning interferometry (FSI). As in the traditional time-lapse method the waves add linearly so that a number of reflecting surfaces may be imaged simultaneously. This method requires the source and imaging transducers to operate over a wide bandwidth. Probably for technical reasons FSI has not been used much in ultrasound imaging to date. However, it is used widely in laser metrology.[3]

In FSI the frequency reflected by moving objects depends on their range and their velocity through the Doppler shift. By analysing signals with the frequency ramping down as well as up, such objects may be located in both range and radial velocity.

Range resolution

The spatial resolution is limited by the wavelength, both transversely in angle and radially in range. Using higher frequencies (20–100 kHz) bats achieve high resolution.[4] They employ combinations of pulse emission and FSI. In the technical and medical use of ultrasound, to get the benefit of the best resolution and directional information frequencies

[1] A parabolic shape is appropriate to a source at infinity. For a single source at a finite distance the shape should be an ellipse. The Winston cone is a shape optimised for a distribution of distant sources. The intuitive conical horn or ear trumpet is a qualitative solution.

[2] This is the type of scan used with RF waves in radar.

[3] The Michelson interferometer is the archetype.

[4] This is about the highest frequency that can be achieved with a transducer without the use of piezoelectric materials which are not available in living tissue, one presumes.

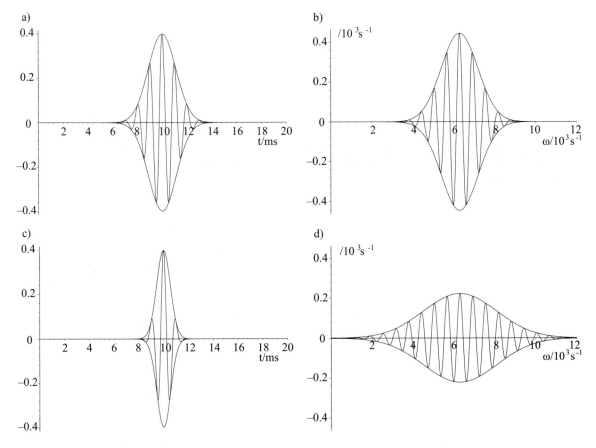

Fig. 9.1 a) A broad Gaussian 1 kHz pulse with $\Delta t = 1.1$ ms plotted in time t.
b) The frequency spectrum of a) showing $Q = \omega/\Delta\omega = 7$.
c) A narrower Gaussian pulse with $\Delta t = 0.55$ ms plotted in time t.
d) The frequency spectrum of c) showing $Q = 3.5$.

up to 10 MHz or beyond may be used. At 1 MHz the wavelength is 1.5 mm in water and this sets the scale of spatial resolution in medical applications. The angular resolution is λ/D where D is the width of the transmitter (or receiver). Although the resolution improves if the frequency is increased further, the range over which ultrasound can be used becomes shorter due to increased absorption.

Let us see how this works. For illustration, Fig. 9.1a shows a pulse $f(t)$ of frequency 1 kHz emitted at time $t = 10$ ms with a width $\Delta t = 1.1$ ms.[5] The reflected signal $F(t')$ observed at time t' is the convolution of f with the reflection coefficient as a function of range, $R(r)$,

$$F(t') = \int f(t' - 2r/c)R(r)\mathrm{d}r. \qquad (9.1)$$

By taking Fourier transforms the radial profile $R(r)$ may be extracted from the observations. Using equation 5.14 the Fourier transform of the object can be written down. However, for frequencies for which

[5]We show Gaussian pulses for which the width Δt is conveniently defined as the standard deviation σ. Then, since the frequency spectrum, the Fourier transform of the time pulse, is also Gaussian, the width of the spectrum $\Delta\omega$ may be defined in the same way.

the PSF is very small, the inversion becomes singular. In other words, spatial detail can only be imaged for those spatial frequencies $k = 2\omega/c$ for which spectral components are present in the pulse. Figure 9.1b shows the values of ω for which this is the case. At other values of ω, and thence k, the reconstruction involves dividing by a small number, thereby amplifying noise in the image.

A pulse narrower in time may be used, as shown in Fig. 9.1c, giving a broader spread in frequency, Fig. 9.1d. The range of spatial information $\Delta k = \Delta\omega/c$ is greater than with the pulse broader in time. So it is the bandwidth and noise that determine spatial resolution. Deconvolution may be used to boost resolution but noise ultimately prevents the amplification of signal at spatial frequencies that are effectively absent.

Imaging of objects in three dimensions

This treatment may be extended to imaging in three dimensions. As well as the range information, the observed signal involves the convolution of the angular distribution of the transmitted pulse, the reflectivity at different angles and the response of the detector as a function of angle. Suppose that, when the detector is pointed at an angle θ', the sensitivity to signals at angle θ is $g(\theta' - \theta)$. Then the observed signal

$$F(t', \theta') = \iint f(t' - 2r/c) g(\theta' - \theta) R(r, \theta) \mathrm{d}r \mathrm{d}\theta. \qquad (9.2)$$

As in optics, the angular pattern of a radiator or receiver with well-defined aperture d involves an angular resolution of λ/d with sidelobes at angles $\pm 1.5 \times \lambda/d$. A similar angular resolution is involved in passive imaging. Angular resolution good at the level of a degree or so therefore requires radiators or detectors with sizes on the scale of 50λ. At acoustic frequencies in the animal kingdom this would be very cumbersome. Nature's solution is to use an array of discrete detectors, a pair of ears. For ultrasound such arrays may be conveniently small.

In general the signal received is the convolution of the object and the 3-D PSF. To reconstruct an image of the object, measurements are taken that sample as much of $\mathbf{k}$-space as possible, as with other modalities. This may be achieved by sweeping the centre spot of the PSF both radially and azimuthally in time as discussed. To complete the 3-D process it should be scanned in polar angle also.

9.1.2 Material testing and medical imaging

Non-destructive testing

The improved reliability of everyday items and technical products in the second half of the twentieth century depended on the development and application of quality control testing of the integrity of materials. This is known as 'non-destructive testing'. While other modalities in this book such as X-rays are also used, sound is often the preferred method of searching for cracks, scratches and other hidden flaws. The changes in density and elastic properties at such discontinuities give rise to large reflection coefficients for ultrasound. For instance, major impedance changes occur at gaseous or liquid occlusions that do not support transverse waves. The high resolution required to resolve small blemishes means that frequencies in excess of 10 MHz are usually used. This is because of the high velocity of sound in metallic and other rigid materials for which wavelengths are otherwise correspondingly longer.

Medical imaging

There are difficulties in applying ultrasound to medical imaging.

The density and compressibility variations in biological tissue are small. Table 9.1 shows the variations in longitudinal sound velocity and impedance. Apart from bone and tooth on the one hand, and air on the other, materials are rather similar. Table 9.2 shows that reflected energies are also small from most interfaces of possible interest. Tissue materials vary in composition from sample to sample so that the numbers in tables 9.1 and 9.2 are rather imprecise as well as being close in value to one another. Further, the tissue boundaries that exist are not sharply defined and are not planar. One might conclude from these arguments that ultrasound medical imaging would not show effective contrast.

There are six reasons why ultrasound is an effective imaging modality in spite of these difficulties.

1. The random thermal noise level in ultrasound is low, so that small signals can be detected. MRI is troubled by the low signal-to-noise ratio that accompanies the small signal from 1 in 10^5 magnetised protons. Ionising radiation modalities are troubled by the large statistical fluctuations that accompany the use of large energy quanta. Relative to the competition, ultrasound images have good signal-to-noise ratio, at least in principle.

2. Reflected ultrasound signals come from incoherent scattering, not from specular reflection by sharply defined planar surfaces. Such scattering is caused by the fact that the materials of living tissue are not homogeneous. Materials are distinguished by the extent to which they are inhomogeneous and generate scattering, rather than by differences in their mean properties. In fact some materials such as blood are remarkably inhomogeneous.

3. Inert contrast agents that increase scattering may be introduced. Because such agents do not rely for their efficacy on any chemical effect or biological action, they are non-toxic and present no hazard beyond their physical bulk.

4. If these inhomogeneities are moving longitudinally, ultrasound can measure their velocity through the Doppler shift in the frequency of the scattered waves.

Table 9.1 Values of the speed of longitudinal sound (c_P, in m s^{-1}) and longitudinal impedance (Z_P, in units of 10^6 kg m^{-2} s^{-1}) for various tissues and substances. [Quoted from Evans & McDicken, table 3.1]

Material	c_P	Z_P
Air (NTP)	330	0.0004
Amniotic fluid	1510	
Aqueous humour	1500	1.50
Blood	1570	1.61
Bone	3500	7.80
Brain	1540	1.58
Cartilage	1660	
Castor oil	1500	1.43
Cerebral spinal fluid	1510	
Fat	1450	1.38
Kidney	1560	1.62
Eye lens	1620	1.84
Liver	1550	1.65
Muscle	1580	1.70
Perspex	2680	3.20
Polythene	2000	1.84
Skin	1600	
Average soft tissue	1540	1.63
Tendon	1750	
Tooth	3600	
Vitreous humour	1520	1.52
Water	1480	1.48

Table 9.2 Calculated reflection coefficients at normal incidence for the amplitude R_A and for the energy R_E. [Quoted from Evans and McDicken, table 3.4]

Boundary	R_A	R_E
Fat/muscle	0.10	0.0108
Fat/kidney	0.08	0.0064
Muscle blood	0.03	0.0007
Bone/fat	0.70	0.4891
Bone/muscle	0.64	0.4123
Lens/aqueous humour	0.10	0.0104
Soft tissue/water	0.05	0.0023
Soft tissue/air	0.9995	0.999
Soft tissue/PZT	0.89	0.800
Soft tissue/polyvinylidene difluoride	0.47	0.0022
Soft tissue/ castor oil	0.07	0.0043

5. In addition to the small variations in longitudinal sound velocity and impedance shown in table 9.1, there are variations in shear modulus between different tissues. At normal incidence these do not affect the reflection of longitudinal P waves. However, as seen in chapter 4, significant polarisation changes in reflected and transmitted waves are to be expected at other angles of incidence. The shear modulus is sensitive to fluidity, muscle tension and other properties of clinical interest. Indeed, in so far as palpation involves a linear process, it is a procedure for sensing a local variation of shear modulus, due for instance to a tumour felt as a lump.

6. Contrast may be generated through the non-linear response of tissue materials. At moderate amplitudes this is seen as harmonic frequency generation, a process characteristic of non-linear phenomena in general. Materials differ more in their non-linear properties than they do in their linear ones, specifically sound velocity and density. It follows that imaging by harmonic generation usually exhibits better contrast than fundamental imaging. In addition contrast agents may be introduced which enhance harmonic generation.

Because the mean acoustic properties of tissue do not vary much, image distortion through refraction is quite mild and can be corrected. In this respect at least the homogeneity of tissue materials represents a useful benefit.

9.2 Generation of ultrasound beams

The generation of ultrasound depends on the use of piezoelectric resonators, made of a material with a non-centrosymmetric crystal structure. Quartz with an internal tetrahedral structure is the classic example. A number of other materials are now commercially available. A resonator consists of a slice of thickness $\lambda/2$ so that the resonator has antinodes at front and back boundaries. In the far-field region (distance $> d^2/\lambda$) the wavefront from such a transducer of width d is approximately spherical. A focusable and steerable ultrasound beam may be emitted by a phased array of such transducers. Such an array can also act as a receiver. The PSF of the beam is determined by the geometry of the array and the wavelength.

9.2.1 Ultrasound transducers

Problems at high frequency

The choice of high frequencies to optimise image resolution has two major drawbacks. Although there are a number of different mechanisms of sound absorption, they all increase with frequency.[6] Ultrasound signals have a short range, especially in materials of low rigidity such as soft tissue. Hence lower frequencies are used in medical imaging than in non-destructive testing. The penetration depth of the required image determines the choice of frequency.

The second disadvantage of the choice of high frequency concerns generation and detection. Effective sources and detectors need to be high frequency mechanical resonators. For a high frequency mechanical resonance a low inertial mass and a high restoring force are required with strong coupling to both electrical input and acoustic output. An ultrasound detector can employ the same resonance as the generator, excited by its coupling to the incident pressure wave and losing energy through coupling to an electrical output signal, instead of the other way around.

Thus for high frequencies we need materials that couple electrical and mechanical behaviour of matter directly and firmly. Piezoelectric materials have this property.

Piezoelectricity is the strain generated in a material by an applied electric field, or vice versa. If the material is constrained a stress will be involved also. The coupling between the stress and strain in the piezoelectric resonator and the stress and strain in the medium of propagation depend on their respective impedances and the mechnical boundary conditions. The coupling on the electrical side depends on the electrical input impedance that the resonator presents to the external circuit. An additional process is that a medium can lose energy internally through mechanical hysteresis. Together these couplings, shown schematically in Fig. 9.2, determine the Q of the resonator and the efficiency with which incident acoustic energy is transformed to electrical energy, and vice versa.

Ferroelectricity and piezoelectricity

An ultrasound transducer that produces acoustic output transforms a voltage signal, a longitudinal E_z, into $\varepsilon_{33} = \partial u_z / \partial z$, the longitudinal strain. Such a transducer is shown in Fig. 9.3. The strain of the active element ε, assumed linear, depends on the third rank[7] piezoelectric tensor d_{ijk}. In general

$$\varepsilon_{ij} = d_{kij} E^k + S_{ij}^{lm} T_{lm} \tag{9.3}$$

where E^k is the electric field, S_{ij}^{lm} is the elastic compliance tensor, and T_{lm} is the stress tensor. When coupled to a low impedance acoustic medium, the piezoelectric material is not subjected to a mechanical stress and the second term nearly vanishes. This means that there is an antinode at the surfaces of the piezoelectric material.[8] We write 'nearly' as a reminder that the medium driven by the transducer does have a small impedance, otherwise no energy would be transmitted at all.

Only certain materials have the lack of a centre of symmetry in their crystal structure required for a non-zero piezoelectric effect. There are two classes, those with pre-existing electric dipole moments, and those

[6]Low frequencies are often preferred for passive imaging. Fog horns operate at the lower limit of audible frequencies in order to maximise the range of signals under foggy conditions. Fog scatters strongly at short wavelengths.

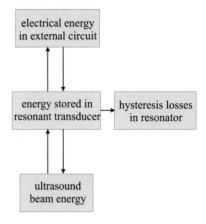

Fig. 9.2 The energy stored in a piezoelectric resonator is supplied by the external circuit and transmitted to the ultrasound beam. Acting as a detector the energy flow is reversed but is also much smaller. The resonator also loses energy to heat internally by hysteresis.

[7]For a discussion of tensors see chapter 4.

[8]The surface is nearly free, hence the material has a thickness $\lambda/2$.

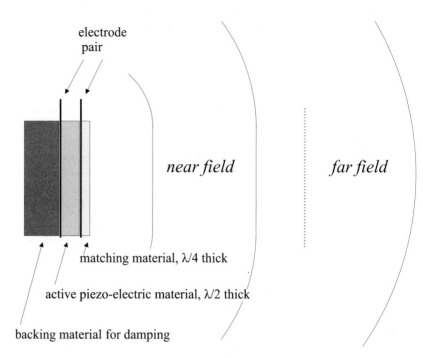

electrode
pair

near field

far field

matching material, λ/4 thick

active piezo-electric material, λ/2 thick

backing material for damping

Fig. 9.3 An individual piezoelectric transducer with its generated near and far wave field regions.

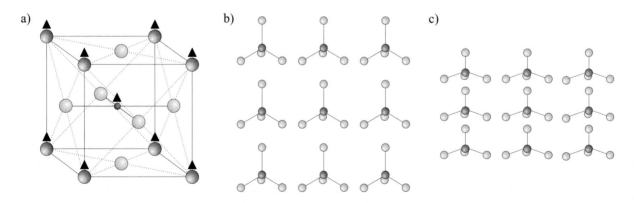

a)   b)   c)

Fig. 9.4 a) An example of a ferroelectric crystal. The crystal structure of barium titanate shows the lattice of positive ions (Ba and Ti) spontaneously displaced by 10^{-11} m relative to the negative oxygen ions, as indicated by small arrows.
b) An illustration of the regular tetrahedral quartz structure, and c) what happens to this under longitudinal strain, vertical as depicted.

Table 9.3 Some values of piezoelectric properties for quartz and proprietary ceramic materials based on lead zirconium titanate (PZT) and barium titanate (BT). The dissipation is the reciprocal of the electrical Q measured at 1 kHz. The density ρ in 10^3 kg m^{-3}. [Data from various websites.]

Material	c m s^{-1}	ρ	ε_r	d_{333} 10^{-12} m V^{-1}	Dissipation $1/Q$
Quartz	5660	6.82	4.5	−2.3	
K300	4060	7.7	1450	350	0.004
PZT-4	4600	7.6	1470	315	0.003
BT301	5640	5.8	1140	127	0.006
PZT5J1		7.4	2600	500	0.02
PZT5H1	4560	7.4	3250	620	0.018

without a centre of symmetry but no pre-existing inherent moment. We consider an example of each.

Ferroelectrics have an inherent charge asymmetry in the crystal unit cell. They include materials such as barium titanate, in which a lattice of barium and titanium ions is displaced with respect to the lattice of oxygen ions by about 10^{-11} m, as illustrated in Fig. 9.4a. As with ferromagnetism and paramagnetism, there is usually a temperature below which the material is spontaneously polarized in this way. Above this temperature the material appears symmetric but shows a large linear polarisability. This is the para-electric state.[9] Such a material exhibits first order piezoelectric behaviour.

There are also materials that show a second order and weaker effect. Unlike ferroelectrics they have no spontaneous electric polarisation, even at low temperature. An example is quartz. This has a non-centrosymmetric crystal, a regular tetrahedral structure which when strained develops a separation of charge in the direction of the longitudinal displacement gradient. This is illustrated in Figs 9.4b and c.

Some examples of piezoelectric materials are given in table 9.3. As well as quartz and barium titanate there are new composite materials. Other important properties are the velocity of sound, the density and the hysteresis loss. The ferroelectric materials have the largest values of ε_r and the piezoelectric coefficient d_{333}. This is the coefficient of interest which relates the longitudinal strain along the polar axis and the electric field in that same direction. Large amplitude behaviour is accompanied by some hysteresis. This heating at high frequencies may be a concern.[10]

Piezoelectric materials may also be used to generate transverse ultrasound waves as shown in Fig. 4.18. These waves are particularly important for non-destructive testing. They can also be generated using the transverse E-field of an RF wave.

[9] In these materials the instability of the unit cell may be described in terms of an optical phonon band that reaches down to zero frequency at $k = 0$, compare Fig. 4.23. However, such dispersion curves do not tell the full underlying stories of non-linearity and temperature dependence.

[10] In medical imaging, to avoid reflective energy losses, transducers are sometimes used in direct contact with the patient. This is a potential safety hazard if the piezoelectric material has a high hysteresis loss so that the transducer runs hot.

[11]Without backing material a transducer would actually radiate a dipole field with waves of opposite phase coming from the rear side. The backing material should also absorb these.

[12]Books and papers on ultrasound often use words like *insonate*, meaning to illuminate with sound. To those outside the field this may seem unfamiliar and unhelpful. In this book *illuminate* and similar words will be used in connection with sound where the analogy with light is unambiguous.

[13]Many transducer arrays are curved rather than being flat. This gives a curved wavefront with default focusing. These shapes are formed for use as internal probes or to follow body curvature.

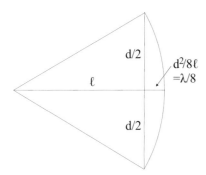

Fig. 9.5 The geometry of the wavefront at the boundary between the near and far wave zones. At this distance ℓ the phase difference between the plane wavefront width d and the spherical wave approximation is $\pi/4$.

[14]The same applies to radar aerials which are composed of an array of elements that may both transmit and receive. Earlier designs involved mechanical rotation and may still be seen at airports and on ships. These are being replaced progressively by phased arrays.

[15]The array may be curved or planar. If it is curved this acts as a focusing element in the signal transit times.

A typical longitudinal ultrasound transducer is shown in Fig. 9.3 with its longitudinal electric field and piezoelectric slice of thickness $\lambda/2$. The backing material is designed to give some damping to the resonance so that short pulses can be generated.[11] Once the ultrasound wave has been emitted by the transducer it is important that it proceeds to illuminate the object with as little reflection on the way as possible.[12] This is achieved by eliminating impedance and density changes, for example by eliminating air gaps and by using water or gel between the transducer and the clinical patient.

9.2.2   Ultrasound beams

Ultrasound beams are generated by arrays of individual transducer elements. As the wave leaves the flat surface,[13] width d, of a transducer, it behaves as a plane wave parallel to the surface with cylindrical waves at the edges. This is the near-field zone. As shown in Fig. 9.3, as it progresses further the wave becomes indistinguishable from a spherical wave. This is the far-field region. The far-field region begins at a distance ℓ where the phase difference between the plane wave and the spherical approximation becomes smaller than, say, $\pi/4$. From the geometry of a circle the sagitta is $d^2/(8\ell)$, as shown in Fig. 9.5. Equating this to $\lambda/8$ we get

$$\ell \approx d^2/\lambda. \qquad (9.4)$$

Individual transducers are small with $d \approx 1\text{–}3$ mm, so that the far-field approximation applies for distances greater than 10 mm, depending on the wavelength.

An array of transducers may be used to steer a beam in a specific direction by introducing a delay for each element of the array. This is shown in Fig. 9.6 in one dimension. If the delay varies linearly across the array, a parallel beam is emitted. By adjusting the delays the beam may be swept in angle. This is faster and more robust than the earlier method which involved mechanically rotating the array.[14] By varying the delay quadratically across the array the beam may be focused. By using a 2-D array the beam may be steered and focused in both planes.[15] When the array acts as a detector the signals reflected from material are collected and summed, again using the delays so that signals from an object in the desired direction and focal region add coherently to the summed signal. This is illustrated in Fig. 9.6d.

The object of interest is scanned by sweeping a directed beam across it and recording the time of the reflected signals as a function of angle. The image data may then be plotted showing a spread of angle and distance. An image is shown in Fig. 9.7 and a transducer in Fig. 9.8. Note how the resolution deteriorates and the signal becomes noisier at the bottom of the image where the signals have suffered the greatest absorption.

Many 3-D clinical images are recorded by mechanically scanning 1-D arrays over the patient's skin. Then, one dimension comes from the

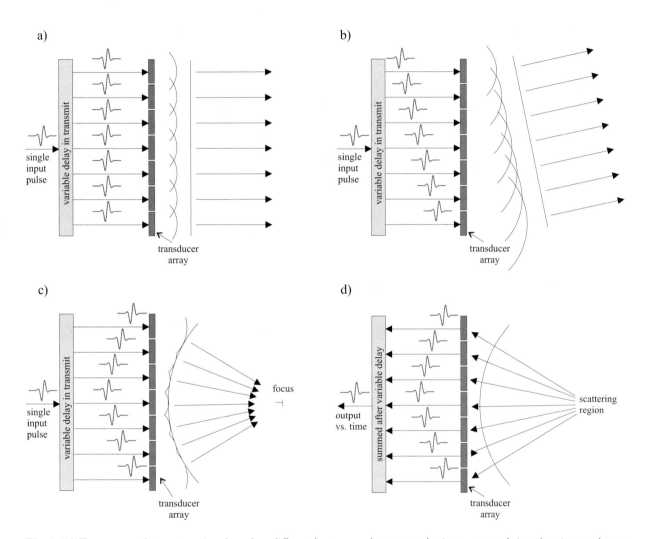

Fig. 9.6 A Huygens wavelet construction shows how different beams may be generated using an array of piezoelectric transducers driven by inputs with variable delays. a) A parallel beam in the forward direction. b) A parallel beam at an angle. c) A beam focused onto a particular region. In the same way, when the array acts as a detector of ultrasound, the same delays applied to the received signals act to refocus the summed signal, as illustrated in d). [See the Duke University website given at the end of the chapter.]

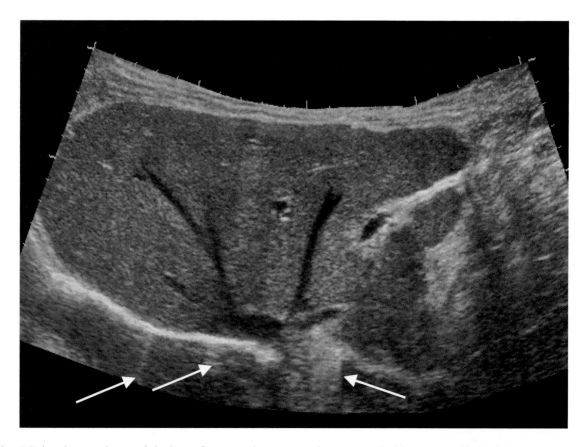

Fig. 9.7 An ultrasound scan of the liver. Some reverberation artefacts are marked by arrows. Much of the image contrast derives from the variation in the density of scattering rather than reflection at boundaries. [Image used with kind permission from GE Healthcare.]

Fig. 9.8 A modern ultrasound transducer for imaging and guided biopsy. The acoustic surface is 50mm × 4mm, the frequency range 5–12 MHz and it can be used with Doppler and harmonic techniques. [Picture reproduced by kind permission of B-K Medical]

echo time, one comes from the angle encoded by the phase between piezoelectric elements, and one comes from the mechanical motion. Since 2000 the availability of 2-D arrays has enabled other faster imaging methods to be developed. In special cases data for volume images may be recorded in times as short as 1 ms for off-line processing.

9.2.3   Beam quality and related artefacts

It is interesting to examine Fig. 9.7. Most of the features in the image are delineated by a change in texture. In a few cases, such as the lower edge of the liver as shown, there is a reflection from a surface. Even where that does occur it seems poorly defined.

In practice beams have significant imperfections which give rise to artefacts and signal loss of various kinds.

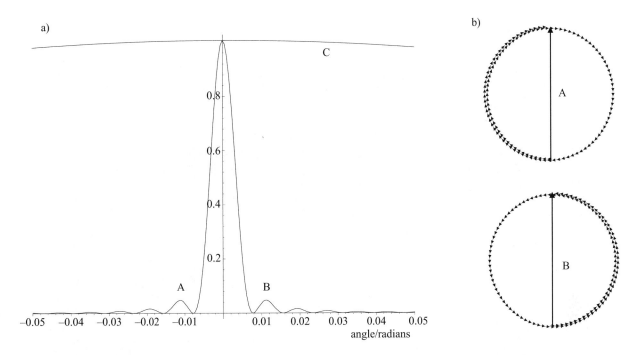

Fig. 9.9 a) The energy diffraction pattern showing the angular distribution in one dimension, in this case for a detector of width $d = 128\lambda$. The sidelobes at A and B at an angle $1.5\lambda/d$ have an intensity $4/(9\pi^2) = 0.045$ relative to the principal peak. b) Vector diagrams showing how the elements add up to give an amplitude maxima of $2/(3\pi)$ at A and B. The segmentation of the detector width into 128 elements has a negligible effect (C).

Effect of absorption

As discussed earlier, the frequency is chosen by considering the depth of material to be imaged and its absorption. To maximise spatial resolution the highest frequency that is consistent with achieving adequate signal-to-noise ratio at the full depth of the image is chosen. Signal amplitudes from the shallow and deep regions of the image then differ by a large factor. While this may be corrected by software in the image processing stage, optimal use of dynamic range is often achieved by ramping the gain of the detector amplifiers in hardware. Then signals from deeper layers that have suffered the greatest absorption arrive late and get boosted by a larger amplification factor.

Sidelobe artefacts

With a linear array of a finite number of elements N and total width D the definition of the beam direction is limited. As well as the PSF peak in angle with a width at base of $\pm\lambda/D$ there are secondary peaks at $\pm1.5\lambda/D$. These are shown in Fig. 9.9 and arise from the sinc function associated with the top-hat-shaped position dependence of the sensitivity/brightness of the detector/emitter array. They are also seen in Fig. 9.10 where an example of a complete 3-D PSF is plotted. These secondary peaks form artefacts in raw data. However, these may be re-

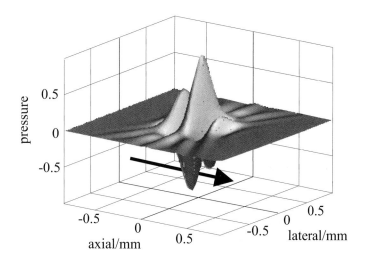

Fig. 9.10 The pressure amplitude of an acoustic pulse from a typical array shown at the acoustic focus as a function of lateral and axial position. Medium water, frequency 7.5 MHz, 60% bandwidth, 128 elements of width equal to the wavelength. [The image of the pulse is reproduced from the Duke University website given at the end of the chapter with permission.]

moved quite simply during the signal filtering given the prior knowledge of the PSF.

Reverberation, refraction and shadowing effects

A surface with a significant reflection coefficient acts as a sonic mirror. The net result is that other sources of reflected waves are imaged in them. Such artefacts are termed reverberations. There may be multiple reverberations but any such reflection must arrive late in the image compared with the signal from the scattering source itself. Examples are to be seen in Fig. 9.7 and are marked by arrows.

Refraction and absorption by layers of material can affect the amplitude of the wave incident on material at greater range, and may then affect the reflected signals on their passage back to the detector. Shadows are cast behind absorbing materials, and images appear displaced by refraction. Regions of bone and air are responsible for the largest effects since they involve the largest velocity and impedance changes. As with reverberations and PSF artefacts, the image can be filtered of such effects iteratively using reconstructions of material in the field of view, provided noise levels are low enough.

9.3   Scattering in inhomogeneous materials

Sound is scattered by a region of different density or elasticity within a material. When the spatial size a is small compared with λ, Rayleigh scattering is observed with cross section $\sim \lambda^{-4}$. Blood is highly inhomogeneous and gives rise to scattering with a very high density of centres.

The noise associated with the interference of amplitudes from a high density of scatterers is termed 'speckle'. This is responsible for the limited effective spatial resolution in ultrasound images. At short wavelengths there may be resonances with very large cross sections. Contrast agents based on tiny bubbles with resonant cross sections can be used. Moving scattering centres may be imaged by their Doppler shift. This is used clinically to give blood-flow images with good time resolution.

9.3.1 A single small inhomogeneity

Source of scattering

Sound may be scattered by a localised inhomogeneity in a material. The inhomogeneity may concern the value of an elastic modulus or the density, or equivalently impedance or velocity. If a, the size of the inhomogeneous region, is large compared with a wavelength, the various reflected and refracted waves are determined by matching boundary conditions, as discussed in chapter 4. If the region is small compared with a wavelength, there are simplifications which may be used to describe the scattering, and these are discussed below.

When a wave is incident on such a small inhomogeneity, in a first approximation each part of the scattering volume may be considered as emitting a secondary wave. The phase differences between these depend on a/λ and the scattering angle θ.[16] The higher angular terms in the angular distribution relate to higher powers of a/λ. Thus the scattered amplitude $\psi(\theta)$ may be written as an expansion of the form

$$\psi(\theta) = \sum_\ell A_\ell \left(\frac{a}{\lambda}\right)^\ell P_\ell(\cos\theta), \tag{9.5}$$

where $P_\ell(\cos\theta)$ are the usual Legendre polynomials. The details need not concern us because, if $a \ll \lambda$, the power series converges rapidly and only the first term, the scattering associated with an equivalent spherical inhomogeneity, is significant. In effect, all the scattered wavelets emanating from the region are then in phase with one another.

The weakness of the inhomogeneity implies that the phase shift suffered by the incident wave crossing the scattering volume must be small. If the density and the modulus of the region differ from the bulk material by $\Delta\rho$ and ΔK respectively, we can write the conditions

$$\frac{a}{\lambda} \ll 1, \quad \frac{a}{\lambda}\sqrt{\frac{\Delta\rho}{\rho}} \ll 1, \quad \frac{a}{\lambda}\sqrt{\frac{\Delta K}{K}} \ll 1. \tag{9.6}$$

The scattering mechanisms relevant to density and elasticity variations are quite different, as illustrated in Figs 9.11a and b respectively.

Small density inhomogeneity

A density difference (in the absence of an elasticity change) implies that the inertia of the sphere is different from the spherical region of homogeneous bulk matter that it displaces. In first order the forces acting on

[16]This is only true for a weak inhomogeneity. For a stronger case these secondary waves may be rescattered by the inhomogeneity. We keep to the simple first order perturbation picture in which this is neglected. This is a Huygens secondary wave description of the scattering process.

Today this is known as the Born approximation, although it was introduced by Lord Rayleigh in 1871, eleven years before Born was born.

a)

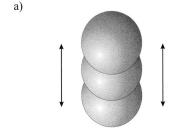

b)

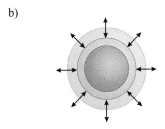

Fig. 9.11 Exaggerated illustrations of the response of a small inhomogeneity, assumed spherical, to an incident wave. a) A density inhomogeneity which moves back and forth relative to the bulk medium because its inertia is different to the bulk medium and therefore has a different acceleration. b) An elasticity inhomogeneity which expands differently to the bulk for the same stress or pressure change.

the sphere are the same as they would be acting on the equivalent homogeneous bulk matter. This force is then either too large or two small to accelerate the sphere in synchronism with its surroundings. The result is that the sphere undergoes simple harmonic motion driven by the wave, as the bulk matter does, but with a different amplitude. The scattering is caused by this difference in amplitude which is inversely proportional to mass. The relative motion excites a longitudinal compression wave in the medium. As the diagram indicates, this P wave has opposite sign in the forward and backward directions, and at 90° the wave vanishes. The scattered wave therefore has a factor $\cos\theta$ in its amplitude. If the incident displacement amplitude is u_0, the amplitude of the scattered wave at large distances for each Huygens wavelet must be proportional to factors $u_0 \cos\theta \exp(\mathrm{i}kr)/r$ and to $\Delta\rho/\rho$. The integration over wavelets brings in the volume of the sphere $V = \frac{4}{3}\pi a^3$. The scattered amplitude is therefore the product of these factors and a constant, C:

$$CV \times \frac{\Delta\rho}{\rho} \times u_0 \cos\theta \frac{\exp(\mathrm{i}kr)}{r}.$$

[17]For a discussion of the meaning of a differential cross section, see section 4.

The differential cross section is the ratio of the scattered intensity multiplied by r^2 and divided by the incident intensity.[17] This gives a cross section

$$\frac{\mathrm{d}\sigma}{\mathrm{d}\Omega} = C^2 V^2 \left(\frac{\Delta\rho \cos\theta}{\rho}\right)^2.$$

The left hand side of this equation has the dimensions of area. This can only be reconciled with the factor V^2 if C has dimensions of inverse length squared. The only candidate length, not already accounted for, is the wavelength. So the final cross section is

$$\frac{\mathrm{d}\sigma}{\mathrm{d}\Omega} = \frac{\pi^2 V^2}{\lambda^4} \left(\frac{\Delta\rho \cos\theta}{\rho}\right)^2, \tag{9.7}$$

where we have slipped in a numerical factor π^2 as revealed in a full derivation.

Small elasticity inhomogeneity

The effect of a small change in elasticity is quite different. It causes the spherical region to expand and contract relative to the bulk medium as the pressure wave passes, as illustrated in Fig. 9.11b. If the modulus had been the same, the rise in pressure would have matched the relative volume change in the surrounding medium. So it is only the difference that causes a change in volume relative to the volume that the sphere would have had under homogeneous conditions. This causes a further compression wave to be propagated, this time isotropic and determined by $\Delta K/K$. Similar general arguments as for the density variation give a differential cross section,

$$\frac{\mathrm{d}\sigma}{\mathrm{d}\Omega} = \frac{\pi^2 V^2}{\lambda^4} \left(\frac{\Delta K}{K}\right)^2. \tag{9.8}$$

The two scattering effects are coherent and we need to consider their relative phase. In a travelling wave the pressure and the particle velocity are in phase and related by the impedance, assumed real. The pressure and displacement are therefore 90° out of phase. Since the scattering by density variation is synchronised to the displacement, and the scattering by elasticity variation is synchronised to the pressure wave, the two scattered waves are always 90° out of phase. At any value of θ the scattering amplitude is the sum of the two processes in quadrature. The combined scattering differential cross section is

$$\frac{\mathrm{d}\sigma}{\mathrm{d}\Omega} = \frac{\pi^2 V^2}{\lambda^4} \left[\left(\frac{\Delta K}{K} \right)^2 + \left(\frac{\Delta\rho \cos\theta}{\rho} \right)^2 \right]. \tag{9.9}$$

This scattering is responsible for many ultrasound signals. For example, the reflection of ultrasound by shoals of fish is not caused by a difference of the *mean* density or elasticity of the fish relative to water (see Fig. 1.1). The density of fish must be very close to the density of water, otherwise fish would just float or sink! The signal detected in fish-finder technology comes from scattering by small gas bubbles within the fish.

In medical imaging with ultrasound the highest frequencies are used to maximise the scattering that comes from the $1/\lambda^4$ characteristic of the Rayleigh cross section, equation 9.9.

Progress in using and applying these results quantitatively, at least in a medical context, has been slow. For many years the piezoelectric technology was not available to enable exploitation of the field. Recently the qualitative use of imaging by scattering has given satisfactory results without any need for quantitative study. The real problem impeding rigorous study has been the intrinsic variability of tissue and its inhomogeneities, and the lack of a clear method for obtaining and quantifying reliable measurements. But these variations are arguably the very parameters likely to be of clinical interest. If he returned today, Lord Rayleigh might be surprised at the lack of progress in this area.

Shear waves and scattering

Such elemental motion also generates a shear S wave in a medium that can support it. The motion shown in Fig. 9.11b is purely radial and generates only outgoing P waves. However, that shown in Fig. 9.11a, while generating a outgoing P wave with a $\cos\theta$ dependence, will also couple to a cylindrical outgoing S wave with a $\sin\theta$ dependence. The particle velocity of such a transverse wave at $\theta = \pi/2$ will be similar to the longitudinal wave at $\theta = 0$ or π. The energy density for this S wave is similar to the P wave, but the flux, and thence the cross section, will be reduced by a factor c_S/c_P. This suggests a differential cross section

$$\frac{\mathrm{d}\sigma_S}{\mathrm{d}\Omega} = \frac{c_S}{c_P} \frac{\pi^2 V^2}{\lambda^4} \left(\sin\theta \frac{\Delta\rho}{\rho} \right)^2. \tag{9.10}$$

The motion in this S wave is at right angles to the P wave, so that there are no interference effects. The smaller the shear modulus, the smaller the cross section. In a viscous fluid such waves are rapidly absorbed and the mechanism is one of absorption rather than scattering.

For solid materials there is a difference between the modulus for longitudinal plane waves $K = \lambda + 2\mu$ and the bulk modulus that describes the spherical expansion at small r, $K_{\mathrm{B}} = \lambda + \frac{2}{3}\mu$. This shows that P waves should be scattered by inhomogeneities of either shear modulus or bulk modulus, or equivalently of either λ or μ. At small r a spherical P wave involves a degree of distortion which is absent in a plane P wave.

Incident S waves are scattered by discontinuities in either density or shear modulus. This makes them particularly suitable for the non-destructive testing of solids where the discontinuity in shear modulus is most pronounced at microscopic bubbles or other fluid micro-inclusions. The choice of S waves also maximises specular reflection in solids at cracks and surfaces bounded by fluids.

Thermodynamics and scattering

Small bubbles of vapour in a liquid do not behave isentropically under the influence of an incident P wave. In the expansion cycle energy is exchanged with the bulk medium as evaporation and condensation occur. These thermodynamic changes are driven fast, not necessarily under conditions of thermodynamic equilibrium. This is the kind of delayed response which gives attenuation of sound through irreversible cyclic change, as discussed in section 43. The details depend critically on the thermodynamics and are not simply described. Similar considerations apply to liquid drops in a vapour. The absorption and scattering of sound in fog is an interesting application. It is common experience that low frequency sound penetrates fog most effectively and that, even at 1 kHz, fog dampens general noise and gives an isolated eerie impression exploited in many a ghost story.

9.3.2   Regions of inhomogeneity

Materials with a concentration of scatterers

Because in tissue the reflection of ultrasound at extended boundaries is small and does not increase with frequency, the f^4 frequency dependence of scattering is important and determines the choice of high frequency. The limitation is absorption which increases at least linearly with frequency by any of the mechanisms discussed in section 4.4.2. The choice also makes for high spatial resolution. Taking this a step further, the spectrum of the size of scattering centres can be measured by observing the frequency dependence of the scattering cross section corrected for absorption. At $a \sim \lambda$ this cross section changes over from a λ^{-4} dependence to being independent of λ.

In a material composed of a large number of scatterers we need to consider the interference between the waves from different centres. Scat-

Table 9.4 Constituents of blood. [Quoted from Evans & McDicken.]

	Number mm^{-3}	Size μm	Density $\times 10^3$ kg m^{-3}	Compressibility m^2 N^{-1}	Volume %
Plasma			1.021	4.09×10^{-10}	~ 50
Erythrocytes	5×10^6	7.2×2.2	1.091	3.41×10^{-10}	45
Leukocytes	8×10^3	9–25			~ 0.8
Platelets	2.5×10^5	2–3			~ 0.2

tered waves emitted by centres that are imaged into different spatially resolved pixels do not interfere. However, when the number of scattering centres contributing to each spatially resolved pixel in the image is greater than 1, there are interference effects.[18] The relative phase comes from the fact that scattering centres are dispersed in depth. The depth probability distribution is likely to be uniform within any depth slice of thickness $\lambda/2$ so that the actual value of this phase for a given pixel is apparently random, though determined by the actual difference in depth of the scatterers for that pixel. The spatial resolution of the optics determines the effective pixel size and therefore which scattering centres can interfere in this way. The result of this interference is called speckle.

> [18]If there are n such scatterers, then there are $n - 1$ phases.

Many biological tissues have such a high concentration of inhomogeneities that their mutual interference has a large effect on scattering. An important example is blood. It consists of a dense suspension of cells in a clear fluid, the plasma. The various constituents are enumerated in table 9.4. The erythrocytes, or red blood cells, predominate. They consist of flexible discs and their concentration is so high that their average separation is only about 10% of their diameter. The conditions for pure Rayleigh scattering are therefore not met. Nevertheless it has been shown experimentally that the f^4 frequency dependence only starts to flatten off above 15 MHz. Note that blood may become anisotropic when the disc-shaped red blood cells become orientated for any reason.

Speckle noise

The amplitude R of an image pixel is the superposition of the scattering amplitudes from a number of scattering centres within the object, each with a phase determined by the depth of the centre. The fluctuation in the resultant R has two sources:

⬦ the number of scatterers, described by the mean and Poisson statistics;

⬦ the phases of the amplitudes varying between 0 and 2π and depending on the depth, as discussed above.

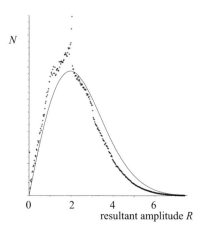

Fig. 9.12 A numerically calculated distribution of the resultant amplitude when the mean number of unit vectors is six. The curve is the corresponding Rayleigh distribution, equation 9.11.

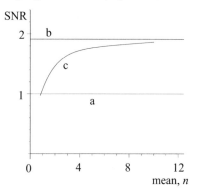

Fig. 9.13 The amplitude resolution, mean over standard deviation, plotted against the mean number of unit vectors $\bar{n}$. Curve a, the value for a linear Poisson distribution which is relevant at $\bar{n} \sim 1$ when the phase is not important. Line b, constant value 1.91 independent of $\bar{n}$ for fluctuations described by the Rayleigh distribution, valid at large $\bar{n}$. Envelope of points c, the result of a simple numerical simulation.

[19] As discussed in the book by Hill *et al.*

The mean resultant R can be drawn as the summation of the contributing amplitudes in an Argand diagram. This increases as $\sqrt{n}$, and the mean energy as $\bar{n}$. If the value of $\bar{n}$ in each pixel is less than or of order 1, phases are irrelevant and the signal and its fluctuations are described by Poisson statistics. Otherwise the phase becomes important and the SNR is determined by speckle noise.

This is simply studied. In Fig. 9.12 a numerical calculation of the distribution of the resultant amplitude R with both Poisson and phase fluctuations is shown for a mean $\bar{n} = 6$ scatterers per pixel. Some structure associated with the contribution of small values of n remains but the distribution is already quite close to the smooth Rayleigh distribution,

$$P(R)\, \mathrm{d}R = \frac{\pi R\, \mathrm{d}R}{2\bar{n}} \exp\left(-\frac{\pi R^2}{4\bar{n}}\right), \qquad (9.11)$$

which obtains for general large $\bar{n}$. Note that this is a function of the ratio $R^2/\bar{n}$ only. It is normalised, its mean is $\sqrt{\bar{n}}$ and its standard deviation is $\sqrt{\bar{n}}/1.91$. The SNR is therefore 1.91 at large $\bar{n}$. Results of the simulation are shown in Fig. 9.13 as curve c. As the mean number of scatterers per pixel is increased from 1 the value of the SNR rises from 1 towards 1.91. As Fig. 9.12 also shows, the Rayleigh distribution is applicable for $\bar{n}$ above 6 or so.

So measurements of the scattered amplitude remain noisy and are not improved as the number of scatterers increases. If the spatial resolution is improved the resolved pixel gets smaller but the fluctuations persist, albeit on a smaller spatial scale. The same phenomenon also occurs in the Rayleigh scattering of light. If the scattering centres move, the speckle pattern changes. By measuring the rate of change of the speckle pattern rates of diffusion or flow may be deduced.

As a result of speckle, medical ultrasound images appear mottled. A typical example is shown in Fig. 9.7. The apparent scale of the mottling is related to the spatial resolution, that is to the transfer function of the imaging. It does not relate to a change in the scattering density in the object. As shown, the amplitude of the mottling is half the mean scattered amplitude, or a quarter in terms of intensity. This has a serious effect on images. It has been shown[19] that speckle causes a reduction in lesion detectability of approximately a factor 8. This radical reduction in contrast resolution is responsible for the poor effective resolution of ultrasound compared with X-ray CT and MRI.

Resonant scattering

If larger values of the scattering phase shift are considered, there is the possibility of resonance. For example, the fundamental oscillation mode of a bubble of perfect gas of radius a in a fluid of density ρ at ambient pressure p is

$$\omega_R = \frac{1}{a}\sqrt{\frac{3\gamma p}{\rho}}. \qquad (9.12)$$

At resonance the cross section is of order λ^2 in the absence of damping. Such a figure is many orders of magnitude above the Rayleigh value.

Typically 4 μm bubbles resonate at 2 MHz with a cross section broadened and reduced by damping. At frequencies above resonance the cross section of a bubble falls rapidly and tends towards its geometrical area.

Other inhomogeneities of the same size resonate at much higher frequencies because they have lower elasticity.

Streaming, mixing and shaking

By driving large amplitude mechanical waves through inhomogeneous materials, at low as well as high frequency, the differential motion causes high rates of shear and thence turbulence. Such irreversible behaviour is familiar in the everyday effects of shaking and mixing. These phenomena have important applications in ultrasonic cleaning baths and industrial homogenisation. In ultrasound the localised flow around inhomogeneities is called streaming.

Ultrasound contrast agents

Contrast agents in ultrasound are provided by administering concentrations of artificial centres of inhomogeneity with large scattering cross sections. In practice these bubbles are small compared with a wavelength with large fractional changes in elasticity and density.

Small gas bubbles may be injected into the bloodstream but tend to dissolve rather too quickly for clinical use. However, air-filled artificial encapsulated bubbles are commercially available with diameters in the range 2–5 μm. These are small enough to pass through the finest blood vessels. These are non-toxic and may be excreted from the body without harmful side effects. They have a spectrum of sizes, and the quantification of this is important, given the size dependence of the resonant cross section. This is the main difficulty in the quantitative use of such agents.

The presence of such contrast agents is detected by scattering and their motion determined using Doppler methods which are discussed below.

9.3.3 Measurement of motion using the Doppler effect

Doppler shift from a moving scatterer

Suppose that the scatterer is moving with a velocity v at angle θ relative to the direction of the ultrasound waves emanating from the source, as depicted in Fig. 9.14a. If the phase velocity of the wave is c, then the relative velocity of waves and scatterer is $c - v\cos\theta$. If the frequency emitted by the stationary source is f, the frequency seen by the scatterer is

$$f' = f \times \frac{c - v\cos\theta}{c}. \tag{9.13}$$

This is the usual Doppler shift for a moving detector.[20]

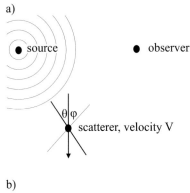

a)

source   observer

$\theta|\varphi$

scatterer, velocity V

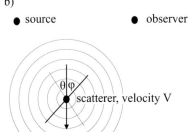

b)

source   observer

$\theta|\varphi$

scatterer, velocity V

Fig. 9.14 a) A source emits a wave towards a scatterer which is moving away from it with speed v and angle θ. b) The scattered wave is detected by an observer at an angle ϕ.

[20]It assumes implicitly that the (apparent) values of the angles do not change rapidly with time.

a)

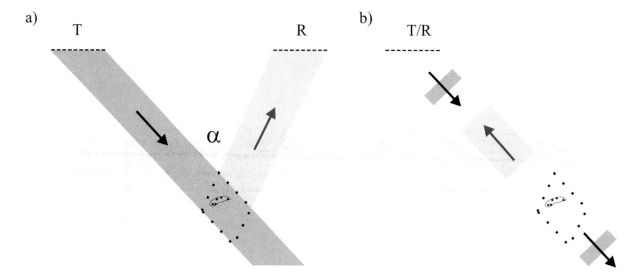

b)

Fig. 9.15 Doppler geometries. Target scattering centres in region illuminated by transmitter T and within the sensitivity of receiver R are detected with a frequency shift. a) Continuous wave (CW) with angular separation α. b) Pulsed wave (PW).

If this scattered echo signal is detected by a stationary observer in the direction ϕ, as shown in Fig. 9.14b, the velocity of the reflected signal relative to the scatterer (as source) is $c + v \cos \phi$. At the frequency f', that makes for a wavelength $\lambda' = (c + v \cos \phi)/f'$. The frequency of the signal detected by the observer as a result of this double Doppler shift is therefore

$$f'' = f \times \frac{c - v \cos \theta}{c + v \cos \phi}. \tag{9.14}$$

If the ratio v/c is small, the frequency shift may be approximated

$$\Delta f = -\frac{v}{c}(\cos \theta + \cos \phi). \tag{9.15}$$

This is the double Doppler shift for the frequency of a stationary source seen by a stationary observer on reflection by a reflector or scatterer moving away at a velocity v.[21]

The value of $\alpha = \theta + \phi$ is determined by the geometry of the source and detector. The smaller the value of α, the larger the sensitivity to Doppler. In medical applications of ultrasound the ratio v/c is always small and the velocity c is effectively constant at 1540 m s^{-1}.

Doppler detection

The Doppler signal from a specific region may be detected in two different ways, termed continuous wave (CW) and pulsed wave (PW). These are shown diagrammatically in Fig. 9.15. In CW the transmitter and receiver are distinct and separated. Using its angular resolution the transmitter illuminates a well-defined column of tissue with a continuous beam. The angular resolution of the detector is used to restrict

[21]This is an approximation. Although the relative velocity of the reflected image of the source and the observer is $v(\cos \theta + \cos \phi)$, the observer would see additionally that half of the medium between him or her and the image of the source is stationary and half is moving. This breaks the mirror symmetry so that equation 9.15 is not the exact result.

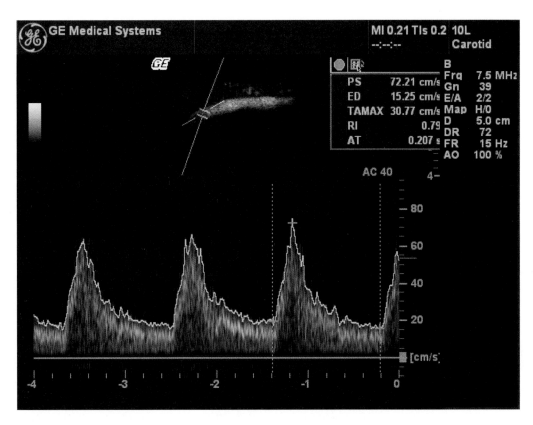

Fig. 9.16 Data from a CW doppler scan of a vascular feature. The geometry is shown at the top. The trace at the bottom shows the velocity spectrum as a function of time. Thus the envelope of the spectra is the maximum blood velocity component at any time. This shows a flow varying regularly from 20–70 cm s^{-1} during the cardiac cycle. [Image used with kind permission from GE Healthcare]

recorded signals to those reflected by a short section of this column, as shown in Fig. 9.15a.

In the PW mode the transmitter and receiver are combined, as shown in Fig. 9.15b with $\alpha = 0$. The position information comes from the echo time of the received signal relative to the transmitted pulse. In PW the length of the transmitted pulse determines both the longitudinal spatial resolution and the bandwidth of the pulse frequency. If the pulse is lengthened, the spatial resolution deteriorates and the velocity resolution improves. The spatial information may be improved but at the expense of the velocity resolution.

In medical ultrasound Doppler signals are generated by scattering at small moving inhomogeneities. Thus the Doppler frequency shift for scattering from red blood corpuscles in the bloodstream is a measure of the longitudinal speed of the blood. Signals from a region are linearly superposed. Therefore by spectral analysis the spectrum of velocity within each image volume can be determined.[22]

Figure 9.16 shows an example of the detailed spectrum that may be obtained. In the upper part of the display is shown the anatomical

[22]This can be applied also to the detection of natural gaseous bubbles and emboli. The latter are minute clots or coagulations in the blood which are difficult to detect by other means. It is of great clinical importance to detect the presence of these before they precipitate an actual embolism or blockage with its serious consequences for the patient.

scan indicating the CW geometry like Fig. 9.15a. The lower part of the diagram shows a plot of velocity (vertically) against time (horizontally). At each time there is a velocity spectrum. The peak velocity has been picked out by a line marking the envelope. This shows clear quantitative measurements of maximum radial blood velocity varying from 20 to 70 m s^{-1} during the cardiac cycle in the targeted region.

Such Doppler measurements only refer to a component of the velocity vector. However, additional components of flow may be measured by adding further detectors R at different angles but aimed at the same region in the CW geometry. With three or more such detectors the complete velocity vector can be measured.

9.4   Non-linear behaviour

Strain and stress changes relative to a reference state may be related by a power series expansion. For small changes the relation is linear. For larger changes all materials show non-linearity and therefore expand when heated. A heated material behaves linearly relative to its quiescent heated state. Non-linearity generates harmonics from an incident monochromatic wave, and a wave–wave interaction picture provides insight. Medical imaging with harmonics gives improved resolution, better penetration and less artefact. However the effects of high power levels are not unpredictable with current knowledge. Higher order terms may be studied only in constituent models. They show how large amplitude progressive waves approach a singular state where harmonic generation and frequency-dependent absorption give irreversible energy deposition; this has applications to homogenising, ultrasonic cleaning and plastic welding. These sound waves can give focussed energy deposition for cancer therapy and other clinical procedures. But constituent models are only illustrative and there is a lack of data for real biological tissue.

9.4.1   Materials under non-linear conditions

Phenomenology

Hooke's law, equation 4.1, states that the change in internal stress is proportional to the strain in a material. This is a statement about two states of the material, the initial or reference state and a changed state. The relation is only true for small strains, that is for pairs of states related by a small strain and a small stress change. We may consider Hooke's law as the approximation involved in taking the first term of a power series which expresses a change in the stress as a Taylor expansion relative to the reference state. Thus in one dimension generally we write the stress change ΔT in terms of the strain $\partial u_z / \partial z$,

$$\Delta T = A \frac{\partial u_z}{\partial z} - \frac{B}{2!} \left(\frac{\partial u_z}{\partial z} \right)^2 + \frac{C}{3!} \left(\frac{\partial u_z}{\partial z} \right)^3 \dots \qquad (9.16)$$

where A is the linear modulus that we have discussed previously.[23] The ratio B/A is a measure of the non-linearity of the relation, but higher orders are often important too. This non-linearity is not peculiar to particular materials, as is the case for the non-linearity of magnetic or electric properties. All materials exhibit mechanical non-linearity to a greater or lesser extent at large strains, and therefore it is a mistake to refer to 'non-linear materials'. Rather one should speak of 'non-linear strains'.

[23]When discussing 3-D materials we take A to be the bulk modulus K_B. Otherwise we discuss non-linear behaviour in one dimension only, ignoring the additional complexities that come with three dimensions.

Reference state

In chapter 4 the displacement of each point in the material u_x, u_y, u_z was described by the difference between the actual position of the point and its reference position x, y, z as a function of this reference position. Then the strain was defined as the spatial derivatives of these displacements with respect to the reference positions. Similarly the change in stress $\Delta T(z)$ at each point was defined as the difference between the actual stress and the reference stress for that point in the material.[24] Also the density ρ_0 was defined as the density in the reference state.

The question of how this reference state should be chosen was not discussed. It was implied perhaps that the reference state should be chosen as the quiescent state. The question of what happens if a different reference state is chosen was not addressed.

Provided that linearity is maintained a different choice would change nothing substantial. All points would have a reference position stretched relative to other choices and the zero of energy would be shifted, but there would be no consequences for waves or the elastic moduli on which they depend.

If, however, non-linear deformations are considered, that is no longer true. In equation 9.16 if the coefficients $B, C, \dots$ are non-zero, a different selection of expansion point changes the coefficient A mathematically, and with it the elastic modulus, wave velocity, impedance and so on. So the choice of reference state matters when behaviour is non-linear but it does not in so far as the behaviour is linear.

[24]Some care is needed here. The 'point' co-moves with the material in the way that we have described it. Thus the change in stress is the difference in T at the point that was at z in the reference state.

Temperature and non-linearity

An important application of these ideas arises in considering the effect of a temperature rise. An atom subject to the interatomic potential such as the Lennard–Jones potential, Fig. 4.21, would sit at the bottom of the well at 0 K. If the material were perfectly linear the curve would be a parabola. However, all such curves must be asymmetric on account of the 'brick wall' at a strain $\varepsilon = -1$ due to electron degeneracy, and the absence of such a wall for positive strain.

At a higher temperature an atom oscillating in such a potential will spend more time at larger separation and less at smaller. In consequence the material expands and the mean strain is positive. We can define a new reference point, the thermal equilibrium state at the raised temperature. Strains will be redefined, there will be a new density and a different modulus related to the changed value of the second derivative of the potential, ϕ''_ℓ. The velocity of small amplitude sound waves at that temperature will be described better by the new parameters at the new reference point with respect to which typical strains are small. They would not be small and linear if referred to the original reference state defined at the lower temperature. We conclude that by changing the reference state to reduce the value of the strain, the appearance of linear behaviour can be recovered.

Data for materials

How large is the dimensionless non-linearity parameter B/A for real materials? A perfect gas obeys the relation $PV^\gamma =$ constant for adiabatic changes. This determines that the ratio[25]

$$\frac{B}{A} = 1 + \gamma. \tag{9.17}$$

This ratio ranges from 2.67 to 2.11 depending on whether the gas is monatomic or polyatomic.[26] For an isothermal change $B/A = 2$. So the adiabatic condition for a perfect gas increases both the non-linearity and the bulk modulus $K_B = \gamma P$.

For liquids and solids the value of B/A may be derived from data on thermal expansion and the change of sound velocity with temperature. Values are shown in table 9.5 together with other parameters of interest. The table is divided into separate parts for solids, liquids and physiological materials. The characterisation of solids is more complicated as there are more possible non-linear parameters. The value of B/A shown is derived from the coefficient of expansion and sound velocity shown. While there is a wide spread in sound velocity there is a smaller range of values for B/A from 4 to 10. Gases are less non-linear.

Since we suppose that the non-linear behaviour of all materials arises from the underlying asymmetric potential curve between atoms and molecules, it is not surprising to find similar values for B/A. Some increase for elements with Z may be understandable in terms of an increasing electron degeneracy. But generally the various values for chemical compounds are harder to understand.

The data shown in table 9.5 include the linear parameters, sound velocity and density, as well as the non-linear ones. These illustrate an important point for physiological materials. Many entries describe large variations. This is because measurements are difficult to make, and conditions difficult to define precisely, particularly *in vivo*. But there may be a genuine wide variation between samples according to the age, background and clinical condition of the patient. This variation is what

[25] See question 9.4.

[26] See table 4.1 for values of γ.

Table 9.5 Examples of data on non-linearity for different materials from various sources. [Data quoted from Hill *et al* and various websites.]

	Thermal exp. $\times 10^{-6}$ K^{-1}	c_P m s^{-1}	dc_P/dT	Density $\times 10^3$ kg m^{-3}	B/A	T °C	Attenuation m^{-1} at 1 MHz
Beryllium	11.3	12890		1.85	4.1		
Sodium	71	2520	−0.5	0.97	2.7	110	0.012
Iron	11.8	5950		7.87	5.6		
Tin	22.0	2471	−0.3	7.26	4.4	240	0.006
Tungsten	4.5	5220		19.60	5.4		
Mercury	60.4	1450	−0.5	13.53	6.1	25	0.006
Lead	28.9	2160		11.34	6.7		
Bismuth	13.4	1651	−3.0	9.79	7.1	280	0.008
Water	206	1482		1.00	5.0	20	0.025
Acetone	1460	1170	−4.5	0.78	9.2	25	0.035
CCl$_4$	1140	921	−3.0	1.59	11.5	25	0.535
Blood		1540–1600		1.06	6.3		0.014–0.018
Fat		1412–1487		0.95	11.1		0.09
Liver		1578–1640		1.06	6.8–7.8		0.041–0.070
Muscle		1529–1629		1.08	7.4–8.1		0.12
Skin		1729			7.9		0.21
Spleen		1567–1635			7.8		0.036–0.062

needs to be measured and imaged for clinical purposes. Present practice in clinical ultrasound is still concentrating on making recognisable pictures and is a long way from being able to deliver quantitative maps of measured properties. As a first step, measurements need to be made for sets of definable and reproducible calibration standards. Relative to these, *in vivo* readings for pathological states of interest could be made which could then be deployed for clinical diagnosis.

Given such quantitative information, proper use could be made of simulation to map the energy flux and to calculate the scattering images expected. This is particularly important when it comes to non-linear behaviour where harmonic imaging, therapy dose and safety are much more sensitive to the energy flux and tissue interaction than would be the case for a purely linear process. There is a real challenge here. Sooner or later such information must become available.

9.4.2   Harmonic imaging

Frequency doubling in first order

Consider the acceleration of an element in a non-linear wave, equation 4.7

$$\rho_0 dz \frac{\partial^2 u_z}{\partial t^2} = \frac{\partial \Delta T}{\partial z} dz.$$

The stress is now given by equation 9.16,

$$\Delta T = K_B \frac{\partial u_z}{\partial z} \left(1 - \frac{1}{2} \frac{B}{A} \frac{\partial u_z}{\partial z} \right)$$

where we have included just the extra B-term at this stage. This gives us an explicit non-linear equation of motion,

$$\frac{\partial^2 u_z}{\partial t^2} = \left(\frac{K_B}{\rho_0} \right) \frac{\partial^2 u_z}{\partial z^2} \left(1 - \frac{B}{A} \frac{\partial u_z}{\partial z} \right), \tag{9.18}$$

in place of equation 4.9.

The effect of this non-linearity on a harmonic wave is to generate frequency components of double the frequency.[27] The solutions of interest are those for which a harmonic sound wave of amplitude u_0 and frequency ω is incident on the material. The effect of the non-linearity then appears as the growth of the harmonic as a function of the distance z into the material. We substitute a trial solution

$$u_z(z,t) = u_0 \exp\left[\mathrm{i}(kz - \omega t)\right] + \beta z \exp\left[2\mathrm{i}(kz - \omega t)\right], \tag{9.19}$$

where $\beta z / u_0$ describes the relative amplitude of the harmonic, assumed small for simplicity. Equating the coefficient of terms in $\exp[\mathrm{i}(kz - \omega t)]$ to zero requires that

$$\frac{\omega^2}{k^2} = \frac{K_B}{\rho_0}, \tag{9.20}$$

giving the expected value of c^2. Doing the same for the coefficient of terms in $\exp[2\mathrm{i}(kz - \omega t)]$ requires that

$$\frac{\beta}{u_0} = \frac{1}{4} \frac{B}{A} k^2 u_0, \tag{9.21}$$

so that the ratio of the displacement amplitudes in the first harmonic and fundamental increases with z,

$$R = \frac{u_1}{u_0} = \frac{1}{4} \frac{B}{A} k^2 u_0 \, z. \tag{9.22}$$

R may be expressed in terms of the incident frequency and strain ε as

$$R = \frac{1}{4} \frac{B}{A} \frac{\omega}{c} \varepsilon z. \tag{9.23}$$

The important result is that this amplitude ratio increases linearly with z, the incident strain amplitude and frequency. The scale length of this initial growth is known as the discontinuity length

$$Z_D = \frac{2\lambda}{\pi\varepsilon} \left(\frac{B}{A} \right)^{-1}. \tag{9.24}$$

[27] Here, by considering just the B-term with B/A small, we restrict attention to the growth of the first harmonic only. In general all harmonics are generated, including zero frequency. Similar effects are responsible for the generation of harmonics of electromagnetic waves using laser beams incident on non-linear dielectric materials.

The ratio for the particle velocity and strain of the harmonic wave to the fundamental is $2R$ because the frequency and wavenumber in the first harmonic are double.

The fraction of energy flux in the first harmonic rises as $4R^2$, that is proportional to z^2 and ω^2. This dependence, combined with the frequency dependence of absorption and the higher order terms in the expansion 9.16 which have been neglected so far, suggest that as non-linear propagation progresses energy is shifted on shorter distance scales into ever higher harmonics which are then rapidly absorbed.

Calculated rates of harmonic generation

In table 9.6 we show some estimated results for Z_D for a number of materials. The values are for an incident frequency of 1 MHz and a power flux 2×10^5 W m^{-2}. For values of z approaching Z_D the assumptions we have made break down progressively.

⋄ Further terms in the Taylor expansion, equation 9.16, generate higher harmonics direct from the fundamental.

⋄ As harmonics grow they start to generate their own harmonics. Since the harmonic pumping mechanism is proportional to ω, these secondary harmonics grow faster proportionately than the parent process.

⋄ As more and more energy is transferred to higher harmonics, it is attenuated by the frequency-dependent absorption.

⋄ Energy conservation ensures that, as the energy flux in the harmonics rises, the strength of the incident fundamental harmonic falls, thereby choking off this harmonic pumping mechanism.

⋄ As the rate of energy deposition rises, local heating occurs which gives rise to inhomogeneity in c and ρ that may then focus and reflect subsequent waves.

The values show that harmonic generation is particularly pronounced in low density materials with low sound velocity. The length varies inversely with frequency and the square root of the power density.

Harmonic generation and wave–wave scattering

If conditions are linear there is no wave–wave interaction. Under non-linear conditions the wave–wave interaction can be calculated in first order by the Born approximation in terms of the anharmonicity of the potential. The space and time integrals involved generate the momentum and energy conservation conditions on the incident and scattered wavenumbers

$$\boldsymbol{k}_1 + \boldsymbol{k}_2 \to \boldsymbol{k}_3 + \boldsymbol{k}_4 \tag{9.25}$$

and frequencies

$$\omega_1 + \omega_2 \to \omega_3 + \omega_4. \tag{9.26}$$

Table 9.6 Values of the discontinuity length Z_D (in metres) as calculated from B/A for different substances at 1 MHz for an ultrasound energy flux of 2×10^5 W m^{-2}. This value of energy flux is typical of that in use in many diagnostic instruments. See also the book by Hill *et al.*

	B/A	Z_D m
Beryllium	4.1	99.60
Sodium	2.7	1.85
Iron	5.6	21.77
Tin	4.4	2.96
Tungsten	5.4	25.70
Mercury	6.1	0.77
Lead	6.7	1.73
Bismuth	7.1	0.78
Water	5.0	0.25
Acetone	9.2	0.07
CCl$_4$	9.0	0.06
Benzene	9.0	0.10
Blood	6.0	0.27
Fat	10.0	0.12
Muscle	7.8	0.21

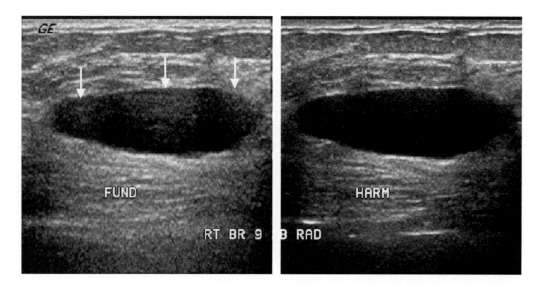

Fig. 9.17 Two ultrasound images showing a large cyst in a breast. The image on the left, scanned with fundamental imaging, shows prominent reverberation echoes in the near field of the cyst (arrows). The harmonic image on the right shows a reduction in artefactual echoes, allowing a more confident diagnosis of a simple breast cyst. [Rapp and Stavros, Journal of Diagnostic Medical Sonography. Image downloaded from GEhealthcare website and reprinted by permission of Sage Publications, Corwin Press Inc.]

Such scattering may be seen as the mechanism of the thermalisation and absorption of an acoustic wave through scattering with low energy waves in thermal equilibrium in the material.

The same mechanism may be used to describe the scattering of an incident wave with itself. The momentum and energy constraints are only satisfied in the forward direction, with

$$\omega_1 + \omega_1 \rightarrow \omega_2 \text{ and } \boldsymbol{k}_1 + \boldsymbol{k}_1 \rightarrow \boldsymbol{k}_2, \tag{9.27}$$

This provides another view of harmonic generation by an incident monochromatic wave, although still in a small amplitude and perturbative approximation.

Use of harmonic generation in diagnostic imaging

In the practical application of harmonic imaging the object is illuminated at one frequency, 5 MHz for example, and then imaged at twice the frequency by filtering the received signal. This has a number of advantages over imaging at the fundamental frequency, as exemplified in Figs 9.17 and 9.18. The resolution is improved in depth and angle. By focusing the incident beam the illumination may be arranged to reduce harmonic generation except within a small region at the depth of interest. As these images show, artefacts are also reduced.

There is an additional benefit of harmonic imaging connected with absorption. The illuminating flux at the lower frequency suffers less absorption on the outward journey. Only the scattered wave on the return

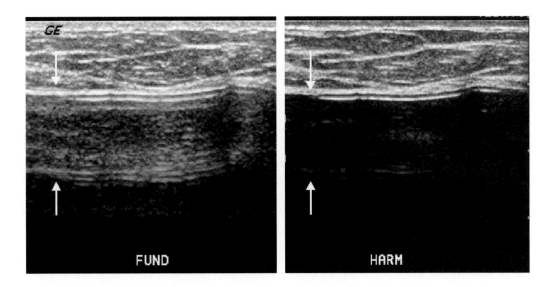

Fig. 9.18 On the left the fundamental image shows reverberation, scatter and clutter artefact in the near field of this silicone implant. This could be mistaken for a pathological condition. On the right in a harmonic image most of the artefactual signals are suppressed, showing that there was no condition. [Rapp and Stavros, Journal of Diagnostic Medical Sonography. Image downloaded from GEhealthcare website and reprinted by permission of Sage Publications, Corwin Press Inc.]

journey suffers the attenuation characteristic of the higher frequency.

Harmonic imaging is in widespread clinical use.

Safety and related concerns

To make harmonic images the power level has to be raised to the threshold of non-linear behaviour. With proper monitoring this should not be a cause for concern because images can be taken rapidly and exposure times kept short. However, the perception that ultrasound is a safe modality means that monitoring is taken, perhaps, less seriously and scanning beams may be left free-running. The physiological data and modelling of energy deposition by ultrasound are not well developed. This is made more serious on account of the non-linearity of harmonic generation.[28] What is the safety margin for ultrasound? Diagnostic doses are considered acceptable if the thermal load raises the temperature of the tissue by less than 1°C, and this is well established. At the other extreme doses that raise the tissue temperature above 56°C for 1 second result in coagulative necrosis.[29] These margins of safety for clinical diagnostic use might be acceptable if the local monitoring of tissue temperature in tissue *in vivo* were more highly developed and if reliable data were available as a basis for simulation and planning.

[28]In a non-linear process the sensitivity to the conditions is increased. For example, in a quadratic process a 10% error in the fundamental amplitude makes for a 20% change in the first harmonic.

[29]Typical doses in ultrasound therapy raise the temperature of the targeted tissue to 80°C.

9.4.3 Constituent model of non-linearity

To understand what happens as higher order terms in the power series, equation 9.16, become important, we need accurate data that describe

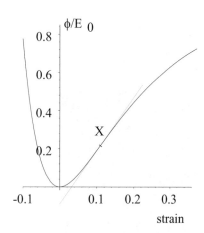

Fig. 9.19 The Lennard–Jones interatomic potential of Fig. 4.21 replotted as a function of strain relative to the reference state chosen to be the equilibrium state at 0 K. Note that at the point marked 'X' the curvature of the potential changes sign.

[30]See question 9.5.

[31]This model calculated rate of decrease of sound velocity with temperature may be compared with some actual values shown in table 9.5.

[32]Given by γP_0.

the materials involved. Since these are not available, we resort to the 1-D constituent model with the Lennard–Jones potential discussed in chapter 4. This will help us understand the phenomena qualitatively, athough quite inadequate to predict actual responses. A feature of such a model is that the higher order terms are all defined and there is no need to consider a truncated series.

Reference state and non-linearity in a constituent model

The change in stress in a material described by the Lennard–Jones potential comes from differentiating twice the potential ϕ given by equation 4.76. We may pick up the story from equation 4.79 where we took the derivative $\phi''(\varepsilon)$ to be constant. Here in the non-linear regime we take on board its dependence on the strain ε and on the choice of reference state.

The absolute value of the potential ϕ is arbitrary. Only relative values are significant for dynamics. The first derivative is absorbed into the definition of T_0, the stress at the reference point. It is only the second derivative of the potential and its variation that need concern us. The wave equation is

$$\rho_\ell \frac{\partial^2 u}{\partial t^2} = \frac{\partial^2 u}{\partial z^2} \frac{1}{\ell^2} \frac{\partial^2 \phi}{\partial \varepsilon^2}(\varepsilon), \tag{9.28}$$

where ρ_ℓ is the density and ℓ the constituent separation in the reference state. This equation is valid for all strains and any choice of reference point but is linear only within any small window of ε for which ϕ'' is effectively constant. By centring the reference point on the chosen window, non-linear effects can be minimised.

At $T = 0$ K the Lennard–Jones potential model (Fig. 9.19) gives the ratio $B/A = 21$ for the reference point at the bottom of the well, $r = \ell$.[30] Compared with real materials this is an overestimate by a factor between 2 and 4. In the model at higher temperatures the equilibrium value of r increases and the reference point moves. The ratio B/A increases and the value of A falls, so that non-linearity becomes more significant and the velocity of sound decreases. These effects are plotted in Fig. 9.20.[31]

It is interesting to note that for a perfect gas the value[32] of the modulus A is positive for all strains. This cannot be the case for a solid or a liquid. The modulus is proportional to the curvature of the potential ϕ''. There is always a value of positive strain at which the potential starts to flatten off and therefore its curvature passes through zero. For the Lennard–Jones potential this occurs at the point marked 'X' in Fig. 9.19.

What happens at such a point, and what does it mean? The answer is shown illustratively in Fig. 9.21 for neighbouring atoms in one dimension. At strains greater than 'X' the modulus K goes negative. An atom between two others at such a strain as shown in Fig. 9.21c. Although there is still tension on the atomic chain, it is energetically favourable for an atom to move away from its median position between its neighbours. It gains more energy by being closer to one neighbour than it loses from moving away from the other. In other words, the chain is

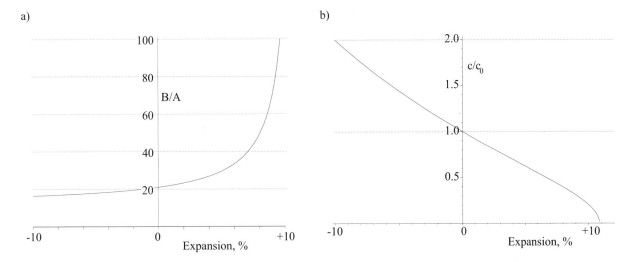

a)

b)

Fig. 9.20 Calculated non-linearity effects in the Lennard–Jones constituent model plotted as a function of the expansion of the reference state relative to the minimum of the potential curve.
a) The first order non-linearity ratio B/A.
b) The relative velocity of sound which reaches zero at a strain of just over 10%.

unstable. Under quasi-static conditions this implies some kind of phase separation. In dynamic conditions the material would suffer cavitation or fracture. With negative K sound waves have an imaginary velocity. Figure 9.20b indicates that this condition happens in a Lennard–Jones material at a strain of just over 10%. In a real material this happens at a much lower strain because of thermal excitation and imperfections in the bonding structure. It is an instance of properties being determined by the proverbial weakest link in the constituent chain.

Superposition principle and non-linear waves

An important aspect of linearity is the superposition principle. This allows that when waves propagate they maintain their identity and do not interact. We can retain some of this picture under certain circumstances even when non-linearity is important. The trick is to redefine the reference state.

Consider two waves. If the strain represented by the first and larger wave defines a new local reference state, the effect of a smaller second wave may be considered as propagating in the compressed (or rarefied) moving medium that the first has provided. The motion of the medium involved is the particle velocity of the first medium, not the velocity of sound. Thus a small wave will propagate quite linearly in a region where the particle velocity is u with a wave velocity, the vector sum, $u + c$. This wave velocity is determined by the compressed (or rarefied) state of the first wave. This only provides a local and short term picture but one that is useful when the waves are moving in the same direction, so that the changed reference state persists for a time.

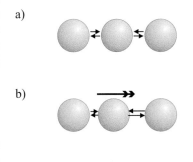

a)

b)

c)

Fig. 9.21 a) At strains below X forces between atoms maintain equal spacing.
b) Below X unequal spacing stabilises. (Repulsion of close neighbour + attraction of far neighbour > attraction of close neighbour + repulsion of far neighbour.)
c) Above X unequal spacing is unstable and ruptures. (Repulsion of close neighbour + attraction of far neighbour < attraction of close neigbour + repulsion of far neighbour.)

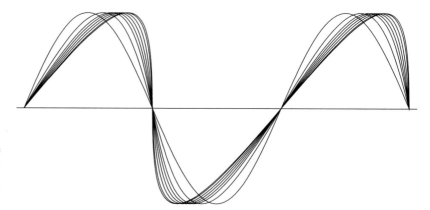

Fig. 9.22 The shape of a harmonic wave, initially a sine wave of frequency 3 MHz, changes shape as it propagates through a Lennard–Jones material at the density of water.

9.4.4   Progressive non-linear waves

General harmonic generation

This view of the reference state gives a useful interpretation of harmonic generation. A small pulse (of either sign) on the peak of a large amplitude travelling wave will move forwards at a greater speed than the bulk of the wave because it is a wave in a more highly compressed medium which is itself moving forward. And by the same argument a small pulse in the trough will move forward more slowly.

What is true for a pulse is true for the wave itself. One may think of each part of a large amplitude wave carrying its own reference state with it. The peak of the wave propagates fast by reference to its own compressed reference state, and the trough slowly. The conclusion is that different parts of the wave propagate at different speeds according to their amplitude. This results in distortion and harmonic generation.

Can we calculate this? In the case of the Lennard–Jones potential, the second derivative of ϕ is known as a function of ε and we may go back to the original wave equation for which the wave velocity is

$$c(\varepsilon) = \sqrt{\frac{1}{\rho_\ell}\frac{1}{\ell^2}\frac{\partial^2\phi(\varepsilon)}{\partial\varepsilon^2}}. \tag{9.29}$$

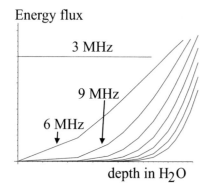

Energy flux

3 MHz

9 MHz

6 MHz

depth in H$_2$O

Fig. 9.23 The growth in harmonic energy as the wave of Fig. 9.22 propagates. As noted in the text, this simplified calculation does not conserve energy.

Consider a wave that is monochromatic in time for all time at $z = 0$. As the wave progresses in z its rising edge becomes steadily steeper as shown in Fig. 9.22. At greater values of z it is still periodic in time but its harmonic content grows as z increases. The illustration is for a 3 MHz wave of 4 W cm^{-2} in a material of the density of water assumed to be described by the Lennard–Jones potential. The curves represent the evolution of the wave shape over a distance of 10–20 cm. The harmonic content grows rapidly as shown in Fig. 9.23. This qualitative illustration obviously does not conserve energy as the harmonics build, but the important point is that a non-linear wave progresses the peaks catch up with the troughs and something dramatic has to occur.

One may see this phenomenon on the sea shore where the deep water harmonic waves have a wave velocity proportional to the square root of the depth for all frequencies.[33] The wave pattern therefore is not dispersive and does not change shape. As the waves reach shallower water where the wave amplitude becomes a significant fraction of the depth the crests gain on the troughs. Eventually the waves break in the familiar way. What happens when a sound wave 'breaks'?

Non-linear absorption, delayed response and irreversibility

The above discussion suggests that at some value of z the wave catches up with itself. For a transverse wave, such as an ocean wave at a beach, this is possible in some sense. For a longitudinal wave it cannot occur because of the implied density singularity. As the curve becomes steeper the rate of change of strain becomes progressively larger at the rising edge of the wave. Eventually the assumption that strain responds without delay to applied stress must fail. A time-independent relation such as the Lennard–Jones potential makes no allowance for response time. The finite delay in the change of strain associated with a very rapid change in stress becomes important. This causes absorption by hysteresis, as illustrated in Fig. 4.24. It is a mistake to try to think quantitatively in terms of separate harmonics as the harmonic waves become strongly coupled in non-linear conditions. It is better to visualise cycling round the stress–strain plot with a strain delayed in time which then generates significant hysteresis when the stress changes very rapidly.

In practice therefore the wave starts to suffer strong attenuation before it reaches the singular condition of catching up with itself. However, the energy deposited by such localised absorption can raise the temperature of the material sufficiently that the properties of the medium themselves are changed, and a shock wave results. Alternatively the absolute stress may be raised above the tensile strength of the material which may then rupture. If the material is a liquid the absolute stress may rise to zero (or to minus the vapour pressure), such that phase separation or cavitation occurs.[34] These processes are thermodynamically irreversible and give rise to heating and absorption. The fracture, heating and cavitation or phase separation represent inhomogeneities that can build with time and precipitate further scattering and absorption in a catastrophic progression. The result is a region of highly localised energy deposition. The sharpness of the edge to the region of energy deposition is related to the high order of non-linearity.

9.4.5   Absorption of high intensity ultrasound

Welding of layered materials

Ultrasound is used for homogenising mixtures of materials and for cleaning solid surfaces immersed in fluids. These are irreversible processes involving energy deposition which we discussed earlier in terms of shear motion and streaming. More generally, ultrasound energy may be deposited selectively at any loose boundary in a material, such as between the layers of plastic polymer sheet in loose contact with one another. The

[33] See section 4.3.3.

[34] Since the absolute stress in a fluid is $-P$, the condition is that this rises to about zero. The effects of the vapour pressure and the surface tension modify the point at which separation occurs.

resulting rise in temperature these may result in them being efficiently welded together.

Let us visualise how this happens. If the density and modulus of the materials at the boundary differ significantly there will be large reflection, leading to standing wave patterns. To match the boundary conditions the displacement amplitudes in the less rigid low density material will be high relative to the rigid dense material. There will therefore be antinodes at the boundary in the more rigid material. These boundary regions will therefore experience higher energy density, larger amplitude and more non-linear absorption than other regions of the dense material. At suitable power levels these regions may fuse with other such layers.

Kidney stone ablation and acoustic surgery

Ultrasound is used to break up and disperse kidney and gall stones. This non-invasive procedure is known as lithotripsy and is now in widespread use. The relative density and rigidity of the stone causes the surface to absorb energy in the same way as discussed for ultrasonic welding. As a result the stone breaks into pieces that may then be excreted in the normal way.

Focused ultrasound may be used to stop bleeding and for minor surgical operations, simply by depositing highly localised energy. This is effectively cauterising a wound. Sources of ultrasound are mobile, and such procedures are cheap and convenient in principle. This is described as acoustic surgery but is not yet developed for general use. It might be suitable for use at the scene of an accident if sufficient monitoring and safeguards are developed.

Cancer therapy

High intensity focused ultrasound (HIFU) has been used for cancer therapy, particularly for the prostate, liver and breast. Much of the clinical work has been pioneered in China. A schematic view of a treatment procedure is shown in Fig. 9.24. The patient is immersed in a water bath to match the ultrasound field to the impedance of the tissue. The treatment plan must be designed so that the beam avoids bones that would disrupt the production of a clean focus. The first stage of the integrated procedure is to provide an image of the organ and tumour with a low power ultrasound beam. This provides registration of the target volume in real time. The therapeutic dose is delivered by a high power beam from a transducer of 12–15 cm aperture and 9–15 cm focal length. A frequency of about 1 MHz is used depending on the tumour depth. The beam is focused to a small cylindrical volume, typically 1.5 mm wide by 15 mm long. The focus is stepped through the region of the tumour raising its temperature locally to more than 55°C with resulting coagulative necrosis.

Under these conditions the catastrophic non-linear absorption occurs at the focus. As the temperature of the treated region rises, bubbles and cavities are formed and the general inhomogeneity increases. As a

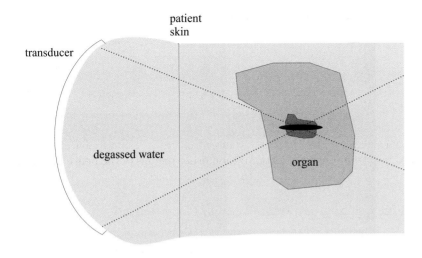

Fig. 9.24 A schematic diagram of cancer therapy treatment using high intensity focused ultrasound.

result the non-linear response and absorption of the region rise further. The location of the energy deposition then moves backwards upstream towards the source, and farther parts of the targeted volume become shadowed by the increased absorption of the nearer parts. Such effects require monitoring and modelling. Outside the focal region there is no energy deposition, and the edge of the region is very sharply defined because of the extreme steepness of the non-linearity at this high power. Such non-linearity may involve higher terms in the expansion 9.16, not just B/A, the first term.[35]

In principle such treatment does not require the small extra non-linear effect of fractionation and is not sensitive to the tissue oxygen content. The only cause for further treatment would be regions of malignancy that were not swept effectively by the therapeutic focal volume. This puts high demands on monitoring and registration that are not easily achieved in the treatment of deep-seated cancers.

Bubble contrast agents are under study as a means of increasing and targeting the non-linear energy deposition of HIFU. In another development bubbles, preloaded with an appropriate chemotherapy pharmaceutical, may be burst by the high intensity ultrasound field precisely at the site of a tumour. This minimises the unfocused general damage and side effects typical of conventional chemotherapy. Figure 9.25 shows a microbubble rupturing in this way. The pulse consisted of one cycle of ultrasound with a frequency centred on 500 kHz. The peak negative acoustic pressure at the region of interest was 0.85MPa. Each frame shows a region 45μm by 27μm at time intervals of 330 ns, except for the time between frames e and f, which was 660 ns. The sequence shows the microbubble: (a) the initial size, 17.6 μm, growing to (c) maximum size, 22.9 μm; (d) shrinking to 20.2 μm before (e) rupture. The collapsed microbubble gets pushed to the lower left side of the frame, apparently by water propelled into the microbubble. A subframe shows the negative of the region of interest. Finally, the deformed microbubble is left as an asymmetric shape (f). These and other ideas are under development.

[35] Alternatively a lower power therapeutic ultrasound beam may be used to raise the temperature of the tumour, voxel by voxel, while imaging it with ultrasound or MRI. In the latter case a temperature map in vivo may be produced using the temperature dependence of the chemical shift of the NMR water peak (this is called MRI thermometry, see page 30). Such instrumentation enables effective doses to be delivered with feedback, achieving coagulative necrosis at 55°C without cavitation.

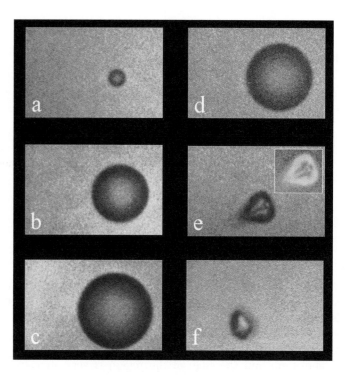

Fig. 9.25 Time-elapsed images of the rupture of a 5.3 μm gas-filled encapsulated microbubble by ultrasound (for details see text). [Images reproduced by kind permission of Nico de Jong, *n.dejong@erasmusmc.nl*.]

The advantages of ultrasound for therapy relative to other treatments may be summarised.

◇ It can be integrated with a real-time imaging capability, thereby minimising registration problems.

◇ It is cheaper, smaller and more portable than the equipment required for radiotherapy.

◇ Unlike γ-radiation it may be focused, and because the energy deposition is very non-linear, the treated volume can be very sharply defined with negligible damage to neighbouring or overlying tissue.

◇ There is a much reduced need for multiple treatment sessions. This reduces costs, shortens the whole elapse period of treatment and improves the patient experience.

The clinical use of ultrasound cancer therapy is not yet widespread. It has a promising future and no adverse public image. Clinical experience is being built up and the track record of performance assessed. A real drawback is the absence of data on the non-linear behaviour of biological tissue. This is highly variable. What is required is the non-linear acoustic properties as a function of position for the patient concerned. Then a reliable treatment plan could be made and margins of safety established. It is not clear how to do this. To deliver the therapy, techniques are needed to measure the real-time response, that is the temperature and acoustic properties of the tissue as a function of position during treatment. Progress is being made on this problem.[36]

[36]See ISTU web search recommendation.

Read more in books

The Theory of Sound, Vols 1 and 2, JWS Rayleigh, reprinted Dover (1945), Rayleigh's original texts on sound

The Physics of Medical Imaging, ed. S Webb, IOP (1988), an edited compilation of useful articles, somewhat out of date but still frequently referenced.

Doppler Ultrasound, DH Evans and WN McDicken, Wiley 2nd edn (2000), a useful recent text

Physical Principles of Medical Ultrasonics, ed. CR Hill, JC Bamber and GR ter Haar, Wiley, 2nd edn (2004), a recommended recent text

Look on the Web

Look up the book website at
   www.physics.ox.ac.uk/users/allison/booksite.htm

Find 'A seminar on k-space applied to medical ultrasound', Martin E. Anderson and Gregg E. Trahey, Department of Biomedical Engineering, Duke University, by searching on
   duke ultrasound k-space

Consider the imaging option of using variable frequency sound waves in medical imaging
   frequency scanning interferometry

Find out how ultrasound is used in NDT
   non-destructive testing

Find out more about developments in piezoelectric materials
   piezoelectric

Read more about blood
   blood composition

Learn about progress and prospects for ultrasound contrast agents
   ultrasound contrast agents

Follow developments in the use of harmonic imaging
   ultrasound harmonic imaging

Monitor advances in ultrasound therapy
   ISTU, MRI thermometry or *hifu ter haar*

Look at other uses of ultrasound for therapy
   lithotripsy or *acoustic surgery*

Questions

9.1 Consider a parallel ultrasound beam of 1 MHz and 100 W m^{-2} in water. The phase velocity is 1540 m s^{-1}. Calculate the impedance, wavelength, pressure difference, particle velocity, displacement and acceleration.

At what energy flux would the absolute pressure reach zero in the trough of the wave (assumed linear)? Does this depend on frequency?

9.2 Estimate the Rayleigh scattering cross section from a red blood cell in plasma at 1 MHz using the data given in table 9.4. If the scattering in blood were complete incoherent, what would be the attenuation length? Compare with the observed value of 0.21 m.

Make a more realistic estimate by assuming that the scattering is completely coherent within a volume of side $\lambda/4$, but incoherent between such volumes. How does this estimate scale with frequency?

9.3 Show that the mean and standard deviation of the Rayleigh distribution, equation 9.11, are

$$\sqrt{\bar{n}} \text{ and } \sqrt{\bar{n}(4/\pi - 1)}.$$

Evaluate the SNR due to speckle noise.

9.4 Show that the non-linearity ratio $B/A = 1 + \gamma$ for a perfect gas under adiabatic conditions ($PV^\gamma =$ constant).

9.5 Show that in one dimension the Lennard–Jones potential, equation 4.76, gives a value of 21 for the non-linearity ratio B/A at 0 K, equation 9.16.

9.6 Show that in the simple Lennard–Jones model the ratio B/A increases when materials are heated and expand. Is this generally true?

Forward look and conclusions

10

in which we look forward to predictable improvements in imaging and significant revolutions in therapy. We conclude that the hazards of ionising radiation have been significantly overestimated with important consequences for choices facing society. We have much yet to learn, not least from the animal kingdom. Progress in ensuring the security of human life continues to depend on knowledge of fundamental physics and informed public education.

10.1	Developments in imaging 307
10.2	Revolutions in cancer therapy 312
10.3	Safety concerns in ultrasound 313
10.4	Rethinking the safety of ionising radiation 315
10.5	New ideas, old truths and education 318
Look on the Web	320

10.1 Developments in imaging

Influence of computational power

The advent of almost unlimited computational power and communication is changing the methods that we use to carry out tasks, although the fundamentals on which these are based are not changed. This revolution has only just begun. The timing of its progress is determined as much by the speed at which trust can be established in new procedures as by the appearance of new technical opportunities. This trust grows as a result of positive experience and education.

Computational power itself will grow and new generations at all levels of society will grow up with it. In decades to come confidence will grow in the use of images in particular as quantitative measurements through better analyses, better displays, optimised designs, improved registration, simulation and calibration. With education these longer term benefits are within our grasp.

Combining data

Additional computing power has eased the task of combining data from different sources, thereby optimising both the contrast and the resolution. These are called fused images and examples[1] were shown in Figs 8.15 and 8.20. In the same way, with careful registration and calibration, images taken on different occasions may be compared, and changes with time monitored.

[1] Recall that these are only 2-D grey-scale sections taken from the full 3-D image.

Modalities can be used together in other ways. Two examples are ultrasound elastography and magnetic resonance elastography which combine palpation with ultrasound and MRI respectively. As limbs and muscles are flexed and other bodily tasks undertaken, the real-time motion and strain of constituent tissue is mapped in three dimensions using ultrasound or MRI. Such 3-D motion may be combined with calculations using finite-element analysis to image the full elastic behaviour of muscles. Alternatively tissue may be subjected to externally applied stress, and then the strain measured by MRI or ultrasound. Possible tumours may be detected as regions with anomalous elastic moduli, as in palpation.[2] Such developments should bring better quantitative data, monitoring procedures and simulations that will permit more quantitative uses of ultrasound.

[2] The thesis by Revell gives some examples of the possibilities that are opened up.

Electro- and magneto-encephalography

Activity in muscles and the brain takes place on timescales of a millisecond or less. Ultrasound, MRI and PET have no resolution on such a scale. However, the electric and magnetic fields that are emitted can be sensed at such speeds. The electric fields emitted by the heart have been used clinically to monitor its activity for many years in the form of the electro-cardiogram (ECG). Such signals are large, being associated with the delivery of significant mechanical power. Electro-encephalography (EEG) concerns the measurement of electrical signals generated by activity in the brain. These are smaller and are partially screened by the conductivity of the skull. Time-varying potential differences on a scale of 10 μV may be measured by an array of electrodes in contact with the skull. Modest spatial information comes from the $1/r^2$ dependence of the potential of an electric dipole.[3]

[3] The potential from any higher multipole would fall off even faster.

Magnetic fields generated by brain activity, though even smaller, are not screened. The detected B-field depends on the orientation of the current elements which may be considered as magnetic dipoles. The measurement and study of these signals is called magneto-encephalography (MEG). As in EEG, spatial information is rudimentary. The necessary detection sensitivity has been developed using superconducting quantum interference devices (SQUIDs). These are the most sensitive magnetometers such devices are available commercially. The usefulness of EEG and MEG may be enhanced by combining their time resolution with the spatial resolution of other modalities.

Passive acoustic imaging

The stethoscope is the icon of medical care and its use is the model of non-invasive clinical examination. Vibrations are emitted during any motion of the human body, and every joint, every muscle, every bowel movement emits sound. Some of the loudest are the low frequency emissions of heart, pulse and lung function. Diagnosis of their malfunction in the doctor's surgery is classic medicine in action. But a doctor has only a single pair of ears of limited frequency response, and the time for

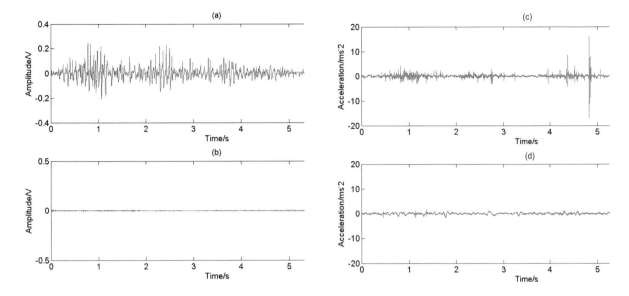

Fig. 10.1 Vibrations emitted from the necks of two people as they move their heads gently from side to side. Traces (a) and (c) are from a person with a 'noisy' neck; traces (b) and (d) are from a person with a 'quiet' neck. Traces (a) and (b) were recorded by a microphone in the person's ear; traces (c) and (d) were recorded with an accelerometer taped to the skull behind the ear. The signals recorded by the microphone show evidence of resonating in the acoustic cavity of the ear. [Waveforms recorded by Lauren McDonald.]

an examination is short.[4] Arrays of modern accelerometers, developed to record every movement of an engineering structure like a car engine, have superior sensitivity, higher frequency range and timing capability than a stethoscope. They hold promise for a useful step forward in simple, affordable and continuous clinical examination of joints and muscles in motion. Significant differences are observed between people. As an example, Fig. 10.1 shows the acoustic signals emitted from the necks of two people, recorded with both a microphone and an accelerometer. The modernisation of the stethoscope is overdue and looks promising.

[4] In addition, the doctor listens to signals that have suffered a large reduction in energy by reflection as they are transmitted from the body into the air.

Non-medical developments

Measurement and simulation have much to offer non-invasive archaeology. As the number and extent of archaeological sites and findings continues to expand, the need to investigate and preserve without excavation becomes more important. Ground-penetrating radar, electrical impedance tomography and acoustic sounding are used to image buried sites and extended to underwater ones too. Ideas and technologies developed for medicine are frequently taken up for geology and security applications.

In parts of the world that have recently experienced war, hidden landmines are a major source of fear for the population, especially in the countryside. These may be located safely by similar imaging and sensing technologies. Greater progress could be made in the development of techniques for land-mine clearance.

In the future other dangerous work such as the decommissioning of nuclear reactors should be eased through the increased use of robotics and autonomous probes on all scales.

Brighter sources and larger signals

The availability of brighter X-ray sources and large area detector arrays with smaller elements gives the prospect of improved resolution with X-ray CT images and XRF. The new types of source, synchrotrons (SR) and X-ray lasers, are still in the physics laboratory. In the future such sources and detectors will make possible cleaner images with reduced scattering and better energy discrimination.

In MRI the signal is limited by the number of spins in the voxel and the small fractional polarisation. Applications are constrained by the maximum acceptable B-field, the small voxel size, the data acquisition time, and the frequency resolution required. Improvement in the acquisition time through further use of multiple detector coils is limited because pickup coils are necessarily external. This suggests that MRI is a mature modality with little prospect of major improvement, but this is not the case. Recent developments involve the use of noble gases hyperpolarised by laser pumping techniques, and the generation of hyperpolarised ^{13}C (and other nuclei) using dynamic nuclear polarisation (DNP).[5] Polarisation may also be enhanced by the use of B_0 fields up to 7T and beyond. Problems with the short wavelength at such fields may be overcome by the use of segmented RF excitation coils. This is difficult because of problems of field and flip-angle uniformity. These are developments to be watched.

For imaging with PET the statistical noise, SNR $= 1/\sqrt{n}$ in the number of quanta per pixel n, limits the contrast. If the safety of ionising radiation were treated in the same way as other hazards, as discussed below, much higher doses would be permitted in PET. If accepted, brighter PET scans with improved large detector arrays are a prospect. The low random background and the high specificity of the take-up suggest a technology for further development. But the spatial resolution of a few millimetres, related to the physics of positron production and annihilation, is a hard limit.

In ultrasound imaging the limit is speckle, and this 'noise' is deterministically related to the 3-D distribution of scatterers contributing to each pixel. Although this looks to be a hard limit, much remains to be done in ultrasound, and there may be surprises. In principle it is cheap and portable. The source technology is not a problem.

Terahertz imaging and infrared spectroscopy

In chapter 1 we dismissed the IR and UV regions of the spectrum as strongly absorbing, and therefore unsuited to imaging in depth. However, neither ultrasound nor ionising radiation is sensitive to the chemical form of matter, and this is a requirement for imaging and probing, for

[5]This is similar to the Overhauser effect discussed in section 2.5.3.

a)

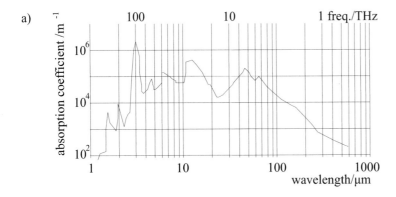

b)

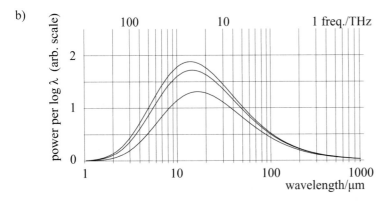

Fig. 10.2 a) The absorption coefficient of liquid water as a function of log wavelength. Note that the vertical scale is logarithmic.
b) The black body radiation spectrum as a function of log wavelength. The lower curve is for $T = 10°$C, the middle curve is for $T = 30°$C, and the upper curve is for blood heat $T = 35°$C. Note that the vertical scale is linear.

example in security applications. Identification is possible exceptionally with NMR by the chemical shift, but this is a second order marginal effect.

In the far IR spectrum warm materials emit thermal radiation, especially at wavelengths for which they are absorbing or black. Indeed thermodynamics requires that, if the absorption coefficient approaches unity, the emission spectrum approaches that of a black body. The black body spectrum at ambient temperatures peaks in the infrared as shown in Fig. 10.2b. Fig. 10.2a shows that water, and materials containing it, are particularly absorptive and black at these wavelengths. By contrast, many dry materials (excluding metals) are quite transparent there. Consequently, by imaging in the terahertz region,[6] the thermal radiation emitted by warm-blooded beings may be detected through colder dry material. Such principles may be used to image hidden survivors and bodies at disaster scenes, and also to find hidden cargoes of immigrants at national ports of entry. The technique depends on the observation of a temperature difference in the field of view, and so, as the ambient temperature approaches blood heat, the sensitivity is reduced.

The strong absorption in the IR is accompanied by strong reflection. In addition, following absorption atoms and molecules may emit characteristic fluorescent radiation. Although the reflected spectrum and the fluorescent radiation may come from a shallow depth, they are useful

[6] Figure 10.2a shows that absorption by water is less than 10^3 m^{-1} at 1 THz.

in the recognition of materials. The same effect is used in everyday life in the optical region when we distinguish things by their colour and reflectivity. In the X-ray region too, fluorescent emission by electronic resonance flags the presence of individual elements (XRF). Similarly, IR fluorescence spectra flag molecules by their individual rotational and vibrational spectrum. When detected in a security context with sufficient resolution, these spectra may indicate the presence of specific drugs or explosives, even through dry packaging in the mail. Technical advances in IR technology suggest that further major developments are expected. For example, the presence of molecules specific to particular cancers may be detected.

The difficulty is that other materials may absorb or reflect bands within the spectrum of interest. In particular the absorption spectrum of water is such that a search can easily be frustrated by a water thickness of a few hundred microns. Materials wrapped in metal foil may escape detection in the same way.

10.2   Revolutions in cancer therapy

Exploitation of non-linearity

Skin cancers can be treated with low energy γ-rays or electron beams without significant peripheral damage. This is because there is no overlying healthy tissue to be damaged. Using a beam of low energy electrons straight from the linac, few photons are created by bremsstrahlung and the electrons do not penetrate much below the superficial volume to be treated.[7]

[7]In the discussion in section 8.1.1 the inefficient production of photons by electrons presented a problem, even in a high Z material like tungsten. In this case with the low Z of tissue, the inefficiency of photon production by an electron beam is beneficial.

For deep-seated cancers treatment by γ-ray therapy must be described as primitive. As the dose contours on Fig. 8.24 show, the dose delivered to the prostate tumour is less than twice as large as the dose delivered to the healthy rectum. The death of the one and the survival of the other is attempted by putting the patient through 20 or more cycles of treatment and recovery in order to exploit the related non-linearity. This is unsatisfactory on four levels.

⋄ The patient experience of this damage-and-recovery iteration is dreadful. The prolonged duration of the treatment is a further element of the adverse patient experience.

⋄ The cost increases with the number of fractions and is therefore high.

⋄ The success rate for the treatment of deep-seated cancers is not good. This is because it is not feasible to give the required dose to the tumour and still ensure the survival of peripheral tissue.

⋄ The healthy tissue is heavily scarred following such treatment such that further courses of treatment may not be possible.

The consequence is that people are frightened. They speak in hushed tones of the 'C word'. This should remind us of where this book started.

In our discussion in chapter 1 we noted that physics has succeeded over the decades and centuries in relieving society of its worst fears. What can be done in this case?

There are two avenues.[8] They are light ion beam therapy and focused ultrasound (HIFU). These were discussed in chapters 8 and 9. The current situation calls for a massive and concerted development in these two areas over the next 20 years, such that in 2025 it can be said that 2005–2025 was an era of great advances in therapy, as 1985–2005 has been in imaging.

Both solutions promise a sharply defined therapeutic spot. Consequently the registration of the treatment with the position of the tumour becomes exacting. This is made difficult by cardiac and respiratory cycles and other differential patient motion. The treatment plan calls for the spot to be raster-scanned through the volume of the tumour. This demands the development of feedback between a real-time image of the tumour and the position of therapeutic beam.

There are new prospects in the design of small high current accelerators that could make the beams of light ions needed for radiotherapy more readily available. Such facilities are already under development in a number of countries. The extension to carbon ions with the extra non-linearity should follow.[9]

In the case of HIFU the application of the therapy is affected by the degree of non-linearity, the early monitoring technology and the considerable variability of tissue properties. These represent a need for serious investment, not just of resources but also of ideas and data. As simulation and monitoring of such therapy develops, confidence will build. Currently prostate therapy is available on an out-patient basis at a number of centres. Much remains to be done before similar treatment is available for other deeper cancers.[10]

[8]We exclude developments in chemotherapy as such which lie outside the physical methods discussed in this book.

[9]See for example the website of the BASROC consortium.

[10]See websites for the International Symposium on Therapeutic Ultrasound (ISTU).

10.3   Safety concerns in ultrasound

Sonoluminescence

The inadequate state of our knowledge of ultrasound in the non-linear regime is indicated by the phenomenon of sonoluminescence in which a small bubble illuminated by ultrasound is observed to emit light consistent with a high temperature black body spectrum.

Figure 10.3 shows a glass vessel of diameter 6.4 cm filled with demineralised water and illuminated by two ultrasound transducers at 25 kHz. Their outline may be seen towards the edge of the picture. At this frequency there is a simple acoustic antinode at the centre of the vessel. An air-filled bubble of initial diameter 75 μm has been placed there by pipette. The observation is that under the influence of the ultrasound field this bubble emits white light in pulses 1 ns long at the point in time when the bubble collapses. This is known as sonoluminescence, which was first discussed in 1934. In the experiment shown here the bubble is the bright point of light indicated by the arrow. The spectrum of the

Fig. 10.3 The observation of sonoluminescence by a bubble marked by the arrow in a water-filled vessel illuminated by an ultrasound standing wave. [Photograph reproduced with thanks to John Cobb, Allan Halliday and Paul Flint]

light decreases with wavelength between 400 and 700 nm and is not inconsistent with black body radiation at a temperature in excess of 12000 K. The number of photons observed was measured with a PMT to be $400 \pm 50 \times 10^3$. Similar results have been found in previous experiments.

The detailed mechanism is not known but the conditions required are not extreme. The frequency and power levels are modest. The point here is that this phenomenon is related to a simple bubble in water, and is not understood. If the local temperature excursion is correct, the safety implications for the clinical use of ultrasound are significant. We have introduced these observations here to show that the physics of ultrasound is still relatively immature, there are unanswered questions and there may yet be further surprises.

Ultrasound safety levels

In chapter 1 guidelines for the safety of energy doses were suggested in terms of four levels; a monitoring level, a background level, a damage level, and a lethal level. In various chapters in this book we have discussed what is known about these levels for different modalities. Relative to the damage level, a sensible safe level may be determined with an extra modest safety factor. For example, in the engineering of a bridge an extra safety factor of two or three in the maximum expected loads might be included in the design. This factor may be increased to take account of uncertainties. The safe level estimated in this way is then available for comparison with other risks, and as a basis for general guidance and regulation. We call the range from the safe level to the damage level the threshold region.

For ultrasound the levels are summarised in Fig. 10.4a on a logarithmic scale in terms of the tissue temperature rise. The threshold region is $\Delta T = 1 - 2°C$; the damage region is $\Delta T = 20°C$. These levels are quite close considering that instrumentation suitable to monitor such temperature rises *in vivo* is a recent development and not yet widely available. The degree of non-linear response of tissue to ultrasound and the paucity of related data is a concern, especially as the advantages of harmonic imaging encourage the use of sound at non-linear intensities. Safety is not improved by the public perception that ultrasound is harmless and particularly suitable for scans of pregnancies. The current unregulated situation is apparent from the website of the UK Health Protection Agency (HPA) where it is stated:[11]

> *There are no specific regulations in the UK for controlling exposure to ultrasound (and infrasound). However, medical products are required to comply with the Medical Devices Regulations 2002 (SI 618/2002). The guidance for medical diagnostic exposures are based on limits specified by the US Food and Drug Administration (FDA) but these do not appear to have a solid scientific basis.[12] Guidance is issued by the professional and membership bodies the British Medical Ultrasound Society, the European Committee for Medical Ultrasound Safety, and the World Federation for Ultrasound in Medicine and Biology.*

We conclude that the situation is a matter of concern.

10.4 Rethinking the safety of ionising radiation

What is a safe radiation dose?

The safety of ionising radiation and ultrasound could hardly be more different. There is no difficulty in monitoring ionising radiation, even down to the level of single photons or less per day. The data from Hiroshima and Nagasaki discussed in chapter 6 show that there is no significant risk of death from solid cancers or leukaemia for a single dose below 100 mSv.[13] This is plotted in Fig. 10.4b. The fatal dose at about 5000 mSv is also shown. Interestingly, the factors that separate the safe, threshold, damage and fatal bands seem much the same as for thermal damage, Fig. 10.4a (although no reason why this should be so is evident).

Also shown is a typical therapy dose per fraction. At 1500 mSv given to healthy tissue every *day* this dose is far from safe; the dose of 3000 mSv to the tumour is inadequately fatal. These figures vary somewhat but are typical of treatment provided in therapy departments on a regular routine basis.

The average annual dose of background radiation in the UK is 2.7 mSv/year.[14] To make a comparison of this dose rate with the above single doses, we need to know the time over which tissue integrates damage. This is the repair time, already known from *in vitro* and therapy experience to vary between several hours and a few days. This is used in

[11] A subgroup was set up at a meeting on 7 December 2004 to consider these concerns. An initial report is expected before the end of 2006.

[12] The underlining emphasis has been added.

[13] A recent report concludes that "*On average, assuming a sex and age distribution similar to that of the entire U.S. population, the BEIR VII lifetime risk model predicts that approximately 1 person in 100 would be expected to develop cancer (solid cancer or leukemia) from a dose of 100 mSv above background, while approximately 42 of the 100 individuals would be expected to develop solid cancer or leukemia from other causes.*" This refers to the whole-life risk and is based in part on a further re-calibration (2002) of the Hiroshima and Nagasaki data. It is an upper limit because the report uses the LNT model. This comes from the Biological Effects of Ionizing Radiation (BEIR) group of the US National Academies (19 June 2005). This conclusion does not differ greatly from that reached in chapter 6.

[14] As discussed in chapter 6.

a) tissue heated above normal temperature (35°C)

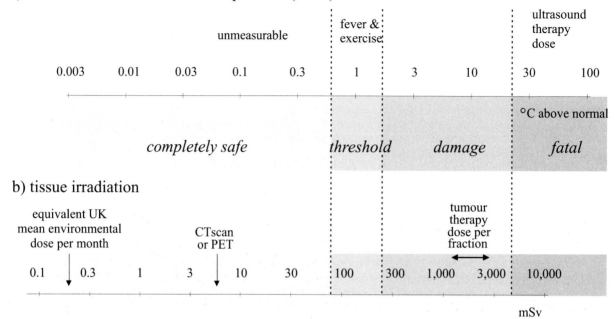

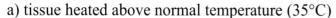

Fig. 10.4 A comparison on a logarithmic scale of ranges of dose separated by dotted lines that, (in turn from left to right) are completely safe, are at the threshold of damage, involve certain damage, and are fatal. These regions are shown for a) tissue heating generated by ultrasound or microwaves, and b) a single dose of ionising radiation. The rise in tissue temperature is in °C and the radiation dose is in mSv.

radiotherapy to set the recovery time between fractions, typically a day. Even if we choose a conservative time of a month, typical environmental radiation exposures are seen to be many orders of magnitude below threshold. With this choice the average UK background dose rate of 2.7 mSv/year is equivalent to a single dose of 0.2 mSv. This is lower than the damage threshold by a factor 1/500. Because of the non-linear repair mechanism this represents not just a proportionately small hazard, it represents no hazard, as was the case for the blood donation discussed earlier.

The dose from a typical PET or CT scan is an order of magnitude below the damage threshold level. This suggests that doses larger by a factor ten should be permitted in cases for which the resulting improved image resolution would give a better clinical diagnosis.

The current public safety limit for any steady extra dose of ionising radiation is 1 mSv/year. This would be equivalent to a single dose of 0.1 mSv, again using the conservative repair time of one month. This is seen to be over-cautious by a factor of about a thousand. Certainly these numbers can be argued up or down by factors of two but the factors of ten are rather firm. Of course room should be left for uncertainties. However, after 50 years these have become rather small.

This is an important conclusion. It means that all safety considera-

tions associated with nuclear radiation need to be revised by a factor of
at least 500. This includes the handling of nuclear material, the decommissioning of reactors and the disposal of waste. The costs of nuclear
power should be very substantially reduced thereby. Given the threat
of global warming this would be an important step. If it is acceptable
for a patient in therapy to risk receiving 15 times the safe dose of 100
mSv *every day*, should not some risks be incurred for the health of the
planet?

Why have we over-reacted to nuclear risks in the past?

A related question is how society came to have the current stringent
safety thresholds for ionising radiation. There are a number of reasons,
some political and some scientific.

◇ As a result of the bombing of Hiroshima and Nagasaki the
public were frightened by the power of nuclear weapons. The
public were told something of the compelling ideas of the fission chain reaction and the power of Einstein's conclusion
that $E = mc^2$. This information did nothing to calm their
fear.[15] Indeed there were political reasons to maintain the
fear of radiation, providing, as it did, the basis for the deter-
ence of the Cold War.

◇ In 1945 and for some time after there was little understanding
of how radiation affected living organisms. Knowledge of
genetics was rudimentary. A cautious approach to safety
seemed wise at that stage.

◇ There was no large scale data on the basis of which it was
possible to limit with confidence any long term damage due
to ionising radiation.

◇ With ample supplies of oil and coal and with no evident
global warming there was no serious disadvantage to a tight
control of radiation safety.

As a result, in the decades that followed World War II wholly new
criteria were set up and applied to ionising radiation safety that were not
applied elsewhere. These became linked to a naive search for absolute
nuclear safety, ignoring the need to balance radiation risks with other
hazards.

◇ A safety regime was set up in which nuclear radiation was to
be set at the lowest possible level. This became the ALARA
principle, discussed in section 1.3.2.

◇ The linear no-threshold (LNT) model was introduced. This
ignored the experience of recovery from radiation dose and
expressed risk in terms of collective dose.

◇ International studies of incidents and accidents (from Hiroshima to Chernobyl) that indicated that risks were exaggerated were ignored by the media. Pressure groups com-
peted with one another in their state of alarm. Even the

[15] We note that neither of these aspects
is relevant to the safety of ionising radiation as such.

label 'nuclear' was dropped from NMR when MRI was introduced. A polarised state of nuclear denial became the norm. To settle practical matters most people followed the prevailing guidelines.

◇ Public authorities confused the need to provide safety from real hazards with the need to make the population feel that they were safe. The latter is an important objective, but education, legislation and advice must be anchored to the former. Politicians having no knowledge of their own and seeking short term advancement, tightened laws and regulations further.

◇ Individual scientists who could have seen what was happening did not have the time to do so.

And then the realisation of global warming dawned – and with it the question of the nuclear power option.

The hazards of global warming need to be weighed against those of ionising radiation. Physicists need to rewrite the description of nuclear safety, for, not only is the current version of it unjustified, but it is obstructing the deployment of nuclear power which is the only large scale 'green' carbon-free solution to the world's power needs. A safety industry entrenched in the old out-dated standards has been built up. It will not be easy to turn this around. Much public education is needed and time is short.

A stumbling block for popular perception is the invisibility of ionising radiation, and it was from the ability to see and its effect on confidence from which this book started. But the public already accepts high levels of ionising radiation[16] in medicine; the need is to achieve an equivalent acceptance of relaxed environmental radiation standards. So the task of public re-education and information should be possible.

Terrorists and rogue states will continue to try to create fear, whether by biological, nuclear or any other means that comes to hand. However, as far as nuclear dangers are concerned the effect of more relaxed safety standards will be to deflate the currency of their threats to some extent. That can only be to the good. However, nothing said here should be construed as suggesting that nuclear power is safe, but it is safer than the alternative and we need it.

[16] Although in radiotherapy the levels accepted by the public should be reduced, as discussed in the previous section.

10.5   New ideas, old truths and education

Nanotechnology probes

Contrast agents have played an important part in the development of imaging. However, these agents have been more or less passive. In the future, miniature probes will become increasingly active. An important task will be the maintenance of communication and supply of power to such probes if they are to undertake sophisticated tasks such as local measurement, surgery and targeted drug delivery. The design and engineering of such miniature devices would harness the tools of DNA self-replication.

Metamaterials and coherent optics

Nanotechnology presents the prospect of making wholly new classes of materials. An example is a metamaterial with a negative refractive index. Samples have already been made. With a negative value of both ε_r and μ_r an electromagnetic pulse may propagate forwards in respect of both group velocity and energy flow, while having a phase velocity in the backward direction.[17]

This negative refractive index causes the incident and transmitted rays at the interface to such a material to be on the same side of the normal, as given by Snell's law. The usual situation is illustrated in Fig. 10.5a, compared with the unconventional result in Fig. 10.5b. A parallel-sided block of such material behaves as a converging lens in the frequency range concerned. Conventional materials are too weakly magnetic at a microscopic level to exhibit these effects. The range of application of such 'metamaterials' remains to be determined, but it seems likely that interesting developments will follow, especially in the RF, microwave and terahertz bands.

Advances in the use of coherent optics, with or without the use of metamaterials, is likely to influence medical imaging in the years ahead.

Smell and spectroscopy

The extent to which dogs and other creatures can distinguish chemicals by smell is surprising. They are able to distinguish different people and to find small quantities of narcotics and explosives on the basis of very low vapour pressures. Substances of molecular weight greater than about 340 cannot be distinguished in this way. Nevertheless dogs can diagnose melanoma by smell. Recently the physiology of smell has been decoded by molecular techniques, and Buck and Axel were awarded the 2004 Nobel Prize for Medicine for their work. Is there a future for animals whose sense of smell has been genetically redesigned for a molecule of interest? Or can the same end be achieved by old-fashioned training?

Navigation in the animal kingdom

While many an ancient mariner could smell land and some birds are known to navigate by following railway lines, on the whole animals, birds and fish have developed their own navigation systems. They are able to recall physical landmarks and features over long periods, they can sense magnetic fields and the polarisation of sunlight, as well as use their powers of smell. Many of their skills appear to be inherited rather than learned. They combine different techniques to able to overcome loss of any one signal. Mankind has still much to learn from understanding how they manage to do this.

[17]We saw in section 3 when discussing the behaviour of a small circuit in a B-field that it may resonate and that, depending on the damping, the magnetic moment that opposes the applied B-field may be negative in the region below resonance. The induced electric dipole moment is opposite to the applied E-field at ω just above ω_p, the plasma frequency. It has been predicted, and recently confirmed, that an artificial material composed of an ensemble of such circuits and conductors on a lattice can be constructed which displays negative values of relative permeability μ_r and relative permeability ε_r in a region of frequency.

a)

b)

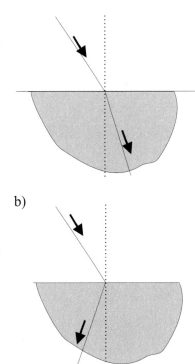

Fig. 10.5 The application of Snell's law. a) A conventional material with transmitted ray on the opposite side of the normal to the incident ray. b) A composite metamaterial with transmitted ray on the same side of the normal to the incident ray.

Gravitational waves and the Universe

The observation of gravitational waves in the first decades of the twenty first century will permit the exploitation of the most penetrating probe in the Universe. The information carried in the form of the polarisation, frequency and direction of gravitational waves will stimulate a whole new branch of physics and astronomy. This new Copernican revolution will ensure that mankind keeps thinking and wondering about new ideas in pure physics. Such stimulation will continue to pay dividends in everyday life even though the benefits will be indirect.

Confidence and progress through open education in pure science

New technology makes further features in pure physics accessible. To survive on the planet for further centuries mankind will need to invest in responsible education and informed confidence. Then, if decisions are taken by those who understand and consider the consequences, there is plenty of scope for optimism. Otherwise, if decisions continue to be driven by greed, secrecy, private gain and sectional interest, prospects are bleak. The way forward relies on decisions being made that are based on pure science and that command the informed confidence of the public. Only in this way may fear be reduced. This should be a major item on the world agenda for the twenty first century.

Look on the Web

Look up the book website at
   www.physics.ox.ac.uk/users/allison/booksite.htm

Study the imaging methods of EEG and MEG
   EEG MEG

Read about infrared spectroscopy
   IR spectroscopy

Consider the competition to IR spectroscopy offered by smell, and the small extent of our knowledge of it
   smell and *Buck Axel Nobel prize*

Read about developments and prospects in radiotherapy (see also chapter 8)
   protontherapy, carbontherapy, BASROC

For progress in ultrasound therapy
   international symposium on ultrasound therapy

Look up elastography
   MRI elastography and *ultrasound elastography*
   and find PhD thesis of James Revell, Bristol (2005)

Watch developments in nanotechnology
   nanotechnology probes
   DNA molecular engineering

Follow new possibilities in MRI using dynamic nuclear polarisation
   DNP enhanced MRI

Read further about sonoluminescence amd metamaterials
   sonoluminescence and *metamaterials*

Look at the links to animal navigation research on the Royal Institute of Navigation website
   animal navigation RIN

Conventions, nomenclature and units

A

Units and nomenclature

Generally we use conventional SI units. In this system the magnetic field is denoted properly by H and measured in A m^{-1}. On the other hand the magnetic flux density or magnetic induction is denoted by B and measured in tesla. In a sense H is fictitious and B is the measurable field. In this book we avoid using H where possible and refer simply to the 'B-field' measured in teslas. This is brief, clear and unambiguous.

In addition we use the following conventional nuclear or atomic units:

Quantity	Unit	SI equivalent	Conversion factor
Cross section	barn		10^{-28} m^2
Energy	eV	e J	1.602×10^{-19} J
Mass	eV/c^2	e/c^2 kg	1.783×10^{-36} kg
Momentum	eV/c	e/c kg m s^{-1}	5.344×10^{-28} kg m s^{-1}

When we refer to 'frequency' we mean angular frequency measured in radians per second, often denoted by ω, unless the frequency is explicitly given in hertz (Hz), meaning cycles per second.

All logarithms are natural logarithms unless another base is explicitly stated.

Decibels

The decibel is used in two different ways in applied physics.

First it is a measure of the ratio of signal energy fluxes on a base-10 logarithmic scale. Thus 1 decibel represents an energy flux ratio of $10^{0.1}$. In general a signal of intensity I and amplitude A is described as n decibels relative to one of intensity I_0 and amplitude A_0, if

$$n = 10 \times \log_{10}\left(\frac{I}{I_0}\right) = 10 \times \log_{10}\left(\frac{A^2}{A_0^2}\right) = 20 \times \log_{10}\left(\frac{A}{A_0}\right). \quad \text{(A.1)}$$

In acoustics the decibel is also used to quantify an absolute level of energy flux relative to a just audible standard. This standard is defined in terms of pressure, and is different in air and water. The decibel is not used in this way in this book.

Lifetimes

The exponential decay of an unstable state may be described, either in terms of a mean life τ such that the population or energy falls with time as $\exp(-t/\tau)$, or in terms of the half-life $t_{1/2}$ such that the energy or population falls by a factor 2 in each period of time $t_{1/2}$. In environmental and radiation contexts the half-life is always used. In quantum mechanics and most of the rest of physics the mean life is used. We explicitly use the terms 'half-life' and 'mean life' to remove ambiguity. The two are related as

$$t_{1/2} = \ln 2 \times \tau. \tag{A.2}$$

Square roots of -1 and time development

It is conventional in electrical circuit theory to denote the root of -1 by j, and to describe time development by $\exp(\mathrm{j}\omega t)$. In quantum theory it is usual to use i for the square root, and to describe time development by $\exp(-\mathrm{i}\omega t)$. It does not really matter which time development is used provided that there is consistency. By flagging the choice by the change of symbol, ambiguity is avoided. In the quantum theory convention a forward-moving progressive wave has an exponent $\exp[\mathrm{i}(\boldsymbol{k} \cdot \boldsymbol{r} - \omega t)]$.

Classical and quantum treatments

Many aspects of the physics that we discuss are adequately described in terms of classical physics, meaning that a description in quantum mechanics gives the same result. If it affords a deeper understanding of what is going on and it is straightforward to do so, we give both quantum and classical descriptions. How it is that classical physics gives the correct answer in the limit, the so-called correspondence principle, is a larger subject beyond the scope of this book.

Glossary of terms and abbreviations

B

The letter in the second column denotes the field to which the term or abbreviation principally belongs: A=archaeology, C=chemistry, E=engineering, G=geology, M=medicine, P=physics, I=imaging.

Term		Meaning
1-D, 2-D, 3-D		one-, two- or three-dimensional
ALARA		as low as reasonably achievable, principle of safety standards
alias	I	an artificially displaced part of an image
angiogram	M	an X-ray image of blood vessels
APD	P	avalanche photodiode
BASROC		British Accelerator Science and Radiation Oncology Consortium
B-field	P	magnetic flux density or magnetic induction, in tesla
bolus	M	concentrated charge of agent introduced into blood or digestive tract
brachytherapy	M	radiotherapy by implanted sources
CCD	E	charge coupled device, a detector array as used in an electronic camera
chiral	P	having handed symmetry, left or right
coronal	M	face-on vertical section of body or head in the upright position
CT scan	M	an X-ray scan analysed in three dimensions by computed tomography
dE/dx	P	mean energy loss with distance along a charged track, excluding bremsstrahlung
DNP		dynamic nuclear polarisation
DTPA	M	di-ethylene-triamine-penta-acetic acid, a chemical 'wrapper' for Gd ions when required as paramagnetic contrast agent in MRI
E1	P	electric dipole resonance. Similarly En, electric multipole resonances
ECG	M	electro-cardiogram
EEG	M	electro-encephalography
erythrocyte	M	red blood cell
ESR	P	electron spin resonance in a laboratory B-field
FDG	M	Fluoro-deoxy-glucose, the carrier used for radioactive fluorine in PET
FFAG	P	(non-scaling) Fixed Field Alternating Gradient, a potential design for a medical therapy accelerator
FGR	P	Fermi's golden rule for the calculation in quantum mechanics of the rate of an interaction in first order time-dependent perturbation theory
FMRIB	I	Functional Magnetic Resonance Imaging of the Brain unit, Oxford University
FOV	I	field of view, the part of the object to be imaged
fractionation	C	change in the relative concentration of isotopes by a mass-dependent process
fractionation	M	delivery of therapy over a period to exploit the non-linearity of response

Term		Meaning
FSI	P	frequency scanning interferometry
fused image	I	an image composed of data from two or more modalities combined
gradient echo	I	condition when all spins in a slice have the same phase due to field gradients
g-factor	P	factor relating a gyromagnetic ratio to its classical value
HPD	P	hybrid photodiode
hypoxic	M	oxygen deficient
infarction	M	blockage, e.g. of a blood vessel
IMRT	M	intensity-modulated radiotherapy
intrinsic	P	pure semiconductor state with carriers arising from thermal excitation alone
in vitro	M	an observation or experiment conducted in a test tube
in vivo	M	a live observation or experiment
IR	P	infra-red region of the spectrum with frequency less than that of light
LET	I	linear energy transfer ($\mathrm{d}E/\mathrm{d}x$) of a charge in tissue
linac	P	linear accelerator powered by RF
LNT		linear no-threshold model of radiation damage
M1	P	magnetic dipole resonance. Similarly Mn, magnetic multipole resonances
magnetisation	P	magnetic dipole moment per unit volume (also called polarisation)
mammography	M	imaging of the breast
MEG	M	magneto-encephalography
metabolite	M	a chemical, an active ingredient of growth or functioning of the body
metastasis	M	the migration of cancer cells from one organ to another
modality	M	group of methods of medical imaging based on a common physical principle
MRI	I	nuclear magnetic resonance imaging
MRS	I	nuclear magnetic resonance spectroscopy with some spatial resolution
muon	P	an unstable heavy version of the electron with 200 times the mass
n-type	P	a semiconductor with impurities so that most charge carriers are negative
NDT	E	non-destructive testing
necrosis	M	death of the cells of an organ or tissue
NICE	M	UK National Institute for Health and Clinical Excellence
NMR	P	nuclear magnetic resonance
non-linear	P	depending more (or less) steeply than the first power
nuclear medicine	M	imaging using administered radioactive nuclei (SPECT and PET)
palpation	M	examination by feeling with fingers and palms
p-type	P	a semiconductor with impurities so that most charge carriers are positive
PET	I	positron emission tomography
phantom	M	a made-up reference sample of known composition
PIXE	P	proton-induced X-ray emission
pixel	I	a discrete element of a 2-D image
PMT	P	photomultiplier tube
polarisation	P	(of a wave) direction of the field vector (not the wavevector)
polarisation	P	(of material) having net electric dipole moment (see also magnetisation)
PSF	I	point spread function
P wave	G	compression wave with longitudinally polarised strain
Q	P	$2\pi \times$ ratio of resonance energy divided by energy lost per cycle
QED	P	quantum electrodynamics

Term		Meaning
radar	E	RAdio Detection And Ranging
RBS	P	Rutherford back scattering with light ion beam
registration	M	matching of position information, both alignment and rotation
RF	P	radiofrequency
RMS		root mean square deviation
RT	M	radiotherapy
sagittal	M	side-on vertical section of body or head in the upright position
SAR	P	specific energy absorption rate
SH wave	G	shear wave with strain perpendicular to the plane of incidence (horiz.)
SI		the international system of metric units
SNR	E	signal-to-noise ratio
speckle	P	a form of statistical noise arising from random phases
SPECT	I	single photon emission computed tomography
spin echo	P	condition in NMR when the precession phase of all spins returns to zero
SPM	P	scanning proton microprobe
SQUID	P	superconducting quantum interference device
SR	P	synchrotron radiation source
standard	A	a made-up reference sample of known composition
SV wave	G	shear wave with strain in the plane of incidence (vertical)
T	P	temperature in K *or* stress in Newtons $/m^{-2}$, according to context
T	P	Tesla, unit of B-field strength
terahertz	P	electromagnetic radiation of frequency around 1 THz
tomography	I	3-D reconstruction of an image
transverse	M	horizontal section of body or head in the upright position
UV	P	ultraviolet region of the spectrum with frequency greater than that of light
vasculature	M	form of the blood vessels
voxel	I	a spatial element of a 3-D image
wavelets	P	a form of wave analysis combining localisation with frequency analysis
XRF	P	X-ray fluorescence
Z	P	the atomic number of an element

Hints and answers to selected questions

2.2

Width 1.8×10^{-30} eV.

In the laboratory the width and lifetime are dominated by collisional de-excitation and the Doppler shift. In inter-stellar space the Doppler shift still dominates the width but the lifetime is the radiative lifetime.

2.3

There are three considerations: a) L comes from the self inductance of the coil element but C can be chosen freely; b) the phase change in each element should be $2\pi/12$ at 120 MHz; c) the input, output and terminating impedances should match for efficiency and to prevent reflected waves generating a field rotating the wrong way around. [For further discussion of such filter circuits refer to the book by Bleaney and Bleaney.]

This ignores the mutual inductances of the coil elements; this is described by a 12×12 inductance matrix. The back emfs are related to the flux linkages of pairs of coil elements. The full set of 12 simultaneous equations should be solved; a consequence is that the C values will not all be the same.

2.4

Consider the Fourier transform of the decay of the resonance energy as a function of time.

2.5

$M = 0.02$ Wb m^{-3}. Frequency 128 MHz.
Consider the maximum possible rate of change of magnetic flux linking the detector coil, and deduce that the mass of water is 0.05 kg.
For the maximum voltage the coil should surround the sample closely so that no return flux passes through the coil. The plane of the coil should contain the B_0 field.

2.6

For an isolated electron the answer is 0.74 MHz. However, the ESR frequency varies too much with the atomic environment of the electron and its damping is too large (Q too low) for this to be a good way to measure small fields.

2.8

The frequency of B_1 must be on resonance at 64 MHz within $\approx 1/50$ MHz. This tolerance is such that the phase will not slip significantly during the 50 μs pulse. During the pulse the moment is rotating about B_1 in its own frame at 2.1×10^4 radians s^{-1}. [See the book by Bleaney and Bleaney for further discussion of this motion.] Answer $60°$.

2.9

In units of 10^{-3} Wb m^{-3}, initially $M = 10$ units in z direction. Immediately after the RF pulse, 5 units in z and 8.7 tranversely. After 100 ms, 5.9 units in z and 1.2 units transversely.

3.1

$$\frac{d\sigma}{d\Omega} = \frac{fr^2}{12 \times 10^3 \times FAmN_a} \text{ m}^2 \text{ steradian}^{-1},$$
where N_a is Avogadro's number.

3.2

Calculate the transverse and longitudinal momentum transfer using the data given. Then use equation 3.9 to show that the target mass is actually that of the electron.

3.4

Form the Lorentz scalar product of the target 4-vector before scattering with that after. In the target frame this is $-mc^2(E + mc^2)$. In the centre-of-mass frame it is

$$-(mc^2)^2 + Q^2/2 = -(mc^2)^2 + Q_4^2/2.$$

3.5

a) Get amplitude a using Newton's second law.
b) Use the impedance Z to get the incident H-field and Poynting's vector to get the incident energy flux, $E_0^2/2Z$.
c) Use Z again.
d) The cross section is the scattered power divided by the incident energy flux.
Answer 6.65×10^{-28} m^2.

3.6

Deuteron range is 11 g cm^{-2} = 110 kg m^{-2}, equivalent to 220 kg m^{-2} for a proton.
Read off from the curve $\beta\gamma = P/mc = 0.4$ so that $P = 800$ MeV/c and $KE = 155$ MeV.
At the same $\beta\gamma$, the proton range is 0.025 m.

3.7

RMS angle 0.017 radians.

3.8

1 mm of copper is 0.896 g cm^{-2}. Assuming that the energy loss is small and that $\beta = 1$, the non-radiative energy loss is 1.228 MeV. The radiative loss is $3.0 \times 0.896/12.25 = 0.22$ MeV. If the incident energy were much less than 3 MeV, the fractional energy loss would not be small and the approximation would fail.

4.1

The isotropic, traceless symmetric and antisymmetric parts:

$$\frac{1}{3}\begin{bmatrix} a+e+i & 0 & 0 \\ 0 & a+e+i & 0 \\ 0 & 0 & a+e+i \end{bmatrix},$$

$$\frac{1}{6}\begin{bmatrix} 4a-2(e+i) & 3(b+d) & 3(c+g) \\ 3(d+b) & 4e-2(i+a) & 3(f+h) \\ 3(g+c) & 3(h+f) & 4i-2(a+e) \end{bmatrix},$$

$$\frac{1}{2}\begin{bmatrix} 0 & b-d & c-g \\ d-b & 0 & f-h \\ g-c & h-f & 0 \end{bmatrix}.$$

For the stress tensor multiply each by the respective modulus $(3\lambda + 2\mu, 2\mu, 0)$ and add back up again.

4.3

Read off c_R from Fig. 4.17. The rest is geometry. Time = 383 s. Distance = 2000 km.

4.4

The wavelengths at resonance are $2\ell/(n+1)$ for the nth harmonic. The open end of the tube is an antinode so that odd n harmonics are missing.

But what are the walls doing? Nothing for the displacement which is longitudinal anyway. But consider the stress (pressure) and why a wave in the tube should be reflected from the open end. What happens if the width, $a \sim \lambda$?

4.5

Use equation 4.68 with $h = \psi_z$ at $z = 0$. Damping of the wave by scouring is clearly a large effect. This is the reason why seabeds are remarkably flat. In the non-linear range the peak of the waves form a forward flow of water which is balanced by a steady backward scouring flow on the seabed. This is the mechanism of coastal erosion under conditions of high waves.

5.5

Let concentrations relative to the references be a, b, c, d, e. The sum-of-squares,

$$S = \Sigma_{i=1}^{100} \left(M(\omega_i) - aA(\omega_i) - bB(\omega_i) \right.$$

$$\left. -cC(\omega_i) - dD(\omega_i) - eE(\omega_i) - X - \omega_i Y \right)^2.$$

Get 7 equations by setting to zero the differential of S with respect to a, b, c, d, e, X, Y in turn. These equations are linear in the unknown concentrations and could be solved by inverting the matrix. A check would then be made that the concentrations and the background levels $X + \omega_i Y$ are everywhere positive for a reasonable fit.

[Such an analysis might also be used to find the concentrations of elements contributing to an observed X-ray emission spectrum taken by XRF or PIXE techniques discussed in chapter 6.]

6.1

a) 470. b) 50×10^6.

6.2

Error on age = mean life × error on count / count. Statistical errors: a) 460 years, 4300 years. b) 1.4 years, 13 years.
Systematic errors due to 1% modern C: for 3000 year old sample is 120 years, for 40,000 year old sample is 10,000 years.

6.3

Assume that 30% of tissue is carbon and that half of the decay energy of ^{14}C is carried away by the neutrino and half deposited by the electron. Sv and Gy are the same for β-decay. This contributes 23 μSv per year to internal dose.

6.4

Suppression of ^{40}K decay relative to neutron decay is 1.6×10^{-14}, ignoring that the energy is 1.5 MeV whereas it is only 0.8 MeV for the neutron. The λ of 1 MeV photon is 1.2×10^{-12} m, and so $a/\lambda \approx 10^{-3}$. For $\Delta J = 4$ the given EM suppression factor comes out at 6×10^{-14}, and this applies approximately to the γ-decay of ^{99}Tc. Both these highly forbidden decays are of great importance, ^{99}Tc in medicine and ^{40}K in the environment.

6.5

Layer of gold (197) on a silicon (28) substrate.

6.6

Apply energy conservation and momentum conservation, longitudinal and transverse.

6.7

a) The KE of the argon is 28 eV.

b) The KE of the radon is 86 keV. Comment: the radon nucleus recoil energy is 3000 times larger and it is probably kicked out of its lattice site which is probably interstitial anyway on account of earlier decays in the radioactive decay sequence. The argon with its small recoil and previously undisturbed lattice site has a good chance to remain trapped. However, I have found no analysis that compares this with the six orders of magnitude difference in containment time.

7.1

Consider the magnetostatic potential due to $\boldsymbol{m}$ and thence its B-field perpendicular to the coil.

7.2

The sensitivity matrix has diagonal elements $8\sqrt{2}$ and off-diagonal elements 1. This would give clear non-singular spatial discrimination.

7.3

Answer: (x, y, z) = (-60, 42, 32) mm relative to the isocentre (the centre of symmetry of the applied gradient fields).

7.5

This chemical shift amounts to increasing the value of the Larmor frequency, 42.55747 MHz T^{-1}, by 4 ppm. The change in calculated position would be (0.0, 1.0, 1.5) mm.

7.6

If the relaxation time is 10 ms, then the maximum power at which energy can be deposited is the full resonance energy every 10 ms. This is the magnetisation energy of 1 in 10^5 protons which is 5.3×10^{-5} J kg^{-1}. Maximum resonant power loading 5.3 mW kg^{-1}. We may conclude that resonant power absorption due to NMR cannot be a significant hazard. (The position is different for ESR.)

7.7

Use energy conservation. The magnetic energy per unit volume is about $-\frac{1}{2}\mu_r B_0^2$ (depending somewhat on the shape as well as the orientation of the screwdriver to the field) and its KE is $+\frac{1}{2}\rho V^2$. These add to zero for a screwdriver starting from rest outside the magnet.

7.8

Ignore any difference between the values of T_2 for A and B, and also the distinction between T_2 and T_2^*. Answer: Before time $t = T'$ there will be no FID from A or B. After time $t = T'$ the ratio of B magnetisation over A magnetisation will be

$$\frac{1 - 2\exp(t/1.3)}{1 - 2\exp(t/0.75)}.$$

The contrast is a maximum at $T' = 0.75/\ln 2$ s, when the amplitude of the FID due to A will be zero.

8.2

KE is 0.1 eV.

8.3

Assume patient has density of water. Then 10^6 voxels per kg and 10^4 counts per voxel. Taking each count to be 60 keV, that makes 6×10^{14} eV per kg. This equals 0.1 mSv at 100% efficiency. This is obviously a significant underestimate for a practical case.

8.5

The range is 200 kg m^{-2}. The range for protons is read off Fig. 3.13; for another particle with the same value of $\beta\gamma$ but charge Z and mass μ the range scales as $Z^{-2} \times \mu/m_p$. A proton with the required incident value of $\beta\gamma$ has a range $200 \times 6^2/12 = 600$ kg m^{-2}. From the curve this $\beta\gamma = 0.85$ and the momentum is 9570 MeV/c. The (relativistic) kinetic energy required is 3517 MeV.

9.1

Impedance 1.54×10^6. Wavelength 1.54 mm.
Pressure amplitude 2.48×10^4 Pa.
Particle velocity 16×10^{-3} m s^{-1}.
Particle displacement 2.6×10^{-9} m.
Acceleration 1.0×10^5 m s^{-2}.
At 1600 W m^{-2} the pressure amplitude would reach atmospheric pressure, 10^5 Pa, independent of frequency.

9.2

Using an oblate ellipsoid for the red blood cell its volume is 6×10^{-17} m^3. Putting this value of V and the differences in density and elasticity from table 9.4 into equation 9.9 and integrating over $\cos \theta$ gives a total cross section for an isolated blood cell $\sigma \sim 2.2 \times 10^{-21}$ m^2. Neglecting the effect of coherence the attenuation length would be 90 km.
The cross section for each coherent volume would be $N^2 \times \sigma = 1.8 \times 10^{-10}$ where $N = 290,000$ is the number of blood cells in a coherent volume. The attenuation length may be estimated from the number of such coherent volumes per unit volume. This estimate of the attenuation length is 0.3 m, varying as 1/frequency and with a fair uncertainty. The observed value is 0.21 m at 1 MHz. The order of magnitude is correct so that we may conclude that the physics is understood.

Index

ablation, 302
absorption
    sound, 126, 301
    X-ray, 66
absorption edge, 236
absorption image, 236
accelerator mass spectrometry, 174
acceptance, 160
acoustic surgery, 302
active sound imaging, 268
acute radiation death, 200
ALARA, 18, 317
aliassing, 213, 220
Alvarez, Luiz, 147
Ampere's law, 22
Anger camera, 248
angiogram, 239
archaeology, 4, 309
artefacts, 15, 139, 211, 212
atmospheric nuclear tests, 197
atomic hydrogen, 9, 31, 32, 35, 41, 54
Auger emission, 158
avalanche photodiode, 167

B-field, 321
    origin, 22
    safety, 19
    strong, 31
    weak, 31
bandwidth, 142
barium titanate, 273
becquerel, 179
beta-light, 187
Bethe–Bloch formula, 75
biological equivalent dose, 179
biological radiation damage, 181, 187
birdcage coil, 44, 54
black body radiation, 311
Bloch equations, 39
blood, 285, 306
blood flow, 223, 287
blood oxygen level dependent scan, 222
body location, 311
Bohr magneton, 25
Boltzmann distribution, 29, 35, 38, 52, 122
Born approximation, 281, 295
boundary
    fluid–fluid, 115
    rock–air, 111

rock–water, 109
    solid–solid, 104
boundary conditions
    displacement, 100
    force, 100
    frequency, 101
    wavevector, 101
Bragg peak, 78, 258
Breit–Rabi plot, 32
bremsstrahlung, 56, 81, 157, 159, 265

calibration, 11
camera, 248
cancer, 197
carbon dating, 174, 175, 205
cavitation, 298
cell reproductive cycle, 189
CERN, 120
chain model, 121, 298, 306
charge conservation, 22
chemical radicals, 29, 159
chemical shift, 41, 218, 225, 231
Cherenkov radiation, 157
Chernobyl, 181, 190, 199, 317
chi-squared method, 146
clean up, 201
clinical MRI scanner, 229
coherent optics, 319
collective dose, 181
combining data, 152
Compton scattering, 8, 65, 84
computation, 307
constituent model, 121
constituent scattering, 60, 84
continuous variables, 133
contrast, 14, 231
contrast agent, 10
    barium and iodine, 239
    bubbles in ultrasound, 271, 287, 303
    paramagnetic, 221
convolution theorem, 137, 156
Copernican revolution, 1
coronal section, 244
cosmic rays, 184
cost, 15
    MRI, 227
    nuclear power, 202, 316
    PET, 255
    radiotherapy, 263
    therapy, 304, 312

cross section, 57
    differential, 58, 84
CT scan, 10
current loop, 22, 23

D'Alembert travelling wave, 89
damage
    biological, 179, 187
    disruptive, 19, 181
    resonant, 19
    thermal, 19
damping, 119
dating, 37
    argon, 176
    carbon-14, 174, 205
    fission track, 177
    thermoluminescence, 178
    uranium series, 177
dE/dx, 74, 76, 82
decibels, 321
deep inelastic scattering, 63
delayed processes, 158
delayed response, 127, 128, 301
delta rays, 157
dephase, 42, 209
diamagnetism, 28, 30
diffusion measurement, 223
Dirac δ-function, 56, 136, 156
discontinuity length, 294
discrete variables, 133
dispersion relation, 91
    chain model, 124
    fluid–fluid waves, 117
    shallow water waves, 118
displacement, 85, 306
distortion, 86
Doppler shift, 10, 14, 271, 287
dynamic nuclear polarisation, 52, 227, 310
dynamics, 59, 69

Earth
    B-field, 22, 48
    crust, 104
    mantle, 104
    normal modes, 120
earthquake, 115
echo planar imaging, 215, 218
education, 16, 203, 307, 317, 318, 320
elasticity, 86

electric
  impedance tomography, 309
  resonance, 9
electro-cardiogram, 308
electro-encephalography, 308
electron spin resonance, 27, 34, 159, 178, 188, 192
element
  analysis, 169
EM radiation detectors, 165
EM shower, 168
energy in seismic waves, 90, 97
  bulk waves, 115
  Rayleigh waves, 115
energy loss, 74
energy per charge carrier, 160
energy transfer, 59
entropy, 88, 131
errors, 145, 152
evanescent wave, 109, 113

far-field imaging, 13
Faraday's law, 22
Fermi Golden Rule, 59, 69
ferroelectricity, 273
ferromagnetism, 28
field of view, 212
filtering, 134, 138
fission reactor, 63
fitting, 144, 156
flow measurement, 223
fluid, 87
fluorescence, 158
flux-gate magnetometer, 49
food sterilisation, 192
forbidden transitions, 205, 247
Fourier reconstruction, 211, 242
Fourier transforms, 135
fracture, 298
frequency, 321
frequency encoding, 210
frequency scanning interferometry, 268
functional imaging, 12
  MRI, 221
  PET, 252
  SPECT, 246
fused image, 14, 152, 255, 307

g-factor, 25, 26, 36
gadolinium, 221
gamma-ray, 7
gas-filled detectors, 163
Gauss's law, 22
Gaussian distribution, 151
Geiger–Müller, 163
geology, 4, 309
geophysics, 4, 102, 104, 130
gradient coils, 207
gradient echo, 324
gradient refocus, 209

gravitational waves, 7, 320
gravity waves, 7, 116, 130
gray, 179
ground penetrating radar, 309
group velocity, 91, 125
gyromagnetic ratio, 24

H-field, 28
Hall probe, 49
harmonic generation, 272, 294, 300
Heaviside stepfunction, 136
high intensity focused ultrasound, 302, 313
Hiroshima and Nagasaki, 182, 196, 315
Hooke's law, 86, 93, 290
hull speed, 118
hybrid photodiode, 167
hyperfine structure, 31, 32, 35

identical particles, 121
incomplete data, 135, 139
information, 11, 131, 156
insonate, 276
internal forces, 85
inversion problem, 259
ionisation detector, 163
IR
  absorption, 9, 311
  fluorescent spectroscopy, 9, 310
irreducible tensor, 93, 130
irreversibility, 301
isotope
  analysis, 172
  separation, 172, 195, 198
isotropic solid, 94

k-space, 215, 217
kinematics, 59
Kramers–Kronig relation, 92

laceration, 16, 181, 190
Lamé constants, 95
Landau fluctuations, 77
Lande g-factor, 26
landmines, 310
Larmor precession, 25
law of reflection, 103
law of transmission, 103
least squares method, 145
Lennard–Jones potential, 122, 291, 297
leukaemia, 197
likelihood method, 149
  extended, 153
linear accelerator, 235
linear energy transfer, 76, 158, 180
linear no-threshold model, 181, 317
linear reconstruction, 242
linear superposition, 210
linearity, 86, 126
liquid drop model, 192

lithotripsy, 302
localisation, 142
Love waves, 115
luminous dials, 187

magnetic
  dipole, 22, 24
    long range field, 33
  field, 47, 321
  induction, 9, 48
  resonance, 9, 23, 34
  torque, 24
magnetic dipole transition, 31, 34
magnetic permeability, 28
magnetic resonance spectroscopy, 225
magnetic susceptibility, 28
magnetisation, 27
  rotation angle, 44, 54
magneto-encephalography, 308
malignant growth, 190
materials testing, 270
Maxwell's equations, 21
mean ionisation potential, 74
mechanical probing, 7
MEG, 52
metabolites, 156, 226
metamaterials, 319
metastasis, 190
microprobe, 170
modelling, 134, 144
moderator, 63
modulus
  bulk, 87, 95
  shear, 87
momentum transfer, 59, 69
monitoring ionising radiation, 37
Monte Carlo, 144, 148
MRI, 4, 13, 207, 310
  safety, 227, 228, 231
  thermometry, 30, 303
multiple collisions, 74
multiple scattering, 78, 265
muon spin resonance, 27
mutations, 190

nanotechnology, 318
navigation, 2, 319
near-field imaging, 13
nearest-neighbour potential, 123
neutron activation analysis, 173
NMR, 27, 30, 50
  B_0 solenoid, 44
  B_1 RF field, 44
  FID pickup coil, 44, 231
  free induction decay, 44, 231
NMR relaxation time
  T_1, 40, 42, 45
  T_2, 40, 42, 46
  T_2^*, 42
noise, 15, 132, 160, 238, 245

non-destructive testing, 270
non-linearity, 19, 312
  biological radiation damage, 187, 189,
      262, 263
  sound, 128, 272, 290, 298
normal modes, 119, 130
  Earth, 120
  piezoelectric crystals, 120
  Sun, 120
nuclear
  abundance, 30
  binding, 192
  fallout, 197
  fission and fusion, 192
  fission products, 193
  fuel, 195
  Oklo reactor, 202
  power, 16, 202
  reactor decommissioning, 310
  record keeping, 203
  shell model, 30
  spin, 30
  surveillance, 198
  waste, 202
  weapon tests, 115, 197
  weapons, 194–196
nuclear EM lifetime, 205, 247
nuclear magnetic resonance, 37
nuclear magneton, 25
nuclear medicine, 246
nuclear-phobia, 318

occupational doses, 187
Overhauser effect, 51
overlapping scans, 215
oxygen effect, 189, 263

P wave, 102, 105
pair production, 8, 68, 159
palpation, 8, 13
paramagnetic resonance, 34
paramagnetism, 28
parity, 104
Paschen–Back effect, 32
passive sound imaging, 267, 308
patient experience, 15
  radiotherapy, 263, 312
  SPECT, 249
  ultrasound therapy, 304
Pauli exclusion principle, 121
pedoscope, 16
perfect gas, 87, 292, 306
permitted annual dose, 316
perturbation theory, 59
PET, 10, 11, 310
  with ^{18}F, 255
phase encoding, 210
phase space, 70
phase velocity, 125
photoabsorption, 66, 74

photoelectric effect, 8
photographic detector, 161, 236
photomultiplier tube, 166
pickup coils, 218, 231
piezoelectric properties, 273
pinhole optics, 57, 233, 248
PIXE, 56
pixel, 14, 161
Planck distribution, 126
point spread function, 14, 137, 270
Poisson distribution, 147
polarisation
  longitudinal, 96
  transverse, 96
portal imaging, 262
positron emission tomography, 252
positron emitters, 253
potassium-40, 183, 205
potential
  inter-atomic, 298
proportional counter, 164
proton magnetometer, 51
pulse sequences, 213
pulse shape, 208
Pyramids, 147

Q, 37, 39, 119
QED, 81
quantisation, 119
quantum magnetometers, 52
quarks, 63
quartz, 120, 273

radar, 10, 13, 268
radiation
  annual exposure, 182
  collective dose, 181
  damage, 20, 64, 189
  detectors, 159
  exposure, 245
  internal exposure, 182
  length, 68, 76, 80, 82
  linear no-threshold model, 181
  medical doses, 185, 265
  memory time, 180, 188, 315
  safety, 181, 314
  weighting factor, 179
radioactive contamination, 201
radioactive decay series, 183
radioactivity, natural, 55, 182
radiotherapy, 181, 256, 315
  brachytherapy, 257
  carbon beam, 258, 264, 265, 313
  cobalt source, 257
  dose monitoring, 261
  electron beam, 257, 312
  fractionation, 262, 263
  IMRT, 260
  oxygen dependence, 263
  photon beam, 257, 312

proton beam, 258, 264, 313
  treatment planning, 259
radon exposure, 184, 205
radon-induced cancer, 182, 190
range, 57, 77, 78, 84
Rayleigh distribution, 286, 306
Rayleigh scattering, 281, 306
  shear motion, 283
  speckle noise, 284, 285
Rayleigh wave, 113, 130
recoil kinematics, 60
reconstruction, 135
reference state, 85, 86, 90, 93, 94, 126,
      291, 298
reflection and transmission
  normal incidence, 99, 130
  polarisation change, 103
reflection at biological boundaries, 271
registration, 14
regulatory websites, 20
resonant circuit, 23
risk levels, 17
rupture, 298
Rutherford back scattering, 63, 171, 205
Rutherford scattering, 72
  classical derivation, 70
  quantum derivation, 69

safety, 16, 17, 227, 228
  B-field, 19
  CT, 316
  ionising radiation, 181, 314, 315
  MRI, 231
  PET, 316
  ultrasound, 297, 314
sagittal section, 244
sampling function, 139
scanning proton microprobe, 170
scattering of sound, 129, 271, 306
scintillation, 158, 162
security, 1, 5, 309, 312
seiches, 120
semi-classical electrodynamics, 81
sensitivity matrix, 220
SH wave, 102, 104
shimming, 225
sievert, 179
signal, 132
simulation of experiments, 144, 153
single photon emission computed
      tomography, 246
slice selection, 208
smell, 319
smoke alarms, 187
Snell's law, 103, 319
SNR, 15
solid state detector, 165
sonoluminescence, 313
spatial encoding, 207, 210, 218
spatial resolution, 12, 14, 161, 172, 215,
      245, 253

specific energy absorption rate, 19
speckle, 135
speckle noise, 285, 306, 310
SPECT, 10
spectroscopic imaging, 225
spin echo, 46
spin–orbit coupling, 26, 75, 247
SQUID, 52, 308
statistical errors, 133
sterilisation, 191
stethoscope, 8, 309
straggling, 77
strain, 85, 87
  tensor, 93
stress, 85, 87
  tensor, 92
structured media, 120, 297
subtraction, 239
sum rule, 67
superposition principle, 89, 299
surface acoustic waves, 115, 275
surface tension waves, 116
susceptibility, 30
SV wave, 102, 107
symmetry, 11
systematic errors, 133

tensile strength, 121
tensors, 92, 130, 223, 273
terahertz imaging, 9, 310
therapy, 4, 302, 312
thermal conductivity, 128
thermal emission, 311
thermal expansion, 291
thermodynamics, 88, 131, 132
thermoluminescence, 178
Thomas–Reiche–Kuhn sum rule, 67
Thomson scattering, 8, 65, 84
Three Mile Island, 199
thyroid, 190, 200
tides, 120
time resolution, 12, 14, 161

time-lapse imaging, 10, 268
tissue temperature rise, 19
transition elements, 29
transverse section, 244
treatment planning, 257
tsunami, 1, 118, 119
Turin Shroud, 175

ultrasonic welding, 301
ultrasound, 8, 267, 310
  absorption, 279, 301
  angular resolution, 270
  artefacts, 278, 279
  beams, 276
  constituent model, 123, 124, 298
  contrast agent, 287, 303
  Doppler, 287
  harmonic generation, 294
  harmonic imaging, 296, 315
  impedance, 271, 306
  inhomogeneous materials, 284
  medical imaging, 271
  near and far field zones, 276
  non-linearity, 290, 293, 295, 306
  range resolution, 268
  refraction and shadowing, 280
  resonant scattering, 286
  reverberation, 280
  safety, 297
  scattering, 281, 306
  shear wave scattering, 283
  streaming, mixing and shaking, 287
  surgery, 302
  therapy, 302, 313
  thermodynamics, 284
  transducers, 272
  wave–wave scattering, 295
  with chemotherapy, 303
underworld, 1
units, 321
unpaired angular momenta, 22, 29, 30,
    32–34, 36, 38, 41, 42, 51, 178

Urals accident, 182

van der Waal's forces, 122
virtual photons, 81
viscosity, 128
voxel, 14, 161

wave
  absorption, 91
  dispersion, 91
  energy density, 90
  equation, 89
  impedance, 90, 271, 306
  longitudinal, 88, 96
  particle velocity, 90, 306
  plane, 88
  transverse, 90, 96
  velocity
    compression, 90, 271
    shear, 90
  water, 116, 117, 130
wavelet transforms, 142
weighted image, 213, 221
Weissäcker–Williams virtual photons, 81
Wigner–Eckart theorem, 94
Windscale, 199
Winston cone, 268
wrapping, 212

X-ray, 7
  bremsstrahlung source, 56
  CT image, 243
  filter, 234
  fluorescence, 169
  image, 4, 10
  source, 56
  synchrotron source, 56, 170, 236, 310
  tube, 233
XRF, 169

Zeeman effect, 31